Structural Dynamics and Vibration
in Practice

Structural Dynamics and Vibration in Practice

An Engineering Handbook

Douglas Thorby

AMSTERDAM • BOSTON • HEIDELBERG • LONDON
NEW YORK • OXFORD • PARIS • SAN DIEGO
SAN FRANCISCO • SINGAPORE • SYDNEY • TOKYO

Butterworth-Heinemann is an imprint of Elsevier

Butterworth-Heinemann is an imprint of Elsevier
Linacre House, Jordan Hill, Oxford OX2 8DP, UK
30 Corporate Drive, Suite 400, Burlington, MA 01803, USA

First edition 2008

British Library Cataloguing in Publication Data
A catalogue record for this book is available from the British Library

Library of Congress Catalog Number: 2007941701

ISBN: 978-0-7506-8002-8

For information on all Butterworth-Heinemann publications visit our
website at http://books.elsevier.com

Printed and bound in Hungary
08 09 10 11 11 10 9 8 7 6 5 4 3 2 1

To my wife, Marjory; our children, Chris and Anne;
and our grandchildren, Tom, Jenny, and Rosa.

Contents

Preface

This book is primarily intended as an introductory text for newly qualified graduates, and experienced engineers from other disciplines, entering the field of structural dynamics and vibration, in industry. It should also be found useful by test engineers and technicians working in this area, and by those studying the subject in universities, although it is not designed to meet the requirements of any particular course of study.

No previous knowledge of structural dynamics is assumed, but the reader should be familiar with the elements of mechanical or structural engineering, and a basic knowledge of mathematics is also required. This should include calculus, complex numbers and matrices. Topics such as the solution of linear second-order differential equations, and eigenvalues and eigenvectors, are explained in the text.

Each concept is explained in the simplest possible way, and the aim has been to give the reader a basic understanding of each topic, so that more specialized texts can be tackled with confidence.

The book is largely based on the author's experience in the aerospace industry, and this will inevitably show. However, most of the material presented is of completely general application, and it is hoped that the book will be found useful as an introduction to structural dynamics and vibration in all branches of engineering.

Although the principles behind current computer software are explained, actual programs are not provided, or discussed in any detail, since this area is more than adequately covered elsewhere. It is assumed that the reader has access to a software package such as MATLAB®.

A feature of the book is the relatively high proportion of space devoted to worked examples. These have been chosen to represent tasks that might be encountered in industry. It will be noticed that both SI and traditional 'British' units have been used in the examples. This is quite deliberate, and is intended to highlight the fact that in industry, at least, the changeover to the SI system is far from complete, and it is not unknown for young graduates, having used only the SI system, to have to learn the obsolete British system when starting out in industry. The author's view is that, far from ignoring systems other than the SI, which is sometimes advocated, engineers must understand, and be comfortable with, all systems of units. It is hoped that the discussion of the subject presented in Chapter 1 will be useful in this respect.

The book is organized as follows. After reviewing the basic concepts used in structural dynamics in Chapter 1, Chapters 2, 3 and 4 are all devoted to the response of the single degree of freedom system. Chapter 5 then looks at damping, including non-linear damping, in single degree of freedom systems. Multi-degree of freedom systems are introduced in Chapter 6, with a simple introduction to matrix methods, based on Lagrange's equations, and the important concepts of modal coordinates and the normal mode summation method. Having briefly introduced eigenvalues and eigenvectors in Chapter 6, some of the simpler procedures for their extraction are described in Chapter 7. Methods for dealing with larger structures, from the original Ritz method of 1909, to today's finite element method, are believed to be explained most clearly by considering them from a historical viewpoint, and this approach is used in Chapter 8. Chapter 9 then introduces the classical Fourier series, and its digital development, the Discrete Fourier Transform (DFT), still the mainstay of practical digital vibration analysis. Chapter 10 is a simple introduction to random vibration, and vibration isolation and absorption are discussed in Chapter 11. In Chapter 12, some of the more commonly encountered self-excited phenomena are introduced, including vibration induced by friction, a brief introduction to the important subject of aircraft flutter, and the phenomenon of shimmy in aircraft landing gear. Finally, Chapter 13 gives an overview of vibration testing, introducing modal testing, environmental testing and vibration fatigue testing in real time.

Douglas Thorby

Acknowledgements

The author would like to acknowledge the assistance of his former colleague, Mike Child, in checking the draft of this book, and pointing out numerous errors.

Thanks are also due to the staff at Elsevier for their help and encouragement, and good humor at all times.

1 Basic Concepts

Contents

This introductory chapter discusses some of the basic concepts in the fascinating subject of structural dynamics.

1.1 Statics, dynamics and structural dynamics

Statics deals with the effect of forces on bodies at rest. *Dynamics* deals with the motion of nominally rigid bodies. The two aspects of dynamics are *kinematics* and *kinetics*. Kinematics is concerned only with the motion of bodies with geometric constraints, irrespective of the forces acting. So, for example, a body connected by a link so that it can only rotate about a fixed point is constrained by its kinematics to move in a circular path, irrespective of any forces that may be acting. On the other hand, in kinetics, the path of a particle may vary as a result of the applied forces. The term *structural dynamics* implies that, in addition to having motion, the bodies are non-rigid, i.e. 'elastic'. 'Structural dynamics' is slightly wider in meaning than 'vibration', which implies only oscillatory behavior.

1.2 Coordinates, displacement, velocity and acceleration

The word *coordinate* acquires a slightly different, additional meaning in structural dynamics. We are used to using coordinates, x, y and z, say, when describing the *location* of a point in a structure. These are *Cartesian* coordinates (named after René Descartes), sometimes also known as 'rectangular' coordinates. However, the same word 'coordinate' can be used to mean the *movement* of a point on a structure from some standard configuration. As an example, the positions of the grid points chosen for the analysis of a structure could be specified as x, y and z coordinates from some fixed point. However, the displacements of those points, when the structure is loaded in some way, are often also referred to as coordinates.

1

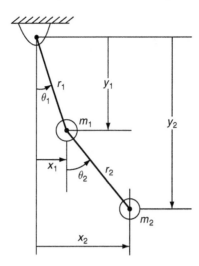

Fig. 1.1   Double pendulum illustrating generalized coordinates.

Cartesian coordinates of this kind are not always suitable for defining the vibration behavior of a system. The powerful Lagrange method requires coordinates known as *generalized coordinates* that not only fully describe the possible motion of the system, but are also independent of each other. An often-used example illustrating the difference between Cartesian and generalized coordinates is the double pendulum shown in Fig. 1.1. The angles θ_1 and θ_2 are sufficient to define the positions of m_1 and m_2 completely, and are therefore suitable as generalized coordinates. All four Cartesian coordinates x_1, y_1, x_2 and y_2, taken together, are not suitable for use as generalized coordinates, since they are not independent of each other, but are related by the two *constraint equations*:

$$x_1^2 + y_1^2 = r_1^2 \quad \text{and} \quad (x_2 - x_1)^2 + (y_2 - y_1)^2 = r_2^2$$

This illustrates the general rule that the number of degrees of freedom, and the number of generalized coordinates required, is the total number of coordinates minus the number of constraint equations. In this case there can only be two generalized coordinates, but they do not necessarily have to be θ_1 and θ_2; for example, x_1 and x_2 also define the positions of the masses completely, and could be used instead.

Generalized coordinates are fundamentally displacements, but can also be differentiated, i.e. expressed in terms of velocity and acceleration. This means that if a certain displacement coordinate, z, is defined as positive upwards, then its velocity, $\dot{z}$, and its acceleration, $\ddot{z}$, are also positive in that direction. The use of dots above symbols, as here, to indicate differentiation with respect to time is a common convention in structural dynamics.

1.3 Simple harmonic motion

Simple harmonic motion, more usually called 'sinusoidal vibration', is often encountered in structural dynamics work.

1.3.1 Time History Representation

Let the motion of a given point be described by the equation:

$$x = X \sin \omega t \tag{1.1}$$

where x is the displacement from the equilibrium position, X the displacement magnitude of the oscillation, ω the frequency in rad/s and t the time. The quantity X is the *single-peak* amplitude, and x travels between the limits $\pm X$, so the *peak-to-peak* amplitude (also known as *double* amplitude) is $2X$.

It appears to be an accepted convention to express displacements as double amplitudes, but velocities and accelerations as single-peak amplitudes, so some care is needed, especially when interpreting vibration test specifications.

Since $\sin \omega t$ repeats every 2π radians, the period of the oscillation, T, say, is $2\pi/\omega$ seconds, and the frequency in hertz (Hz) is $1/T = \omega/2\pi$. The velocity, dx/dt, or $\dot{x}$, of the point concerned, is obtained by differentiating Eq. (1.1):

$$\dot{x} = \omega X \cos \omega t \tag{1.2}$$

The corresponding acceleration, d^2x/dt^2, or $\ddot{x}$, is obtained by differentiating Eq. (1.2):

$$\ddot{x} = -\omega^2 X \sin \omega t \tag{1.3}$$

Figure 1.2 shows the displacement, x, the velocity, $\dot{x}$, and the acceleration $\ddot{x}$, plotted against time, t.

Since Eq. (1.2),

$$\dot{x} = \omega X \cos \omega t$$

can be written as

$$\dot{x} = \omega X \sin \left(\omega t + \frac{\pi}{2} \right)$$

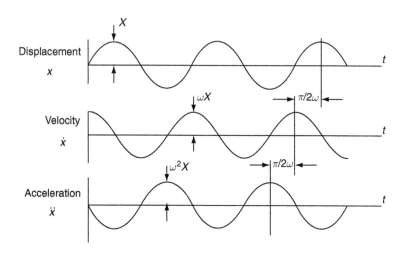

Fig. 1.2   Displacement, velocity and acceleration time histories for simple harmonic motion.

or

$$\dot{x} = \omega X \sin\left[\omega\left(t + \frac{\pi}{2\omega}\right)\right] \tag{1.4}$$

any given feature of the time history of $\dot{x}$, for example the maximum value, occurs at a value of t which is $\pi/2\omega$ less (i.e. earlier) than the same feature in the wave representing x. The velocity is therefore said to 'lead' the displacement by this amount of time. This lead can also be expressed as a quarter-period, $T/4$, a phase angle of $\pi/2$ radians, or 90°.

Similarly, the acceleration time history, Eq. (1.3),

$$\ddot{x} = -\omega^2 X \sin \omega t$$

can be written as

$$\ddot{x} = -\omega^2 X \sin(\omega t + \pi) \tag{1.5}$$

so the acceleration 'leads' the displacement by a time π/ω, a half-period, $T/2$, or a phase angle of π radians or 180°. In Fig. 1.2 this shifts the velocity and acceleration plots to the *left* by these amounts relative to the displacement: the lead being in time, not distance along the time axis.

The 'single-peak' and 'peak-to-peak' values of a sinusoidal vibration were introduced above. Another common way of expressing the amplitude of a vibration level is the root mean square, or RMS value. This is derived, in the case of the displacement, x, as follows:

Squaring both sides of Eq. (1.1):

$$x^2 = X^2 \sin^2 \omega t \tag{1.6}$$

The mean square value of the whole waveform is the same as that of the first half-cycle of $X \sin \omega t$, so the mean value of x^2, written $\langle x^2 \rangle$, is

$$\langle x^2 \rangle = X^2 \frac{2}{T} \int_0^{T/2} \sin^2 \omega t \cdot dt \tag{1.7}$$

Substituting

$$t = \frac{1}{\omega}(\omega t), \quad dt = \frac{1}{\omega}d(\omega t), \quad T = \frac{2\pi}{\omega},$$

$$\langle x^2 \rangle = X^2 \frac{\omega}{\pi} \int_0^{\pi/\omega} \sin^2 \omega t \cdot d(\omega t) = \frac{X^2}{2} \tag{1.8}$$

Therefore the RMS value of x is $X/\sqrt{2}$, or about $0.707X$. It can be seen that this ratio holds for any sinusoidal waveform: the RMS value is always $1/\sqrt{2}$ times the single-peak value.

The waveforms considered here are assumed to have zero mean value, and it should be remembered that a steady component, if present, contributes to the RMS value.

Example 1.1

The sinusoidal vibration displacement amplitude at a particular point on an engine has a single-peak value of 1.00 mm at a frequency of 20 Hz. Express this in terms of single-peak velocity in m/s, and single-peak acceleration in both m/s² and g units. Also quote RMS values for displacement, velocity and acceleration.

Solution

Remembering Eq. (1.1),

$$x = X \sin \omega t \tag{A}$$

we simply differentiate twice, so,

$$\dot{x} = \omega X \cos \omega t \tag{B}$$

and

$$\ddot{x} = -\omega^2 X \sin \omega t \tag{C}$$

The single-peak displacement, X, is, in this case, 1.00 mm or 0.001 m. The value of $\omega = 2\pi f$, where f is the frequency in Hz. Thus, $\omega = 2\pi(20) = 40\pi$ rad/s.

From Eq. (B), the single-peak value of $\dot{x}$ is ωX, or $(40\pi \times 0.001) = 0.126$ m/s or 126 mm/s.

From Eq. (C), the single-peak value of $\ddot{x}$ is $\omega^2 X$ or $[(40\pi)^2 \times 0.001] = 15.8$ m/s² or $(15.8/9.81) = 1.61$ g.

Root mean square values are $1/\sqrt{2}$ or 0.707 times single-peak values in all cases, as shown in the Table 1.1.

Table 1.1
Peak and RMS Values, Example 1.1

	Single peak value	RMS Value
Displacement	1.00 mm	0.707 mm
Velocity	126 m/s	89.1 mm/s
Acceleration	15.8 m/s²	11.2 m/s²
	1.61g	1.14g rms

1.3.2 Complex Exponential Representation

Expressing simple harmonic motion in complex exponential form considerably simplifies many operations, particularly the solution of differential equations. It is based on Euler's equation, which is usually written as:

$$e^{i\theta} = \cos\theta + i\sin\theta \tag{1.9}$$

where e is the well-known constant, θ an angle in radians and i is $\sqrt{-1}$.

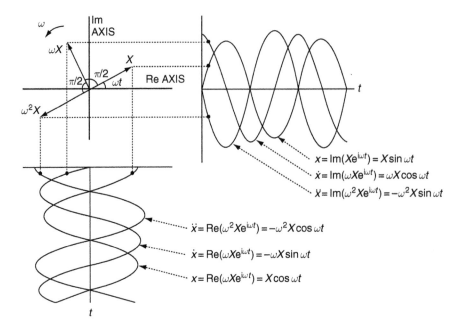

Fig. 1.3   Rotating vectors on an Argand diagram.

Multiplying Eq. (1.9) through by X and substituting ωt for θ:

$$Xe^{i\omega t} = X\cos\omega t + iX\sin\omega t \tag{1.10}$$

When plotted on an *Argand diagram* (where real values are plotted horizontally, and imaginary values vertically) as shown in Fig. 1.3, this can be regarded as a vector, of length X, rotating counter-clockwise at a rate of ω rad/s. The projection on the real, or x axis, is $X\cos\omega t$ and the projection on the imaginary axis, iy, is $iX\sin\omega t$. This gives an alternate way of writing $X\cos\omega t$ and $X\sin\omega t$, since

$$X\sin\omega t = \text{Im}\left(Xe^{i\omega t}\right) \tag{1.11}$$

where Im ( ) is understood to mean 'the imaginary part of ( )', and

$$X\cos\omega t = \text{Re}\left(Xe^{i\omega t}\right) \tag{1.12}$$

where Re ( ) is understood to mean 'the real part of ( )'.

Figure 1.3 also shows the velocity vector, of magnitude ωX, and the acceleration vector, of magnitude $\omega^2 X$, and their horizontal and vertical projections

Equations (1.11) and (1.12) can be used to produce the same results as Eqs (1.1) through (1.3), as follows:

If

$$x = \text{Im}\left(Xe^{i\omega t}\right) = \text{Im}(X\cos\omega t + iX\sin\omega t) = X\sin\omega t \tag{1.13}$$

then

$$\dot{x} = \text{Im}\left(i\omega Xe^{i\omega t}\right) = \text{Im}[i\,\omega(X\cos\omega t + iX\sin\omega t)] = \omega X\cos\omega t \tag{1.14}$$

(since $i^2 = -1$) and

$$\ddot{x} = \text{Im}\left(-\omega^2 X e^{i\omega t}\right) = \text{Im}\left[-\omega^2 (X\cos\omega t + iX\sin\omega t)\right] = -\omega^2 X\sin\omega t \qquad (1.15)$$

If the displacement x had instead been defined as $x = X\cos\omega t$, then Eq. (1.12), i.e. $X\cos\omega t = \text{Re}\left(X e^{i\omega t}\right)$, could have been used equally well.

The interpretation of Eq. (1.10) as a rotating complex vector is simply a mathematical device, and does not necessarily have physical significance. In reality, nothing is rotating, and the functions of time used in dynamics work are real, not complex.

1.4 Mass, stiffness and damping

The accelerations, velocities and displacements in a system produce forces when multiplied, respectively, by mass, damping and stiffness. These can be considered to be the building blocks of mechanical systems, in much the same way that inductance, capacitance and resistance (L, C and R) are the building blocks of electronic circuits.

1.4.1 Mass and Inertia

The relationship between mass, m, and acceleration, $\ddot{x}$, is given by Newton's second law. This states that when a force acts on a mass, the rate of change of momentum (the product of mass and velocity) is equal to the applied force:

$$\frac{\mathrm{d}}{\mathrm{d}t}\left(m\frac{\mathrm{d}x}{\mathrm{d}t}\right) = F \qquad (1.16)$$

where m is the mass, not necessarily constant, $\mathrm{d}x/\mathrm{d}t$ the velocity and F the force. For constant mass, this is usually expressed in the more familiar form:

$$F = m\ddot{x} \qquad (1.17)$$

If we draw a *free body diagram*, such as Fig. 1.4, to represent Eq. (1.17), where F and x (and therefore $\dot{x}$ and $\ddot{x}$) are defined as positive to the right, the resulting inertia force, $m\ddot{x}$, acts to the left. Therefore, if we decided to define all quantities as positive to the right, it would appear as $-m\ddot{x}$.

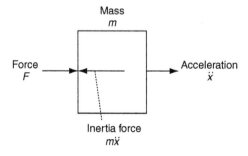

Fig. 1.4   D'Alembert's principle.

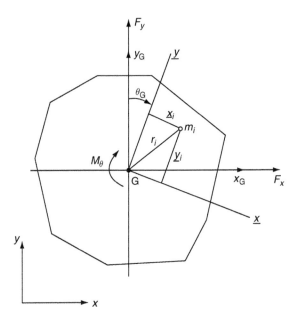

Fig. 1.5   Plane motion of a rigid body.

This is known as *D'Alembert's principle*, much used in setting up equations of motion. It is, of course, only a statement of the fact that the two forces, F and $m\ddot{x}$, being in equilibrium, must act in opposite directions.

Newton's second law deals, strictly, only with particles of mass. These can be 'lumped' into rigid bodies. Figure 1.5 shows such a rigid body, made up of a large number, n, of mass particles, m_i, of which only one is shown. For simplicity, the body is considered free to move only in the plane of the paper. Two sets of coordinates are used: the position in space of the *mass center* or 'center of gravity' of the body, G, is determined by the three coordinates x_G, y_G and θ_G. The other coordinate system, $\underline{x}$, $\underline{y}$, is fixed in the body, moves with it and has its origin at G. This is used to specify the locations of the n particles of mass that together make up the body. Incidentally, if these axes did not move with the body, the moments of inertia would not be constant, a considerable complication.

The mass center, G, is, of course, the point where the algebraic sum of the first moments of inertia of all the n mass particles is zero, about both the $\underline{x}$ and the $\underline{y}$ axes, i.e.,

$$\sum_{i=1}^{n} m_i \underline{x}_i = \sum_{i=1}^{n} m_i \underline{y}_i = 0, \qquad (1.18)$$

where the n individual mass particles, m_i are located at $\underline{x}_i$, $\underline{y}_i$ ($i = 1$ to n) in the $\underline{x}$, $\underline{y}$ coordinate system.

External forces and moments are considered to be applied, and their resultants through and about G are F_x, F_y and M_θ. These must be balanced by the internal inertia forces of the mass particles.

Thus in the x direction, since F_x acts at the mass center,

$$F_x + \sum_{i=1}^{n} -m_i\ddot{x}_i = 0; \tag{1.19a}$$

or

$$F_x = m\ddot{x}_G; \tag{1.19b}$$

where $m = \Sigma_{i=1}^{n} m_i =$ total mass of body.

Note that the negative sign in Eq. (1.19a) is due to D'Alembert's principle. Similarly in the y direction,

$$F_y + \sum_{i=1}^{n} -m_i\ddot{y}_i = 0 \tag{1.20a}$$

or

$$F_y = m\ddot{y}_G \tag{1.20b}$$

For rotation about G, the internal tangential force due to one mass particle, m_i is $-m_i r_i \ddot{\theta}_G$, the negative sign again being due to D'Alembert's principle, and the moment produced about G is $-m_i r_i^2 \ddot{\theta}_G$. The total moment due to all the mass particles in the body is thus $\sum_{i=1}^{n} (-m_i r_i^2 \ddot{\theta}_G)$, all other forces canceling because G is the mass center. This must balance the externally applied moment, M_θ, so

$$M_\theta + \sum_{i=1}^{n} (-m_i r_i^2 \ddot{\theta}_G) = 0 \tag{1.21a}$$

or

$$M_\theta = I_G \ddot{\theta}_G \tag{1.21b}$$

where

$$I_G = \sum_{i=1}^{n} m_i r_i^2 \tag{1.22}$$

Equation (1.22) defines the *mass moment of inertia* of the body about the mass center. In matrix form Eqs (1.19b), (1.20b) and (1.21b) can be combined to produce

$$\begin{bmatrix} F_x \\ F_y \\ M_\theta \end{bmatrix} = \begin{bmatrix} m & 0 & 0 \\ 0 & m & 0 \\ 0 & 0 & I_G \end{bmatrix} \begin{bmatrix} \ddot{x}_G \\ \ddot{y}_G \\ \ddot{\theta}_G \end{bmatrix} \tag{1.23}$$

The (3×3) matrix is an example of an *inertia matrix*, in this case diagonal, due to the choice of the mass center as the reference center. Thus the many masses making up the system of Fig. 1.5 have been 'lumped', and can now be treated as a single mass (with two freedoms) and a single rotational moment of inertia, all located at the mass center of the body, G.

In general, of course, for a three-dimensional body, the inertia matrix will be of size (6×6), and there will be six coordinates, three translations and three rotations.

1.4.2 Stiffness

Stiffnesses can be determined by any of the standard methods of static structural analysis. Consider the rod shown in Fig. 1.6(a), fixed at one end. Force F is applied axially at the free end, and the extension x is measured. As shown in Fig. 1.6(b), if the force F is gradually increased from zero to a positive value, it is found, for most materials, that *Hooke's law* applies; the extension, x, is proportional to the force, up to a point known as the *elastic limit*. This is also true for negative (compressive) loading, assuming that the rod is prevented from buckling. The slope $\delta F/\delta x$ of the straight line between these extremes, where δF and δx represent small changes in F and x, respectively, is the stiffness, k. It should also be noted in Fig. 1.6(b) that the energy stored, the *potential energy*, at any value of x, is the area of the shaded triangle, $\frac{1}{2}Fx$, or $\frac{1}{2}kx^2$, since $F = kx$.

Calculating the stiffness of the rod, for the same longitudinal loading, is a straight-forward application of elastic theory:

$$k = \frac{\delta F}{\delta x} = \frac{\delta \sigma}{\delta \varepsilon} \cdot \frac{a}{L} = \frac{Ea}{L} \tag{1.24}$$

where a is the cross-sectional area of the rod, L its original length and $\delta \sigma/\delta \varepsilon$ the slope of the plot of stress, σ, versus strain, ε, known as *Young's modulus, E*.

The stiffness of beam elements in bending can similarly be found from ordinary elastic theory. As an example, the vertical displacement, y, at the end of the uniform built-in cantilever shown in Fig. 1.6(c), when a force F is applied is given by the well-known formula:

$$y = \frac{FL^3}{3EI} \tag{1.25}$$

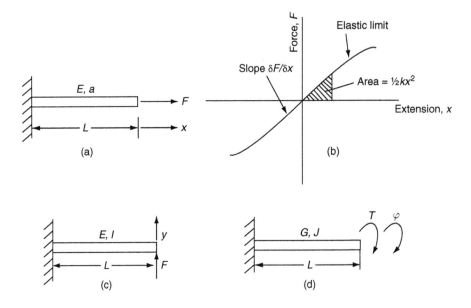

Fig. 1.6

where L is the length of the beam, E Young's modulus and I the 'moment of inertia' or second moment of area of the cross-section applicable to vertical bending.

From Eq. (1.25), the stiffness $\delta F/\delta y$, or k_y, is

$$k_y = \frac{\delta F}{\delta y} = \frac{3EI}{L^3} \tag{1.26}$$

For a rod or beam in torsion, fixed at one end, with torque, T, applied at the other end, as shown in Fig. 1.6(d), the following result applies.

$$\frac{T}{J} = \frac{G\varphi}{L} \tag{1.27}$$

where T is the applied torque, J the polar area 'moment of inertia' of the cross-section, G the shear modulus of the material, φ the angle of twist at the free end and L the length.

From Eq. (1.27), the torsional stiffness at the free end is

$$k_\varphi = \frac{\delta T}{\delta \varphi} = \frac{GJ}{L} \tag{1.28}$$

Example 1.3

Figure 1.7, where the dimensions are in mm., shows part of the spring suspension unit of a small road vehicle. It consists of a circular-section torsion bar, and rigid lever. The effective length of the torsion bar, L_1, is 900 mm, and that of the lever, L_2, is 300 mm. The bar is made of steel, having elastic shear modulus $G = 90 \times 10^9$ N/m² ($= 90$ GN/m² or 90 GPa), and its diameter, d, is 20 mm. Find the vertical stiffness, for small displacements, as measured at point B.

Solution

From Eq. (1.28), the torsional stiffness, k_φ, of the bar between point A and the fixed end, is

$$k_\varphi = \frac{\delta T}{\delta \varphi} = \frac{GJ}{L_1} \tag{A}$$

where δT is a small change in torque applied to the bar at A and $\delta \varphi$ the corresponding twist angle. G is the shear modulus of the material, J the polar moment of inertia of the bar cross-section and L_1 the length of the torsion bar. However, we require the linear stiffness k_x or $\delta F/\delta x$, as seen at point B. Using Eq. (A), and the relationships, valid for small angles, that $\delta T = \delta F \cdot L_2$, and $\delta \varphi = \delta x/L_2$, it is easily shown that

$$k_x = \frac{\delta F}{\delta x} = \frac{\delta T}{\delta \varphi} \cdot \frac{1}{L_2^2} = \frac{GJ}{L_1 L_2^2} = k_\varphi \cdot \frac{1}{L_2^2} \tag{B}$$

Numerically, $G = 90 \times 10^9$ N/m²; $J = \pi d^4/32 = \pi(0.02)^4/32 = 15.7 \times 10^{-9}$ m⁴; $L_1 = 0.90$ m; $L_2 = 0.30$ m giving $k_x = 17\,400$ N/m or 17.4 kN/m.

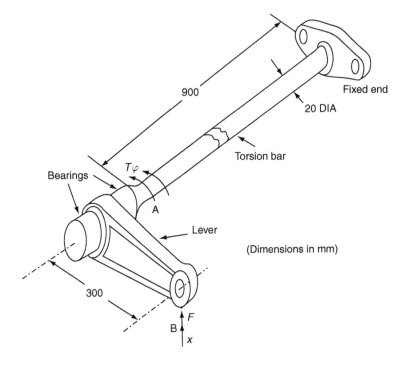

900

Fixed end

20 DIA

Torsion bar

$T\varphi$

Bearings

A

Lever

(Dimensions in mm)

300

B

F

x

Fig. 1.7   Torsion bar discussed in Example 1.3.

1.4.3 Stiffness and Flexibility Matrices

A stiffness value need not be defined at a single point: the force and displacement can be at the same location, or different locations, producing a *direct stiffness* or a *cross stiffness,* respectively. An array of such terms is known as a *stiffness matrix* or a matrix of *stiffness influence coefficients.* This gives the column vector of forces or moments f_1, f_2, etc., required to be applied at all stations to balance a unit displacement at one station:

$$\begin{Bmatrix} f_1 \\ f_2 \\ \vdots \end{Bmatrix} = \begin{bmatrix} k_{11} & k_{12} & \cdots \\ k_{21} & k_{22} & \cdots \\ \vdots & \vdots & \vdots \end{bmatrix} \begin{Bmatrix} x_1 \\ x_2 \\ \vdots \end{Bmatrix} \tag{1.29}$$

or

$$\{f\} = [k]\{x\} \tag{1.30}$$

The stiffness matrix, $[k]$, is square, and symmetric (i.e. $k_{ij} = k_{ji}$ throughout).

The mathematical inverse of the stiffness matrix is the *flexibility matrix* which gives the displacements x_1, x_2, etc., produced by unit forces or moments f_1, f_2, etc.

$$\begin{Bmatrix} x_1 \\ x_2 \\ \vdots \end{Bmatrix} = \begin{bmatrix} \alpha_{11} & \alpha_{12} & \cdots \\ \alpha_{21} & \alpha_{22} & \cdots \\ \vdots & \vdots & \vdots \end{bmatrix} \begin{Bmatrix} f_1 \\ f_2 \\ \vdots \end{Bmatrix} \tag{1.31}$$

or

$$\{x\} = [\alpha]\{f\} \tag{1.32}$$

The flexibility matrix, $[\alpha]$, is also symmetric, and its individual terms are known as *flexibility influence coefficients.*
Since $[\alpha]$ is the inverse of $[k]$, i.e. $[\alpha] = [k]^{-1}$,

$$[\alpha][k] = [k]^{-1}[k] = [I] \tag{1.33}$$

where $[I]$ is the unit matrix, a square matrix with all diagonal terms equal to unity, and all other terms zero.

Since the stiffness matrix (or the flexibility matrix) relates forces (or moments) applied anywhere on a linear structure to the displacements produced anywhere, it contains all there is to know about the stiffness properties of the structure, provided there are sufficient coordinates.

Example 1.4

(a) Derive the stiffness matrix for the chain of springs shown in Fig. 1.8.
(b) Derive the corresponding flexibility matrix.
(c) Show that one is the inverse of the other.

Solution

(a) The (3×3) stiffness matrix can be found by setting each of the coordinates x_1, x_2 and x_3 to 1 in turn, with the others at zero, and writing down the forces required at each node to maintain equilibrium:

when : $x_1 = 1$ $x_2 = 0$ $x_3 = 0$
then : $f_1 = k_1 x_1 + k_2 x_2$ $f_2 = -k_2 x_1$ $f_3 = 0$ (A$_1$)

when : $x_1 = 0$ $x_2 = 1$ $x_3 = 0$
then : $f_1 = -k_2 x_2$ $f_2 = k_2 x_2 + k_3 x_2$ $f_3 = -k_3 x_2$ (A$_2$)

when : $x_1 = 0$ $x_2 = 0$ $x_3 = 1$
then : $f_1 = 0$ $f_2 = -k_3 x_3$ $f_3 = k_3 x_3$ (A$_3$)

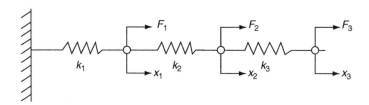

Fig. 1.8   Chain of spring elements discussed in Example 1.4.

Equations (A₁) now give the first column of the stiffness matrix, Eqs (A₂) the second column, and so on, as follows:

$$\begin{Bmatrix} f_1 \\ f_2 \\ f_3 \end{Bmatrix} = \begin{bmatrix} (k_1 + k_2) & -k_2 & 0 \\ -k_2 & (k_2 + k_3) & -k_3 \\ 0 & -k_3 & k_3 \end{bmatrix} \begin{Bmatrix} x_1 \\ x_2 \\ x_3 \end{Bmatrix} \tag{B}$$

The required stiffness matrix is given by Eq. (B).

(b) The flexibility matrix can be found by setting f_1, f_2 and f_3 to 1, in turn, with the others zero, and writing down the displacements x_1, x_2 and x_3:

$$\begin{aligned} f_1 = 1 \qquad & f_2 = 0 \qquad & f_3 = 0 \\ x_1 = 1/k_1 \qquad & x_2 = 1/k_1 \qquad & x_3 = 1 k_1 \end{aligned} \tag{C$_1$}$$

$$\begin{aligned} f_1 = 0 \qquad & f_2 = 1 \qquad & f_3 = 0 \\ x_1 = 1/k_1 \qquad & x_2 = 1/k_1 + 1/k_2 \qquad & x_3 = 1/k_1 + 1/k_2 \end{aligned} \tag{C$_2$}$$

$$\begin{aligned} f_1 = 0 \qquad & f_2 = 0 \qquad & f_3 = 1 \\ x_1 = 1/k_1 \qquad & x_2 = 1/k_1 + 1/k_2 \qquad & x_3 = 1/k_1 + 1/k_2 + 1/k_3 \end{aligned} \tag{C$_3$}$$

Equations (C₁) give the first column of the flexibility matrix, Eqs (C₂) the second column, and so on:

$$\begin{Bmatrix} x_1 \\ x_2 \\ x_3 \end{Bmatrix} = \begin{bmatrix} 1/k_1 & 1/k_1 & 1/k_1 \\ 1/k_1 & (1/k_1 + 1/k_2) & (1/k_1 + 1/k_2) \\ 1/k_1 & (1/k_1 + 1/k_2) & (1/k_1 + 1/k_2 + 1/k_3) \end{bmatrix} \begin{Bmatrix} f_1 \\ f_2 \\ f_3 \end{Bmatrix} \tag{D}$$

Equation (D) gives the required flexibility matrix.

(c) To prove that the flexibility matrix is the inverse of the stiffness matrix, and vice versa, we multiply them together, which should produce a unit matrix, as is, indeed, the case.

$$\begin{bmatrix} (k_1 + k_2) & -k_2 & 0 \\ -k_2 & (k_2 + k_3) & -k_3 \\ 0 & -k_3 & k_3 \end{bmatrix} \times \begin{bmatrix} 1/k_1 & 1/k_1 & 1/k_1 \\ 1/k_1 & (1/k_1 + 1/k_2) & (1/k_1 + 1/k_2) \\ 1/k_1 & (1/k_1 + 1/k_2) & (1/k_1 + 1/k_2 + 1/k_3) \end{bmatrix}$$

$$= \begin{bmatrix} 1 & 0 & 0 \\ 0 & 1 & 0 \\ 0 & 0 & 1 \end{bmatrix} \tag{E}$$

1.4.4 Damping

Of the three 'building blocks' that make up the systems we deal with in structural dynamics, mass and stiffness are *conservative* in that they can only store, or conserve, energy. Systems containing only mass and stiffness are therefore known as *conservative systems*. The third quantity, damping, is different, in that it dissipates energy, which is lost from the system.

Some of the ways in which damping can occur are as follows.

(a) The damping may be inherent in a structure or material. Unfortunately, the term 'structural damping' has acquired a special meaning: it now appears to mean 'hysteretic damping', and cannot be used to mean the damping in a structure, whatever its form, as the name would imply. Damping in conventional jointed metal structures is partly due to hysteresis within the metal itself, but much more to friction at bolted or riveted joints, and pumping of the fluid, often just air, in the joints. Viscoelastic materials, such as elastomers (rubber-like materials), can be formulated to have relatively high damping, as well as stiffness, making them suitable for the manufacture of vibration isolators, engine mounts, etc.

(b) The damping may be deliberately added to a mechanism or structure to suppress unwanted oscillations. Examples are discrete units, usually using fluids, such as vehicle suspension dampers and viscoelastic damping layers on panels.

(c) The damping can be created by the fluid around a structure, for example air or water. If there is no relative flow between the structure and the fluid, only *radiation damping* is possible, and the energy loss is due to the generation of sound. There are applications where this can be important, but for normal structures vibrating in air, radiation damping can usually be ignored. On the other hand, if relative fluid flow is involved, for example an aircraft wing traveling through the air, quite large aerodynamic damping (and stiffness) forces may be developed.

(d) Damping can be generated by magnetic fields. The damping effect of a conductor moving in a magnetic field is often used in measuring instruments. Moving coils, as used in loudspeakers and, of particular interest, in vibration testing, in electro-magnetic exciters, can develop surprisingly large damping forces.

(e) Figure 1.9 shows a discrete damper of the type often fitted to vehicle suspensions. Such a device typically produces a damping force, F, in response to closure velocity, $\dot{x}$, by forcing fluid through an orifice. This is inherently a square-law rather than a linear effect, but can be made approximately linear by the use of a special valve, which opens progressively with increasing flow. The damper is then known as an *automotive damper*, and the one shown in Fig. 1.9 will be assumed to be of this type. Then the force and velocity are related by:

$$F = c\dot{x} \qquad\qquad (1.34)$$

where F is the external applied force and $\dot{x}$ the velocity at the same point. The quantity c is a constant having the dimensions force/unit velocity. Equation (1.34) will apply for both positive and negative values of F and $\dot{x}$, assuming that the device is double-acting.

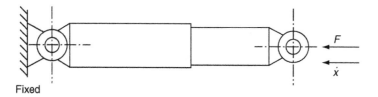

Fixed

Fig. 1.9   Common form of damper used in vehicle suspensions.

Example 1.5

An automotive damper similar to that shown in Fig. 1.9 is stated by the supplier to produce a linear damping force, in both directions, defined by the equation $F = c\dot{x}$, where F is the applied force in newtons, $\dot{x}$ is the stroking velocity, in m/s, and the constant c is 1500 N/m/s. A test on the unit involves applying a single-peak sinusoidal force of ± 1000 N at each of the frequencies, $f = 1.0$, 2.0 and 5.0 Hz. Calculate the expected single-peak displacement, and total movement, at each of these frequencies.

Solution

The applied force is $F = c\dot{x}$, where $F = 1000 \sin(2\pi ft)$. The expected velocity, $\dot{x}$, is therefore:

$$\dot{x} = \frac{F}{c} = \frac{1000 \sin (2\pi ft)}{1500} \tag{A}$$

The displacement x is given by integrating the velocity with respect to time:

$$x = -\frac{1000 \cos (2\pi ft)}{1500(2\pi f)} + x_0 \tag{B}$$

where the initial displacement x_0 is adjusted to mid-stroke. The single-peak displacement, $|x|$, measured from the mid-stroke position, is therefore:

$$|x| = \frac{1000}{1500 \times 2\pi f} = \frac{0.106}{f} \text{ m} \tag{C}$$

The expected single-peak displacement and total movement at each test frequency are given in Table 1.2.

Table 1.2

Frequency f (Hz)	Single-peak displacement $|x|$ (m)	Total movement of damper piston (m)
1	0.106	0.212
2	0.053	0.106
5	0.021	0.042

1.5 Energy methods in structural dynamics

It is possible to solve some problems in structural dynamics using only Newton's second law and D'Alembert's principle, but as the complexity of the systems analysed increases, methods based on the concept of energy, or work, become necessary. The terms 'energy' and 'work' refer to the same physical quantity, measured in the same units, but they are used in slightly different ways: work put into a conservative system, for example, becomes the same amount of energy when stored in the system.

Three methods based on work, or energy, are described here. They are (1) *Rayleigh's energy method*; (2) *the principle of virtual work* (or *virtual displacements*), and (3) *Lagrange's equations*. All are based on the principle of the conservation of energy. The following simple definitions should be considered first.

(a) Work is done when a force causes a displacement. If both are defined at the same point, and in the same direction, the work done is the product of the force and displacement, measured, for example, in newton-meters (or lbf. -ft). This assumes that the force remains constant. If it varies, the *power*, the instantaneous product of force and velocity, must be integrated with respect to time, to calculate the work done. If a moment acts on an angular displacement, the work done is still in the same units, since the angle is non-dimensional. It is therefore permissible to mix translational and rotational energy in the same expression.

(b) The *kinetic energy, T*, stored in an element of mass, m, is given by $T = \frac{1}{2}m\dot{x}^2$, where $\dot{x}$ is the velocity. By using the idea of a mass moment of inertia, I, the kinetic energy in a rotating body is given by $T = \frac{1}{2}I\dot{\theta}^2$, where $\dot{\theta}$ is the angular velocity of the body.

(c) The *potential energy, U*, stored in a spring, of stiffness k, is given by $U = \frac{1}{2}kx^2$, where x is the compression (or extension) of the spring, not necessarily the displacement at one end. In the case of a rotational spring, the potential energy is given by $U = \frac{1}{2}k_\theta\theta^2$, where k_θ is the angular stiffness, and θ is the angular displacement

1.5.1 Rayleigh's Energy Method

Rayleigh's method (not to be confused with a later development, the *Rayleigh–Ritz method*) is now mainly of historical interest. It is applicable only to single-DOF systems, and permits the natural frequency to be found if the kinetic and potential energies in the system can be calculated. The motion at every point in the system (i.e. the mode shape in the case of continuous systems) must be known, or assumed. Since, in vibrating systems, the maximum kinetic energy in the mass elements is transferred into the same amount of potential energy in the spring elements, these can be equated, giving the natural frequency. It should be noted that the maximum kinetic energy does not occur at the same time as the maximum potential energy.

Example 1.6

Use Rayleigh's energy method to find the natural frequency, ω_1, of the fundamental bending mode of the uniform cantilever beam shown in Fig. 1.10, assuming that the vibration mode shape is given by:

$$\frac{y}{y_T} = \left(\frac{x}{L}\right)^2 \tag{A}$$

where y_T is the single-peak amplitude at the tip; y the vertical displacement of beam at distance x from the root; L the length of beam; m the mass per unit length; E the Young's modulus and I the second moment of area of beam cross-section.

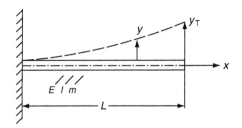

Fig. 1.10   Uniform cantilever beam used in Example 1.6.

Solution

From Eq. (A) it is assumed that

$$y = y_T \frac{x^2}{L^2} \sin \omega t \qquad (B)$$

Then,

$$\dot{y} = \omega_1 y_T \frac{x^2}{L^2} \cos \omega t \qquad (C)$$

The maximum kinetic energy, T_{max}, occurs when $\cos \omega t = 1$, i.e. when $\dot{y} = \omega_1 y_T x^2/L^2$, and is given by:

$$T_{max} = \int_0^L \tfrac{1}{2} m \dot{y}^2 = \tfrac{1}{2} \frac{m}{L^4} \omega_1^2 y_T^2 \int_0^L x^4 \cdot \mathrm{d}x = \tfrac{1}{10} m L \omega_1^2 y_T^2 \qquad (D)$$

To find the maximum potential energy, we need an expression for the maximum curvature as a function of x.

From Eq. (A),

$$y = y_T \frac{x^2}{L^2} \sin \omega t$$

so

$$\frac{\mathrm{d}y}{\mathrm{d}x} = y_T \frac{2x}{L^2} \sin \omega t$$

and

$$\frac{\mathrm{d}^2 y}{\mathrm{d}x^2} = \frac{2y_T}{L^2} \sin \omega t$$

therefore

$$\left(\frac{\mathrm{d}^2 y}{\mathrm{d}x^2} \right)_{max} = \frac{2y_T}{L^2} \qquad (E)$$

This is independent of x in this particular case, the curvature being constant along the beam. The standard expression for the potential energy in a beam due to bending is

$$U = \tfrac{1}{2} \int_0^L EI \left(\frac{\mathrm{d}^2 y}{\mathrm{d}x^2} \right)^2 \cdot \mathrm{d}x \qquad (F)$$

The maximum potential energy U_{max} is given by substituting Eq. (E) into Eq. (F), giving

$$U_{max} = \frac{2EI \cdot y_T^2}{L^3} \tag{G}$$

Now the basis of the Rayleigh method is that $T_{max} = U_{max}$. Thus, from Eqs (D) and (G) we have

$$\frac{1}{10}mL\omega_1^2 y_T^2 = \frac{2EI \cdot y_T^2}{L^3},$$

which simplifies to

$$\omega_1 = \frac{\sqrt{20}}{L^2}\sqrt{\frac{EI}{m}} = \frac{4.47}{L^2}\sqrt{\frac{EI}{m}} \tag{H}$$

The exact answer, from Chapter 8, is $\frac{3.52}{L^2}\sqrt{\frac{EI}{m}}$, so the Rayleigh method is somewhat inaccurate in this case. This was due to a poor choice of function for the assumed mode shape.

1.5.2 The Principle of Virtual Work

This states that in any system in equilibrium, the total work done by all the forces acting at one instant in time, when a small virtual displacement is applied to one of its freedoms, is equal to zero. The system being 'in equilibrium' does not necessarily mean that it is static, or that all forces are zero; it simply means that all forces are accounted for, and are in balance. Although the same result can sometimes be obtained by diligent application of Newton's second law, and D'Alembert's principle, the virtual work method is a useful time-saver, and less prone to errors, in the case of more complicated systems. The method is illustrated by the following examples.

Example 1.7

The two gear wheels shown in Fig. 1.11 have mass moments of inertia I_1 and I_2. A clockwise moment, M, is applied to the left gear about pivot A. Find the equivalent mass moment of the whole system as seen at pivot A.

Solution

Taking all quantities as positive when clockwise, let the applied moment, M, produce positive angular acceleration, $+\ddot{\theta}$, of the left gear. Then the counter-clockwise acceleration of the right gear must be $-(R/r)\ddot{\theta}$. By D'Alembert's principle, the corresponding inertia moments are

Left gear: $-I_1\ddot{\theta}$ Right gear: $+I_2\dfrac{R}{r}\ddot{\theta}$

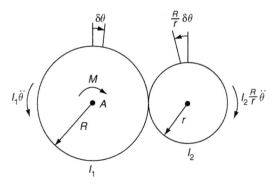

Fig. 1.11

A virtual angular displacement $+\delta\theta$ is now applied to the left gear. This will also produce a virtual angular displacement $-(R/r\delta\theta)$ of the right gear. Now multiplying all moments acting by their corresponding virtual displacements, summing and equating to zero:

$$M \cdot \delta\theta - I_1 \ddot{\theta} \cdot \delta\theta - \frac{R}{r}\delta\theta \cdot I_2 \frac{R}{r}\ddot{\theta} = 0 \qquad \text{(A)}$$

or

$$M = \left[I_1 + \left(\frac{R}{r}\right)^2 I_2 \right] \ddot{\theta} \qquad \text{(B)}$$

This is the equation of motion of the system, referred to point A. The effective mass moment at this point is $I_1 + (R/r)^2 I_2$.

In Example 1.7, the mass moment of inertia, I_2, was scaled by $(R/r)^2$, i.e. by the square of the velocity ratio. This result can easily be generalized for any linear devices, such as masses, springs and dampers, connected via a mechanical velocity ratio. It does not work for non-linear devices, however, as shown in the following example.

Example 1.8

Figure 1.12 shows part of an aircraft landing gear system, consisting of a lever and a double-acting square-law damper strut. The latter produces a force $\overline{F}$, given by $\overline{F} = C\dot{z}^2 \text{sgn}(\dot{z})$, where $\dot{z}$ is the closure velocity of the strut, and C is a constant. The expression 'sgn$(\dot{z})$', meaning 'sign of $(\dot{z})$', is simply a way to allow the force $\overline{F}$ to change sign so that it always opposes the direction of the velocity. Thus, upward velocity, $\dot{z}$, at point A, produces a downward force $\overline{F}$ on the lever, and vice versa.

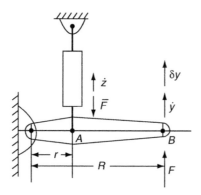

Fig. 1.12   Part of a landing gear system discussed in Example 1.8.

If *F* is an external force applied at point *B,* and $\dot{y}$ is the velocity at that point, find an expression for the equivalent damper, as seen at point *B*, in terms of *F* and $\dot{y}$.

Solution

Defining the forces and displacements at *A* and *B* as positive when upwards, let a virtual displacement $+\delta y$ be applied at *B*. Then the virtual displacement at *A* is $+(r/R)\delta y$. The forces acting on the lever are $+F$ at *B*, and $-C((r/R)\dot{y})^2$ at *A*.

Multiplying the forces acting by the virtual displacements at corresponding points, and summing to zero, we have

$$\delta y.F + \frac{r}{R}\delta y\left[-C\left(\frac{r}{R}\dot{y}\right)^2\right] = 0 \tag{A}$$

or

$$F = \left(\frac{r}{R}\right)^3 C\dot{y}^2. \tag{B}$$

So the effective square-law damping coefficient at point *B* is the actual value *C*, multiplied by the *cube* of the velocity ratio $(r/R) = (\dot{z}/\dot{y})$, a useful result when dealing with square-law dampers.

1.5.3 Lagrange's Equations

Lagrange's equations were published in 1788, and remain the most useful and widely used energy-based method to this day, especially when expressed in matrix form. Their derivation is given in many standard texts [1.1, 1.2], and will not be repeated here.

The equations can appear in a number of different forms. The basic form, for a system without damping, is

$$\frac{\mathrm{d}}{\mathrm{d}t}\left(\frac{\partial T}{\partial \dot{q}_i}\right) - \frac{\partial T}{\partial q_i} + \frac{\partial U}{\partial q_i} = Q_i \tag{1.35}$$

$$i = 1, 2, 3, \ldots, n.$$

where

T is the total kinetic energy in the system;

U is the total potential energy in the system;

q_i are generalized displacements, as discussed in Section 1.2. These must meet certain requirements, essentially that their number must be equal to the number of degrees of freedom in the system; that they must be capable of describing the possible motion of the system; and they must not be linearly dependent.

Q_i are generalized external forces, corresponding to the generalized displacements q_i. They can be defined as those forces which, when multiplied by the generalized displacements, correctly represent the work done by the actual external forces on the actual displacements.

For most structures, unless they are rotating, the kinetic energy, T, depends upon the generalized velocities, $\dot{q}_i$, but not upon the generalized displacements q_i, and the term $\partial T/\partial q_i$ can often be omitted.

Usually, the damping terms are added to a structural model after it has been transformed into normal coordinates, and do not need to appear in Lagrange's equations. However, if the viscous damping terms can be defined in terms of the generalized coordinates, a dissipation function D can be introduced into Lagrange's equations which, when partially differentiated with respect to the $\dot{q}_i$ terms, produces appropriate terms in the final equations of motion. Thus, just as we have $T = \frac{1}{2}m\dot{x}^2$ for a single mass, and $U = \frac{1}{2}kx^2$ for a single spring, we can invent the function $D = \frac{1}{2}c\dot{x}^2$ for a single damper c, where c is defined by Eq. (1.34), i.e. $F = c\dot{x}$. Lagrange's equations, with this addition, become

$$\frac{\mathrm{d}}{\mathrm{d}t}\left(\frac{\partial T}{\partial \dot{q}_i}\right) - \frac{\partial T}{\partial q_i} + \frac{\partial U}{\partial q_i} + \frac{\partial D}{\partial \dot{q}_i} = Q_i, \quad i = 1, 2, 3, \ldots, n. \tag{1.36}$$

Lagrange's equations, without damping terms, may sometimes be seen written in the form:

$$\frac{\mathrm{d}}{\mathrm{d}t}\left(\frac{\partial L}{\partial \dot{q}_i}\right) - \frac{\partial L}{\partial q_i} = Q_i \tag{1.37}$$

where the *kinetic potential*, L is defined as:

$$L = T - U \tag{1.38}$$

By substituting Eq. (1.38) into Eq. (1.37), the latter can be seen to be exactly the same as Eq. (1.35), since $\partial U/\partial \dot{q}_i = 0$.

Many examples showing the practical application of Lagrange's equations appear in Chapters 2, 6 and 8.

1.6 Linear and non-linear systems

The *linearity* of a mechanical system depends upon the linearity of its components, which are mass, stiffness and damping. Linearity implies that the response of each of these components (i.e. acceleration, displacement, and velocity, respectively) bears a straight line or *linear* relationship to an applied force. The straight line relationship must extend over the whole range of movement of the component, negative as well as positive.

Nearly all the mathematical operations used in structural dynamics, such as super-position, Fourier analysis, inversion, eigensolutions, etc., rely on linearity, and although analytic solutions exist for a few non-linear equations, the only solution process always available for any seriously non-linear system is step-by-step solution in the time domain.

Fortunately, with the possible exception of damping, *significant* non-linearity in vibrating structures, as they are encountered in industry today, is actually quite rare. This may seem a strange statement to make, in view of the large number of research papers produced annually on the subject, but this is probably due to the fact that it is an interesting and relatively undeveloped field, with plenty of scope for original research, rather than to any great need from industry. In practical engineering work, it is quite likely that if very non-linear structural behavior is found, it may well be indicative of poor design. It may then be better to look for, and eliminate, the causes of the non-linearity, such as backlash and friction, rather than to spend time modeling the non-linear problem.

1.7 Systems of units

In dynamics work, the four main *dimensions* used are mass, force, length and time, and each of these is expressed in *units*. The unit of time has always been taken as the second (although frequencies in the aircraft industry were still being quoted in cycles/minute in the 1950s), but the unit of length can be feet, meters, centimeters and so on. The greatest cause of confusion, however, is the fact that the same unit (usually the pound or the kilogram) has traditionally been used for both mass and force.

Any system of units that satisfies Newton's second law, Eq. (1.17), will work in practice, with the further proviso that the units of displacement, velocity and acceleration must all be based on the same units for length and time.

The pound (or the kilogram) cannot be taken as the unit of force *and* mass in the same system, since we know that a one pound mass acted upon by a one pound force produces an acceleration of g, the *standard* acceleration due to gravity assumed in defining the units, not of unity, as required by Eq. (1.17). This problem used to be solved in engineering work (but not in physics) by writing W/g for the mass, m, where W is the *weight* of the mass under *standard* gravity. The same units, whether pounds or kilograms, could then, in effect, be used for both force and mass. Later practice was to choose a different unit for either mass or force, and incorporate the factor g into one of the units, as will be seen below. The very latest system, the SI, is, in theory, independent of earth gravity. The kilogram (as a *mass* unit), the meter, and the second are fixed, more or less arbitrarily, and the force unit, the newton, is then what follows from Newton's second law.

1.7.1 Absolute and Gravitational Systems

The fact that the pound, kilogram, etc. can be taken as either a mass or a force has given rise to two groups of systems of units. Those in which mass is taken as the fundamental quantity, and force is inferred, are known as *absolute* systems. Some examples are the following:

(a) the CGS system, an early metric system, used in physics, but now obsolete. Here the gram was the unit of mass, and acceleration was measured in centimeters/s^2. The resulting force unit was called the 'dyne'.

(b) an obsolete system based on British units, also used in physics, where the pound was taken as the unit of mass, acceleration was in ft/s^2, and the resulting force unit was called the poundal.

(c) the current SI system (from *Système International d'Unités*), in which the mass unit is the kilogram, acceleration is in m/s^2, and the force unit is the newton.

The other group, *gravitational* systems, take force as the fundamental quantity, and infer mass. The acceleration due to gravity, even at the earth's surface, varies slightly, so a standard value has to be assumed. This is nominally the value at sea level at latitude 45° north, and is taken as 9.806 65 m/s^2, or about 32.1740 ft/s^2. In practice, the rounded values of 9.81 m/s^2 or 32.2 ft/s^2 will do for most purposes. Some gravitational systems are the following:

(a) A system using the pound force (lbf) as the fundamental quantity, with acceleration in ft/s^2. The unit of mass is given by Eq. (1.17) as $m = F/\ddot{x}$. Since a unit of F is one lbf, and a unit of $\ddot{x}$ is one (ft/s^2), then a unit of m is one (lbf/ft/s^2), more correctly expressed as one (lbf ft^{-1}s^2). This is known as the 'slug'. Its weight, W, the force that would be exerted on it by standard gravity, is equal to the value of F given by Eq. (1.17), with $m = 1$ and $\ddot{x} = g$, the acceleration due to gravity, about 32.2 ft/s^2. Therefore, $W = F = mg = (1 \times 32.2) = 32.2$ lbf.

(b) A system, as above, except that the inch is used instead of the foot. The mass unit is one (lbf in.$^{-1}$ s^2) sometimes known as the 'mug'. This has a weight of about 386 lbf, as can be confirmed by applying the method used in (a) above.

(c) A system, formerly used in some European countries, taking the kilogram as a unit of *force* (kgf), and measuring acceleration in m/s^2. Using the same arguments as above, the unit of mass, the kilopond (kp), is seen to be one (kgf m^{-1} s^2), and has a weight of 9.81 kgf. In this system the kilogram-force (kgf) is very close to one decanewton (daN) in the SI system, a fact sometimes used (confusingly) when presenting data.

Table 1.3 summarizes the six systems of units mentioned above. Some are now completely obsolete, but are included here not only because they can be found in old books, and are of historical interest, but also because they help to illustrate the principles involved.

Although the SI system will eventually replace all others, some engineering work still appears to be being carried out in the two British gravitational systems, and many engineers will have to use both SI and British units for some time to come. Also, of course, archived material, in obsolete units, may be kept for many years. The best way to deal with this situation is to be familiar with, and able to work in, all systems. The examples in this book therefore use both SI units and the British units still in use.

Table 1.3
Some Current and Obsolete Systems of Units

	Absolute systems			Gravitational systems		
	SI	CGS	British (1)	British (2)	British (3)	European
Where used (STATUS)	International (Current)	International (Obsolete)	English speaking countries (Obsolete)	English speaking countries (Becoming obsolete)	English speaking countries (Becoming obsolete)	European mainland (Obsolete)
Mass unit	kilogram (kg)	gram (g)	pound mass (lbm)	slug ($\mathrm{lb\,ft^{-1}\,s^2}$)	'mug' ($\mathrm{lb\,in.^{-1}\,s^2}$)	kilopond (kp) ($\mathrm{kg\,m^{-1}\,s^2}$)
Force unit	newton (N)	dyne (dyn)	poundal (pdl)	pound force (lbf)	pound force (lbf)	kilogram force (kgf)
Length unit	meter (m)	centimeter (cm)	foot (ft)	foot (ft)	inch (in.)	meter (m)
Time unit	second (s)	second (s)	second (s)	second (s)	second (s)	second (s)
Acceleration due to gravity (approx)	$9.807\ \mathrm{m/s^2}$	$980.7\ \mathrm{cm/s^2}$	$32.17\ \mathrm{ft/s^2}$	$32.17\ \mathrm{ft/s^2}$	$386.1\ \mathrm{in./s^2}$	$9.807\ \mathrm{m/s^2}$

To summarize, the rules, when working in an unfamiliar system, are simply the following:

(a) The mass and acceleration units must satisfy Newton's second law, Eq. (1.17); that is, one unit of force acting on a unit mass must produce one unit of acceleration.
(b) Acceleration, velocity and displacement must all use the same basic unit for length. This may seem obvious, but care is needed when the common practice of expressing acceleration in g units is used.

1.7.2 Conversion between Systems

When converting between SI units and British units, it is helpful to remember that one inch $= 25.4$ mm *exactly*. Therefore, if the number of inches in a meter is required, it can be remembered as $(1000/25.4)$ inches exactly. Similarly, a foot expressed in meters is $(12 \times 25.4)/1000$ m. Unfortunately, there is no easy way with force and mass, and we just have to know that one pound force (lbf) $= 4.448\ 22$ N, and one pound mass $= 0.453\ 592$ kg.

Example 1.6

(a) The actual force output of a vibration exciter is being measured by using it to shake a rigid mass of 2 kg, to which an accurately calibrated accelerometer is attached. The exciter specification states that when supplied with a certain sinusoidal current, it should produce a force output of ± 100 N. When this current is actually supplied, the accelerometer reads $\pm 5.07g$. Calculate the actual force output of the exciter. The stiffness of the mass suspension, and gravity, may be ignored.
(b) Repeat the calculation using the British pound–inch–second system.

Solution

(a) It is clear from the data given that the SI system is being used. The required force output is given by Eq. (1.17):

$$F = m\ddot{x}$$

We note that in this equation, the force, F, must be in newtons, the mass, m, must be in kilograms, and the acceleration $\ddot{x}$, in m/s^2. The acceleration, given as $\pm 5.07\,g$, must therefore first be converted to m/s^2 by multiplying by 9.807, giving $\pm(5.07 \times 9.807) = \pm 49.72$ m/s^2. Then the force output of the exciter in newtons is $F = m\ddot{x} = \pm(2 \times 49.72) = \pm 99.4$ N, slightly less than the nominal value of ± 100 N.

(b) In the stated British gravitational system:

The test mass $= 2\,\text{kg} = (2/0.4536) = 4.409\,\text{lb. wt} = (4.410/386.1) = 0.01142\,\text{lb in.}^{-1}\text{s}$.
The measured acceleration $= \pm 5.07g = \pm(5.07 \times 386.1)$ in./s$^2 = \pm 1958$ in./s^2.
The required force output in lbf. is, as before, given by Eq. (1.17):

$$F = m\ddot{x} = \pm(0.01142 \times 1958) = \pm 22.36\,\text{lbf}$$

Converting back to SI units: ± 22.36 ibf $= \pm(22.36 \times 4.448)$ N $= \pm 99.4$ N, agreeing with the original calculation in SI units. The conversion factors used in this case were accurate to four figures in order to achieve three figure accuracy overall.

1.7.3 The SI System

The main features of the SI system have already been described, but the following points should also be noted.

Prefixes are used before units to avoid very large or very small numbers. The *preferred* prefixes likely to be used in structural dynamics work are given in Table 1.4.

Prefixes indicating multiples and sub-multiples other than in steps of 10^3 are frowned upon officially. The prefixes shown in Table 1.5 should therefore be used only when the preferred prefixes are inconvenient.

It should be noted that the centimeter, cm, is not a preferred SI unit.

Only one prefix should be used with any one unit; for example, 1000 kg should not be written as 1 kkg, but 1 Mg is acceptable. When a unit is raised to a power, the power applies to the whole unit, including the prefix; for example, mm^3 is taken as $(\text{mm})^3 = (10^{-3}\text{ m})^3 = 10^{-9}\text{ m}^3$, *not* 10^{-3} m^3.

The SI system, unlike some other systems, has units for pressure or stress. The pascal (Pa) is defined as one N m^{-2}. Young's modulus, E, may therefore be found specified in pascals, for example a steel sample may have $E = 200$ GPa (200 giga-pascals), equal to $(200 \times 10^9)\ \text{N m}^{-2}$. The pascal may also be encountered in defining sound pressure levels.

Table 1.4
Preferred SI Prefixes

Prefix	Symbol	Factor
giga	G	10^9
mega	M	10^6
kilo	k	10^3
milli	m	10^{-3}
micro	μ	10^{-6}

Table 1.5
Non-preferred SI Prefixes

Prefix	Symbol	Factor
hecto	h	10^2
deca	da	10
deci	d	10^{-1}
centi	c	10^{-2}

The bar (b), equal to 10^5 Pa, is sometimes used for fluid pressure. This is equivalent to 14.503 lbf/in^2, *roughly* one atmosphere.

Units generally use lower case letters, e.g. m for meter, *including* those named after people, when spelt out in full, e.g. newton, ampere, watt, but the latter use upper case letters when abbreviated, i.e. N for newtons, A for amperes, W for watts.

A major advantage of the SI system is that the units for all forms of energy and power are the same. Thus, one newton-meter (N m) of energy is equal to one joule (J), and one J/s is a power of one watt. All other systems require special factors to relate, say, mechanical energy to heat or electrical energy.

References

1.1 Scanlan, RH and Rosenbaum, R (1951). *Introduction to the Study of Aircraft Vibration and Flutter*. Macmillan. Reprinted 1968 Dover Publications.
1.2 Meriam, JL (1966). *Dynamics*. Wiley.

2 The Linear Single Degree of Freedom System: Classical Methods

Contents

Single degree of freedom (single-DOF) theory, as outlined in this and in the next two chapters, enables a surprisingly large proportion of day-to-day structural dynamics problems to be solved. This chapter first describes how the equations of motion of single-DOF systems can be set up, using a variety of methods. It will be seen that most systems can be reduced to one of the two basic configurations:
(1) Systems excited by an external force;
(2) Systems excited by the motion of the base or supports.

Having set up the differential equations of motion, their direct solution by 'classical' methods is then discussed. These methods are somewhat tedious in use, and more practical methods of solution are discussed in Chapters 3 and 4. However, they are fundamental to an understanding of vital concepts such as damped and undamped natural frequencies, critical damping, etc., and cannot be omitted from any study of structural dynamics.

2.1 Setting up the differential equation of motion

Several methods can be used to set up the differential equation of motion. The three methods introduced here are the following:
(a) By inspection of the forces involved, using Newton's second law, with D'Alembert's principle, and the fundamental properties of mass, stiffness and damping. This is only suitable for very simple systems, such as that shown in Fig. 2.1.
(b) For single-DOF systems with multiple mass, spring or damper elements connected together by levers or gears, the *principle of virtual work*, introduced in Chapter 1, provides a simple approach.
(c) Using *Lagrange's equations*, introduced in Chapter 1. This is the standard energy method, always used for dealing with multi-DOF systems, but it can also be used with advantage for some of the more complicated single-DOF systems.

2.1.1 Single Degree of Freedom System with Force Input

Figure 2.1 shows a single-DOF system, with viscous damping and external forcing, in schematic form. The reality may look quite different; it may have several springs, masses and dampers, or the motion may be angular.

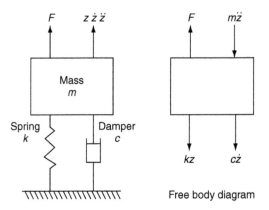

Fig. 2.1   Single-DOF system with applied force in schematic form.

The mass, m, is assumed to be excited directly by an external force, F. The system will be assumed to be in equilibrium when the applied force is zero; any compression or extension of the spring due to gravity, for example, is assumed to have already taken place, and can be ignored as far as the vibration response is concerned.

In the following, the applied force F, and the coordinates z, $\dot{z}$ and $\ddot{z}$ are functions of time, and should strictly be written $F(t)$, $z(t)$, $\dot{z}(t)$ and $\ddot{z}(t)$. However, the (t) will be omitted here for clarity. It should also be noted that if, as in this case, z is defined as positive upwards, then its derivatives $\dot{z}$ and $\ddot{z}$ will also be positive upwards.

The system shown in Fig. 2.1 is simple enough for the first method listed above to be used: simple reasoning, using Newton's second law and D'Alembert's principle. Considering the free body diagram for the mass, on the right, upward displacement, z, will extend the spring, and thus produce a downward force, kz, on the mass. Similarly, upward velocity $\dot{z}$, acting on the damper, will produce a downward force $c\dot{z}$ on the mass. By D'Alembert's principle, upward acceleration of the mass will produce a downward inertia force, $m\ddot{z}$, upon itself. The applied force, F, is positive upward by definition. So, considering the four forces *as they act on the mass*, and regarding them as positive when upwards and negative when downwards, we have, for equilibrium,

$$-m\ddot{z} - c\dot{z} - kz + F = 0$$

or

$$m\ddot{z} + c\dot{z} + kz = F \tag{2.1a}$$

and this is the equation of motion of the system shown in Fig. 2.1.

If the system had, instead, consisted of a moment of inertia, I, rotating about a fixed point, a rotary spring, k_θ, and a rotary damper, c_θ, and had an external applied moment, M, the equation of motion would have become

$$I\ddot{\theta} + c_\theta\dot{\theta} + k_\theta\theta = M \tag{2.1b}$$

This is mathematically equivalent to Eq. (2.1a), however, and exactly the same procedures can be used for its solution.

The following example shows how an apparently more complicated system can be reduced to the schematic system of Fig. 2.1, using either the principle of virtual work or Lagrange's equations.

Example 2.1

The system shown in Fig. 2.2(a) consists of a pivoted, rigid beam, considered massless, with two attached concentrated masses, m_1 and m_2. The spring, k_1, and the viscous damper, c_1, connect the beam to fixed points, as shown. An external force, $\overline{F}$, is applied vertically to m_2.

Derive the equation of motion, for small displacements, of this single-DOF system, using:

(a) the principle of virtual work;
(b) Lagrange's equations.

Solution

(a) *Method 1: using the principle of virtual work*
The system has only one degree of freedom, and its motion *must* be represented by only one coordinate. However, the choice of which coordinate to use is arbitrary.

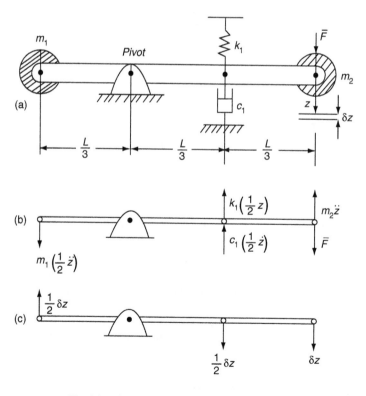

Fig. 2.2 Pivoted beam considered in Example 2.1.

Possible candidates are the angle of rotation about the pivot, or the vertical dis-placement of one of the masses. Let us choose to use the displacement, z, of the mass m_2, which is defined as positive in the downward direction. It follows that the velocity, $\dot{z}$, and the acceleration, $\ddot{z}$, will also be positive downwards at that point. The inertia, stiffness, damping and external forces acting are shown in Fig. 2.2(b). These forces are now assumed not to change when an infinitesimal additional downward virtual dis-placement, δz, is applied to m_2. Figure 2.2(c) shows the resulting virtual displacements at this and other points in the system. The virtual displacement at the left end of the beam, for example, is $\frac{1}{2}\delta z$, and upwards. At the point where the spring and damper are attached, the virtual displacement is $\frac{1}{2}\delta z$, and downwards.

The total work done by the forces in Fig. 2.2(b), acting on the corresponding virtual displacements in Fig. 2.2(c) (counting the work as positive when the force and virtual displacement are in the same direction, and negative when they are in opposite directions), is, by the principle of virtual work, equal to zero, so:

$$-m_1\left(\frac{1}{2}\ddot{z}\right)\left(\frac{1}{2}\delta z\right) - m_2\ddot{z}\cdot\delta z - k_1\left(\frac{1}{2}z\right)\left(\frac{1}{2}\delta z\right) - c_1\left(\frac{1}{2}\dot{z}\right)\left(\frac{1}{2}\delta z\right) + \overline{F}\cdot\delta z = 0 \quad \text{(A)}$$

Dividing Eq. (A) through by δz, and simplifying, gives the required equation of motion:

$$\left(\frac{1}{4}m_1 + m_2\right)\ddot{z} + \left(\frac{1}{4}c_1\right)\dot{z} + \left(\frac{1}{4}k_1\right)z = \overline{F} \quad \text{(B)}$$

It is seen that the above process has 'lumped' the two masses m_1 and m_2 into a single *effective mass* of $\left(\frac{1}{4}m_1 + m_2\right)$, considered to be located at the point where z and $\overline{F}$ are defined, at the right end of the beam. Similarly, the stiffness and damping have changed because they have been referred to the new location.

(b) *Method 2: using Lagrange's equations*

If a single-DOF system has viscous damping, and also the kinetic energy depends only upon the velocities (i.e. not the displacements) of the masses, as in this case, Lagrange's equation is given by Eq. (1.36), Chapter 1, noting that $\partial T/\partial q = 0$:

$$\frac{\mathrm{d}}{\mathrm{d}t}\left(\frac{\partial T}{\partial \dot{q}}\right) + \frac{\partial U}{\partial q} + \frac{\partial D}{\partial \dot{q}} = Q \quad \text{(C)}$$

where q is the single generalized coordinate, taken as equal to z in this case; T the total kinetic energy in the system; U the total potential energy in the system; D a dissipation function, included to represent the energy lost in damping and Q the generalized external force. This last quantity, Q, must be defined so that when it acts upon q, the correct expression for the work done by the external force is obtained, so that in this case, since $q = z$, then $Q = \overline{F}$, since clearly $\overline{F}z$ is the work done by the external force $\overline{F}$, and this must be equal to Qq.

The kinetic energy, T, the potential energy, U, and dissipation energy, D, of the system can now be written down. We know that the kinetic energy of a single mass particle, m, with velocity $\dot{z}$ is equal to $\frac{1}{2}m\dot{z}^2$; the potential energy stored in a spring of stiffness k, with displacement z, is $\frac{1}{2}kz^2$, and the dissipation energy lost in the damper c with closure velocity $\dot{z}$ is $\frac{1}{2}c\dot{z}^2$. Using these simple relationships, we have, by inspection of Fig. 2.2(a):

$$T = \frac{1}{2}m_1\left(\frac{\frac{1}{3}L}{\frac{2}{3}L}\dot{z}\right)^2 + \frac{1}{2}m_2\dot{z}^2 = \frac{1}{2}m_1\left(\frac{1}{2}\dot{z}\right)^2 + \frac{1}{2}m_2\dot{z}^2 \tag{D}$$

$$U = \frac{1}{2}k_1\left(\frac{\frac{1}{3}L}{\frac{2}{3}L}z\right)^2 = \frac{1}{2}k_1\left(\frac{1}{2}z\right)^2 \tag{E}$$

$$D = \frac{1}{2}c_1\left(\frac{\frac{1}{3}L}{\frac{2}{3}L}\dot{z}\right)^2 = \frac{1}{2}c_1\left(\frac{1}{2}\dot{z}\right)^2 \tag{F}$$

Applying Eq. (C), term by term:

$$\frac{\mathrm{d}}{\mathrm{d}t}\left(\frac{\partial T}{\partial \dot{q}}\right) = \frac{\mathrm{d}}{\mathrm{d}t}\left(\frac{\partial T}{\partial \dot{z}}\right) = \left(\frac{1}{4}m_1 + m_2\right)\ddot{z} \tag{G}$$

$$\frac{\partial U}{\partial q} = \frac{\partial U}{\partial z} = \frac{1}{4}k_1 z \tag{H}$$

$$\frac{\partial D}{\partial \dot{q}} = \frac{\partial D}{\partial \dot{z}} = \frac{1}{4}c_1 z \tag{I}$$

$$Q = \overline{F} \tag{J}$$

Substituting Eqs (G), (H), (I), and (J) into Eq. (C) gives the equation of motion:

$$\left(\frac{1}{4}m_1 + m_2\right)\ddot{z} + \left(\frac{1}{4}c_1\right)\dot{z} + \left(\frac{1}{4}k_1\right)z = \overline{F} \tag{K}$$

which is seen to agree with Eq. (B), derived by the principle of virtual work.

By comparing Eq. (B) or (K), in the example above, with Eq. (2.1a), it can be seen that the apparently more complicated system shown in Fig. 2.2(a) can be represented by the basic schematic single-DOF system of Fig. 2.1, if 'equivalent' values are substituted for m, c, k and F.

2.1.2 Single Degree of Freedom System with Base Motion Input

The system discussed in Section 2.1.1, and illustrated in Fig. 2.1, was excited by an external force. The same mass, spring and damper could also be excited by applying motion to the base, which was previously assumed fixed. This important case is illustrated in Fig. 2.3. A translational system is shown, but a rotational system is also possible. 'Lumping', as in the worked example above, may be required to reduce the system to this simplified form.

In Fig. 2.3, the absolute displacement of the base is x, the displacement of the mass, m, relative to the base is y; and z is the absolute displacement of the mass in space.

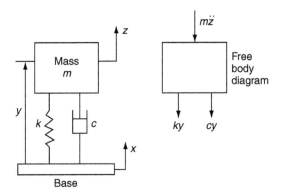

Fig. 2.3   Single-DOF system with base motion input in schematic form.

In the free body diagram, on the right, for positive x and y, there are three downward (negative) forces acting on the mass, so for equilibrium:

$$-m\ddot{z} - c\dot{y} - ky = 0$$

or

$$m\ddot{z} + c\dot{y} + ky = 0 \tag{2.2}$$

but since

$$y = z - x, \quad \dot{y} = \dot{z} - \dot{x}, \quad \text{and} \quad \ddot{y} = \ddot{z} - \ddot{x} \tag{2.3}$$

then

$$m\ddot{y} + c\dot{y} + ky = -m\ddot{x} \tag{2.4}$$

It should be noted that y is a relative, not an absolute, coordinate, and that the inertia force acting on m is $m\ddot{z}$, not $m\ddot{y}$.

Equation (2.4) is seen to be of the same form as Eq. (2.1a), but with the term $-m\ddot{x}$ replacing the force F.

If Eq. (2.4) is required in terms of the absolute coordinates, x and z, Eqs (2.3) can be used to write it in the form:

$$m\ddot{z} + c\dot{z} + kz = c\dot{x} + kx \tag{2.5}$$

2.2 Free response of single-DOF systems by direct solution of the equation of motion

If the system shown in Fig. 2.1, and represented by Eq. (2.1a), has no applied force, the equation of motion becomes

$$m\ddot{z} + c\dot{z} + kz = 0 \tag{2.6}$$

This is a linear, second-order, *homogeneous* (because all terms are of the same kind), differential equation, and it can be solved by trial. Since there is no input, it can only represent the free response of the system; however, it can do this with various *initial conditions*. We try the solution:

$$z = Ce^{\lambda t} \tag{2.7}$$

where C is a constant. Then,

$$\dot{z} = \lambda C e^{\lambda t} \tag{2.8}$$

and

$$\ddot{z} = \lambda^2 C e^{\lambda t} \tag{2.9}$$

Substituting Eqs (2.7) – (2.9) into Eq. (2.6):

$$m\lambda^2 C e^{\lambda t} + c\lambda C e^{\lambda t} + kC e^{\lambda t} = 0$$

and Eq. (2.7) is seen to be a solution. Dividing through by $Ce^{\lambda t}$ gives the *characteristic equation:*

$$m\lambda^2 + c\lambda + k = 0 \tag{2.10}$$

Solving this quadratic equation by the formula method:

$$\lambda = \frac{-c}{2m} \pm \sqrt{\left(\frac{c}{2m}\right)^2 - \frac{k}{m}} \tag{2.11}$$

which consists of two roots, λ_1 and λ_2, say, where

$$\lambda_1 = \frac{-c}{2m} + \sqrt{\left(\frac{c}{2m}\right)^2 - \frac{k}{m}} \tag{2.12}$$

and

$$\lambda_2 = \frac{-c}{2m} - \sqrt{\left(\frac{c}{2m}\right)^2 - \frac{k}{m}} \tag{2.13}$$

so $z = C_1 e^{\lambda_1 t}$ and $z = C_2 e^{\lambda_2 t}$ are also solutions, and their sum is the general solution:

$$z = C_1 e^{\lambda_1 t} + C_2 e^{\lambda_2 t} \tag{2.14}$$

Particular solutions now depend on the constants, C_1 and C_2, which are determined by the *initial conditions*, which are the values of x and $\dot{x}$ at $t = 0$.

If there is no damping, i.e. $c = 0$, then Eqs (2.12) and (2.13) become

$$\lambda_1 = i\sqrt{\frac{k}{m}} \quad \text{and} \quad \lambda_2 = -i\sqrt{\frac{k}{m}}$$

where i is the complex operator $\sqrt{-1}$.

Substituting in Eq. (2.14):

$$z = C_1 e^{i\omega_n t} + C_2 e^{-i\omega_n t} \tag{2.15}$$

where

$$\omega_n = \sqrt{\frac{k}{m}} \tag{2.16}$$

ω_n is the *undamped natural frequency*, which is a useful concept, even in damped systems.

Returning to the system *with* damping, using Eq. (2.16), Eq. (2.11) can be written as:

$$\lambda = \frac{-c}{2m} \pm \sqrt{\left(\frac{c}{2m}\right)^2 - \omega_n^2} \qquad (2.17)$$

The nature of the two roots λ_1 and λ_2 given by Eq. (2.17) depends upon whether the quantity under the square root is positive, exactly zero, or negative. When it is exactly zero, $\omega_n = c/2m$, and this particular value of c is called the *critical* value, c_c, given by:

$$c_c = 2m\omega_n \qquad (2.18)$$

The critical value of damping is so called because it marks the boundary between oscillatory and non-oscillatory behavior.

The non-dimensional viscous damping coefficient, γ, is defined by:

$$\gamma = \frac{c}{c_c} \qquad (2.19)$$

Also known as the 'fraction of critical damping', γ is the most generally used measure of damping. It is sometimes expressed in percentage terms, for example $\gamma = 0.02$ may be referred to as '2% viscous damping'.

It is now helpful to write Eqs (2.6) and (2.17) in terms of the undamped natural frequency, ω_n, and the non-dimensional damping coefficient γ. Using Eqs (2.16), (2.18) and (2.19), Eq. (2.6) is first written in the form:

$$\ddot{z} + 2\gamma\omega_n\dot{z} + \omega_n^2 z = 0 \qquad (2.20)$$

Then, using Eqs (2.18) and (2.19), Eq. (2.17) can be written as:

$$\lambda = -\gamma\omega_n \pm \omega_n\sqrt{(\gamma^2 - 1)} \qquad (2.21)$$

Since $i = \sqrt{-1}$, Eq. (2.21) can also be written in the form:

$$\lambda = -\gamma\omega_n \pm i\omega_n\sqrt{(1 - \gamma^2)} \qquad (2.22)$$

The form of the solution of Eq. (2.20), now depends upon whether γ is less than 1, more than 1 or equal to 1 as follows
(a) Critical damping, $\gamma = 1$:
     From Eq. (2.21):

$$\lambda = \lambda_1 = \lambda_2 = -\gamma\omega_n$$

So the roots are real and equal. This apparently gives only the solution $z = Ae^{-\gamma\omega_n t}$, but it can be shown that $z = Bte^{-\gamma\omega_n t}$ is another solution, so the general solution is

$$z = (A + Bt)e^{-\gamma\omega_n t} \qquad (2.23)$$

(b) More than critical damping, $\gamma > 1$:
     From Eq. (2.21), $\lambda = -\gamma\omega_n \pm \omega_n\sqrt{(\gamma^2 - 1)}$, the roots are real and different, and from Eq. (2.14):

$$z = C_1 e^{\lambda_1 t} + C_2 e^{\lambda_2 t}$$

or:

$$z = e^{-\gamma\omega_n t}\left[C_1 e^{-\omega_n\sqrt{(\gamma^2-1)}\,t} + C_2 e^{\omega_n\sqrt{(\gamma^2-1)}\,t}\right] \tag{2.24}$$

(c) Less than critical damping, $\gamma < 1$:

This is the case of most practical interest, since most engineering structures will have far less than critical damping. In this case it will be more convenient to take the roots from Eq. (2.22), $\lambda = -\gamma\omega_n \pm i\omega_n\sqrt{(1-\gamma^2)}$ which can be written as:

$$\lambda = -\gamma\omega_n \pm i\omega_d \tag{2.25}$$

where

$$\omega_d = \omega_n\sqrt{(1-\gamma^2)} \tag{2.26}$$

and ω_d is the *damped natural frequency*. This is the frequency at which the system vibrates when disturbed. It is lower than the undamped natural frequency, ω_n, but negligibly so for small γ.

The general solution, from Eq. (2.14), is $z = C_1 e^{\lambda_1 t} + C_2 e^{\lambda_2 t}$. Substituting Eq. (2.25) into Eq. (2.14):

$$z = e^{-\gamma\omega_n t}\left(C_1 e^{i\omega_d t} + C_2 e^{-i\omega_d t}\right) \tag{2.27}$$

Using the identities: $e^{i\omega_d t} = \cos\omega_d t + i\sin\omega_d t$ and $e^{-i\omega_d t} = \cos\omega_d t - i\sin\omega_d t$, Eq. (2.27) can be written as:

$$z = e^{-\gamma\omega_n t}(A\cos\omega_d t + B\sin\omega_d t) \tag{2.28}$$

where $A = (C_1 + C_2)$ and $B = i(C_1 - C_2)$.

Example 2.2

The system shown in Fig. 2.1, but without the applied force, F, has the following properties: $m = 1$ kg; $k = 10\,000$ N/m and $c = 40$ N/m/s. Plot the time history of z for the initial conditions: $z = 0.1$ m, $\dot{z} = 0$, at $t = 0$.

Solution

From Eq. (2.16),

$$\omega_n = \sqrt{\frac{k}{m}} = \sqrt{\frac{10000}{1}} = 100\,\text{rad/s}$$

From Eqs (2.18) and (2.19),

$$\gamma = \frac{c}{c_c} = \frac{c}{2m\omega_n} = 40/(2 \times 1 \times 100) = 0.2$$

From Eq. (2.26), $\omega_d = \omega_n\sqrt{(1-\gamma^2)} = 100\sqrt{(1-0.2^2)} = 98.0\,\text{rad/s}$.

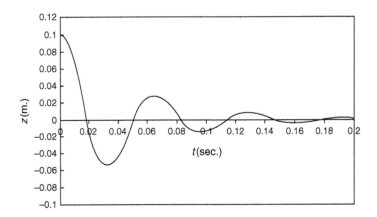

Fig. 2.4   Free response of a single-DOF system (Example 2.2).

Since $\gamma < 1$, the response is given by Eq. (2.28):

$$z = e^{-\gamma\omega_n t}(A\ \cos\omega_d t + B\ \sin\omega_d t) \tag{A}$$

We shall also require an expression for the velocity, $\dot{z}$. Differentiating Eq. (A):

$$\dot{z} = e^{-\gamma\omega_n t}[(B\omega_d - A\gamma\omega_n)\cos\omega_d t - (A\omega_d + B\gamma\omega_n)\sin\omega_d t] \tag{B}$$

The constants A and B can be found from Eqs (A) and (B) by substituting the numerical values of ω_n, ω_d and γ, as calculated above, and the initial conditions, $z = 0.1$, and $\dot{z} = 0$ at $t = 0$, giving $A = 0.1$, and $B = 0.0204$. Then from Eq. (A):

$$z = e^{-20t}(0.1\ \cos 98.0t + 0.0204\ \sin 98.0t) \tag{C}$$

This is easily plotted using a spreadsheet program, as in Fig. 2.4.

2.3 Forced response of the system by direct solution of the equation of motion

If, now, a force, F, is applied to the system, the equation to be solved is

$$m\ddot{z} + c\dot{z} + kz = F \tag{2.29}$$

and so far we have only considered the case where $F = 0$. We found that the solution, or *complementary function* $z = C_1 e^{\lambda_1 t} + C_2 e^{\lambda_2 t}$, Eq. (2.14), made the left side of the equation equal to zero, as required. However, we now wish to make it equal to F, and the complete solution will be, say,

$$z = C_1 e^{\lambda_1 t} + C_2 e^{\lambda_2 t} + X \tag{2.30}$$

where X is whatever is required to achieve this. The function X is known as the *particular integral*. Standard texts on the solution of linear, second-order differential equations list particular integrals that should be tried for various forms of F, and Table 2.1 is an example.

Table 2.1

Form of F	Form of the particular integral
$F = $ a constant	$z = C$
$F = At$	$z = Ct + D$
$F = At^2$	$z = Ct^2 + Dt + E$
$F = A\sin\omega t$ or $A\cos\omega t$	$z = C\cos\omega t + D\sin\omega t$

Example 2.3

(a) Derive an expression for the displacement response, z, when a step force of magnitude P is applied to the single-DOF system shown in Fig. 2.10. Assume that the damping is less than critical, and that the initial conditions are $z = \dot{z} = 0$ at $t = 0$.
(b) Plot the displacement, z, in non-dimensional form, as a multiple of the static displacement for the same load, $x_s = P/k$, with the non-dimensional viscous damping coefficient, γ, equal to 0.1, and the undamped natural frequency, ω_n, equal to 10 rad/s.

Solution

Part (a):
In this case the equation to be solved is

$$m\ddot{z} + c\dot{z} + kz = P \qquad (A)$$

Equation (A) can be written in a form similar to Eq. (2.20), sometimes known as the 'standard form':

$$\ddot{z} + 2\gamma\omega_n\dot{z} + \omega_n^2 z = P/m \qquad (B)$$

Since $\gamma < 1$, the complementary function $z = e^{-\gamma\omega_n t}(A\cos\omega_d t + B\sin\omega_d t)$, Eq. (2.28), is appropriate. For the particular integral, from Table 2.1, we try $z = C$, which makes $\dot{z} = 0$ and $\ddot{z} = 0$. Substituting these into Eq. (B) gives $C = P/m\omega_n^2$, so the complete solution is

$$z = e^{-\gamma\omega_n t}(A\cos\omega_d t + B\sin\omega_d t) + P/(m\omega_n^2) \qquad (C)$$

Differentiating

$$\dot{z} = e^{-\gamma\omega_n t}[(B\omega_d - A\gamma\omega_n)\cos\omega_d t - (A\omega_d + B\gamma\omega_n)\sin\omega_d t] \qquad (D)$$

Inserting the initial conditions $z = 0$ and $\dot{z} = 0$ at $t = 0$ into Eqs (C) and (D), and solving for the constants A and B gives, using Eqs (2.16) and (2.26),

$$A = -P/(m\omega_n^2) = -P/k \qquad (E)$$

and

$$B = -P\gamma\omega_n/(m\omega_n^2\omega_d) = -\frac{P\gamma}{k\sqrt{1-\gamma^2}} \qquad (F)$$

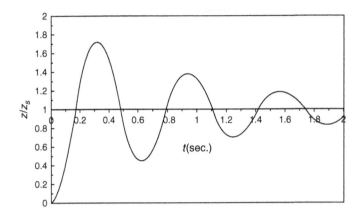

Fig. 2.5   Step response of a single-DOF system (Example 2.3).

Inserting Eqs (E) and (F) into Eq. (C), gives the solution:

$$z = \frac{P}{k}\left[1 - e^{-\gamma\omega_n t}\left(\cos\omega_d t + \frac{\gamma}{\sqrt{(1-\gamma^2)}}\sin\omega_d t\right)\right] \tag{G}$$

Part (b):
Now the displacement of the system when a *static* load, equal in magnitude to P, is applied, is z_s, say, where $z_s = P/k$. Dividing both sides of Eq. (G) by z_s, we have a non-dimensional version:

$$\frac{z}{z_s} = 1 - e^{-\gamma\omega_n t}\left(\cos\omega_d t + \frac{\gamma}{\sqrt{1-\gamma^2}}\sin\omega_d t\right) \tag{H}$$

This is plotted in Fig. 2.5, with $\gamma = 0.1$ and $\omega_n = 10$. It is seen that the displacement approaches twice the static value, before settling at the static value.

Example 2.4

Derive an expression for the displacement response of a single-DOF system excited by a force of the form $F = P\sin\omega t$. Assume that the initial conditions are $z = \dot{z} = 0$ at $t = 0$, and that the non-dimensional damping coefficient, γ, is less than 1.

Solution

In this case the equation to be solved is, from Eq. (2.29),

$$m\ddot{z} + c\dot{z} + kz = P\sin\omega t \tag{A}$$

or in standard form:

$$\ddot{z} + 2\gamma\omega_n\dot{z} + \omega_n^2 z = (P/m)\sin\omega t \tag{B}$$

and the initial conditions are $z = \dot{z} = 0$ at $t = 0$. Since the damping is assumed less than critical, i.e. $\gamma < 1$, the appropriate complementary function is

$$z = e^{-\gamma\omega_n t}(A\cos\omega_d t + B\sin\omega_d t) \tag{C}$$

The particular integral assumed will be as given by Table 2.1, i.e.,

$$z = C\cos\omega t + D\sin\omega t \tag{D}$$

then

$$\dot{z} = -\omega C\sin\omega t + \omega D\cos\omega t \tag{E}$$

and

$$\ddot{z} = -\omega^2 C\cos\omega t - \omega^2 D\sin\omega t \tag{F}$$

Substituting Eqs (D)–(F) into Eq. (B):

$$-\omega^2 C\cos\omega t - \omega^2 D\sin\omega t - 2\gamma\omega_n\omega C\sin\omega t + 2\gamma\omega_n\omega D\cos\omega t + \omega_n^2 C\cos\omega t$$
$$+ \omega_n^2 D\sin\omega t = (P/m)\sin\omega t \tag{G}$$

Sine and cosine terms in Eq. (G) can now be equated:
Sine terms:

$$-\omega^2 D - 2\gamma\omega_n\omega C + \omega_n^2 D = P/m \tag{H}$$

Cosine terms:

$$-\omega^2 C + 2\gamma\omega_n\omega D + \omega_n^2 C = 0 \tag{I}$$

Solving for C and D:

$$C = \frac{P(-2\gamma\omega_n\omega)}{m\left[\left(\omega_n^2 - \omega^2\right)^2 + (2\gamma\omega_n\omega)^2\right]} \tag{J}$$

$$D = \frac{P\left(\omega_n^2 - \omega^2\right)}{m\left[\left(\omega_n^2 - \omega^2\right)^2 + (2\gamma\omega_n\omega)^2\right]} \tag{K}$$

The complete solution is the sum of the complementary function and the particular integral:

$$z = e^{-\gamma\omega_n t}(A\cos\omega_d t + B\sin\omega_d t) + C\cos\omega t + D\sin\omega t \tag{L}$$

where A and B have still to be found. This requires another equation, which is obtained by differentiating Eq. (L):

$$\dot{z} = e^{-\gamma\omega_n t}[(B\omega_d - A\gamma\omega_n)\cos\omega_d t - (A\omega_d + B\gamma\omega_n)\sin\omega_d t] - C\omega\sin\omega t + D\omega\cos\omega t \tag{M}$$

A and *B* are now found from the initial conditions, in this case $z = \dot{z} = 0$ at $t = 0$, using Eqs (L) and (M) as a pair of simultaneous equations, as follows:

From Eq. (L):

$$A = -C = \frac{P(2\gamma\omega_n\omega)}{m\left[\left(\omega_n^2 - \omega^2\right)^2 + (2\gamma\omega_n\omega)^2\right]} \tag{N}$$

and from Eq. (M):

$$B = \frac{-(C\gamma\omega_n + D\omega)}{\omega_d} = \frac{P\omega\left[\left(\omega^2 - \omega_n^2\right) + \left(2\gamma^2\omega_n^2\right)\right]}{m\omega_d\left[\left(\omega_n^2 - \omega^2\right)^2 + (2\gamma\omega_n\omega)^2\right]} \tag{O}$$

where *C* and *D* are given by Eqs (J) and (K) above

Example 2.5

Plot the displacement response of the system shown in Fig. 2.1, for the two cases given in Table 2.2, commenting on the results.

Table 2.2

Case	Forcing function	Peak force P (N)	Mass m (kg)	Forcing frequency f (Hz)	ω (rad/s)	Undamped nat. frequency f_n (Hz)	ω_n (rad/s)	Viscous damping coeff. γ
(a)	$P\sin\omega t$	100	100	10	62.83	10	62.83	0.02
(b)	$P\sin\omega t$	100	100	9	56.55	10	62.83	0.01

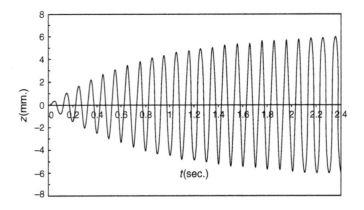

Fig. 2.6   Response of a single-DOF system to a sinusoidal force at the natural frequency (Example 2.5 (a))

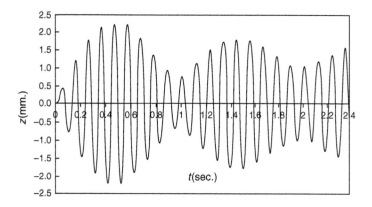

Fig. 2.7 Response of a single-DOF system to sinusoidal excitation close to the natural frequency, showing beating.

Case (a)

In this case the input force is $P \sin \omega t$ and the displacement response is given by Eq. (L), in Example 2.4. The values of the constants A, B, C and D are given by Eqs (J), (K), (N) and (O), of Example 2.4.

The response is shown in Fig. 2.6, which illustrates the surprisingly long time taken for the response to approach the steady-state amplitude when a system is excited at its resonant frequency.

Case (b)

The displacement response is again given by Eq. (L) in Example 2.4, and the values of A, B, C and D are again given by Eqs (J), (K), (N) and (O), in Example 2.4. Fig. 2.7 shows the response, z, for this case, with the forcing frequency, at 9 Hz, close to, but not equal to, the undamped natural frequency, at 10 Hz. This shows a phenomenon known as *beating*, between the forced and the transient responses.

3 The Linear Single Degree of Freedom System: Response in the Time Domain

Contents

Chapter 2 looked at how the differential equations of motion of linear single-DOF systems can be set up, and how the free and forced response can be found by direct solution of those equations, using classical mathematics. This method is not only tedious in use, as has been seen, but also requires the forcing function to be analytic, and the system to be linear. In this chapter, therefore, additional methods for finding the time or 'transient' response of a single-DOF system, more suitable for everyday use, are introduced. All methods fall into one of the three groups listed below.

(a) Other exact analytical methods.

   If the input function can be expressed analytically, *and* the system is linear, there are two main alternatives to the direct solution of the differential equation of motion, as already discussed. These are

   (i) The Laplace transform method;
   (ii) The convolution integral method.

   These, like the classical method, are exact analytical methods.

(b) 'Semi-analytical' methods.

   When the system is linear, but the input function, usually an applied force, is not known analytically, it can be split up into sections, not necessarily short sections, and an analytic, exact, solution applied to each section. These can be called 'semi-analytical' methods, because once the input function has been approximated, the subsequent operations give the exact analytical solution to that input.

(c) Step-by-step numerical methods using approximate derivatives.

   In these methods, the response is always calculated in very small increments. In 'direct' methods, the solution at the next step is found without iteration; 'indirect' methods use iteration between steps. No method of this type is exact, but they can give acceptable accuracy for most purposes, and they can be used for linear and non-linear systems. Four methods of this type are described in this chapter:

   (i) The Euler method;
   (ii) The modified Euler method;

(iii) The central difference method;

(iv) The fourth-order Runge-Kutta method.

3.1 Exact analytical methods

3.1.1 The Laplace Transform Method

The Laplace transform method can produce the time response, from the equation of motion, in exactly the same form as the classical method described in Chapter 2. It requires the input function to be known analytically, and as with the classical method, the effects of both the input and the initial conditions can be found. The following introduction has been simplified by assuming that all initial conditions are zero, as in most practical cases.

In the system shown in Fig. 2.1, suppose that we wish to find the time history of z when various force time histories, F, are applied. The equation of motion was seen to be

$$m\ddot{z} + c\dot{z} + kz = F, \tag{3.1}$$

where z and F should strictly be written $z(t)$ and $F(t)$, to show that they are functions of time, t.

Using Eq. (2.16):

$$\omega_n = \sqrt{\frac{k}{m}}$$

Equation (2.18):

$$c_c = 2m\omega_n$$

and Eq. (2.19):

$$\gamma = \frac{c}{c_c}$$

Equation (3.1) can be expressed in the standard form:

$$\ddot{z} + 2\gamma\omega_n\dot{z} + \omega_n^2 z = F/m \tag{3.2}$$

The transfer function

To express Eq. (3.2), for example, in Laplace notation, provided that the initial conditions (i.e. the values of z and $\dot{z}$ at $t = 0$) are zero, we simply replace d/dt by the Laplace operator, s (and d^2/dt^2 by s^2) so that it becomes

$$\left(s^2 + 2\gamma\omega_n s + \omega_n^2\right)\underline{z} = \underline{F}/m \tag{3.3}$$

where it will be noticed that z and F have been changed to $\underline{z}$ and $\underline{F}$, respectively, to indicate that they are now functions of s rather than of t, and should, strictly, be written as $\underline{z}(s)$ and $\underline{F}(s)$.

Writing Eq. (3.3) in the form

$$\left(\frac{\underline{z}}{\underline{F}}\right) = \frac{1}{m} \cdot \frac{1}{\left(s^2 + 2\gamma\omega_n s + \omega_n^2\right)} \tag{3.4}$$

the quantity $(\underline{z}/\underline{F})$ is an example of a *transfer function*, in this case relating $\underline{z}$ and $\underline{F}$.

Equation (3.4) gives the displacement due to an applied force. Other possibilities are that we may need to know the response in terms of velocity, or acceleration, rather than displacement, or the excitation could be by base motion rather than an external force. These alternate forms are easily found using Laplace notation; for example, if velocity, $\dot{z}$, rather than displacement, z, was required as the response, then, since differentiation corresponds to multiplication by s,

$$\left(\frac{\dot{z}}{F}\right) = \frac{s}{m} \cdot \frac{1}{\left(s^2 + 2\gamma\omega_n s + \omega_n^2\right)} \tag{3.5}$$

If the acceleration, $\ddot{z}$, was required as the response, we simply multiply by s again:

$$\left(\frac{\ddot{z}}{F}\right) = \frac{s^2}{m} \cdot \frac{1}{\left(s^2 + 2\gamma\omega_n s + \omega_n^2\right)} \tag{3.6}$$

If the excitation is by base motion, as in the system shown in Fig. 2.3, rather than by an applied force, its equation of motion is

$$m\ddot{y} + c\dot{y} + ky = -m\ddot{x} \tag{3.7}$$

where x, y, etc. are defined by Fig. 2.3. Equation (3.7) can also be written as:

$$\ddot{y} + 2\gamma\omega_n\dot{y} + \omega_n^2 y = -\ddot{x} \tag{3.8}$$

which, it will be noticed, is independent of m. This means that for a given base motion input, the response is independent of the scale of the system, depending only upon the natural frequency and damping coefficient.

Using the rule, as before, of replacing d/dt by s to find the transfer function, Eq. (3.8) can be written as:

$$\left(\frac{y}{x}\right) = \frac{-s^2}{\left(s^2 + 2\gamma\omega_n s + \omega_n^2\right)} \tag{3.9}$$

The transfer functions $\left(\dot{y}/\dot{x}\right)$ and $\left(\ddot{y}/\ddot{x}\right)$ are also given by Eq. (3.9), since the s or s^2 in both numerator and denominator can be canceled. It is also easily shown that

$$\left(\frac{y}{\dot{x}}\right) = \left(\frac{\dot{y}}{\ddot{x}}\right) = \frac{-s}{\left(s^2 + 2\gamma\omega_n s + \omega_n^2\right)} \tag{3.10}$$

and

$$\left(\frac{y}{\ddot{x}}\right) = \frac{-1}{\left(s^2 + 2\gamma\omega_n s + \omega_n^2\right)} \tag{3.11}$$

The Laplace transform

The Laplace transform, $f(s)$, of a function of time, $f(t)$, is defined as:

$$f(s) = \int_0^\infty e^{-st} f(t) dt \tag{3.12}$$

Equation (3.12) can be written as $f(s) = L[f(t)]$, where $L[\ ]$ indicates 'the Laplace transform of $[\ ]$'.

The Laplace operator, s, is complex and can be written as $s = \sigma + i\omega$. In the special case of $\sigma = 0$, the Laplace transform reduces to the Fourier transform, and if the latter exists, it can therefore be formed from the Laplace transform by substituting $i\omega$ for s. However, the inclusion of the real part, σ, enables the integral to converge for a wider range of functions than the Fourier transform, making the Laplace transform more generally useful for representing transients.

The Laplace transforms of common functions of time are tabulated in standard texts, and a short table is given in Appendix A. It will be noticed that the Laplace transform of a unit impulse is 1, so the Laplace transform of the impulse response of a system is equal to its transfer function.

To find the time history, $z(t)$, when a force time history, $F(t)$, or other input, acts on a single-DOF system, we proceed as follows:

(1) The Laplace transform of $F(t)$ is looked up in Appendix A, and is, say, $L[F(t)]$.
(2) The Laplace transform of the response, $z(t)$, which is $L[z(t)]$, is then given by multiplying the Laplace transform of the force (or other input) by the transfer function of the system:

$$L[z(t)] = L[F(t)] \times \left(\frac{z}{F}\right) \tag{3.13}$$

(3) The inverse transform of the right side of Eq. (3.13) is the required response, $z(t)$, and, in principle, the tables are then used in reverse to obtain it. Unfortunately, some algebra is usually required to achieve this.

Example 3.1

Use the Laplace transform method to solve Example 2.3, i.e. to find an expression for the displacement response, $z(t)$, of the system shown in Fig. 2.1 when the force, $F(t)$, consists of a positive step, of magnitude P, applied at $t = 0$, with the initial conditions: $z = \dot{z} = 0$ at $t = 0$.

Solution

The Laplace transform of the displacement response $z(t)$ is given by Eq. (3.13),

$$L[z(t)] = L[F(t)] \times \left(\frac{z}{F}\right) \tag{A}$$

where in this case, $L[F(t)]$, the Laplace transform of the force, $F(t)$, is that of a unit step function, which from the table is $1/s$, multiplied by the magnitude of the step, P, so

$$L[z(t)] = \frac{P}{s} \tag{B}$$

From Eq. (3.4), the transfer function is

$$\left(\frac{z}{F}\right) = \frac{1}{m} \cdot \frac{1}{\left(s^2 + 2\gamma\omega_n s + \omega_n^2\right)} \tag{C}$$

Substituting Eqs (B) and (C) into Eq. (A),

$$L[z(t)] = \frac{P}{s} \times \frac{1}{m} \cdot \frac{1}{\left(s^2 + 2\gamma\omega_n s + \omega_n^2\right)} = \frac{P}{m}\left[\frac{1}{s} \cdot \frac{1}{\left(s^2 + 2\gamma\omega_n s + \omega_n^2\right)}\right] \quad \text{(D)}$$

Ignoring, for the present, the constant P/m, which simply multiplies the final answer, we now require the inverse transform of

$$\left[\frac{1}{s} \cdot \frac{1}{\left(s^2 + 2\gamma\omega_n s + \omega_n^2\right)}\right]$$

This will not usually appear in standard lists of transforms, and it must be written in the form of *partial fractions*. The rules for forming partial fractions can be found in standard texts, often under 'integration', since this is a common method for breaking expressions down into smaller components before integration. In this case the rules say that the required partial fractions are given by:

$$\left[\frac{1}{s} \cdot \frac{1}{\left(s^2 + 2\gamma\omega_n s + \omega_n^2\right)}\right] \equiv \frac{A}{s} + \frac{Bs + C}{\left(s^2 + 2\gamma\omega_n s + \omega_n^2\right)} \quad \text{(E)}$$

where A, B and C are constants to be found. Equation (E) is an identity, valid for any value of s. Writing it in the form

$$1 \equiv A\left(s^2 + 2\gamma\omega_n s + \omega_n^2\right) + Bs^2 + Cs \quad \text{(F)}$$

the constants, A, B and C can now be found, either by sequentially substituting any three values for s, say the numbers 1, 2 and 3, or by equating the constants, the coefficients of s and those of s^2, on each side of the equation, in turn. Using the latter method:

Equating constants: $1 = A\omega_n^2$ or $A = \frac{1}{\omega_n^2}$

Equating coeffs of s: $0 = 2A\gamma\omega_n + C$ or $C = -\frac{1}{\omega_n^2}\left(2\gamma\omega_n\right)$

Equating coeffs of s^2: $0 = A + B$ so $B = -\frac{1}{\omega_n^2}$.

Substituting for A, B and C into the right side of Eq. (E) we have

$$\frac{1}{\omega_n^2}\left[\frac{1}{s} - \frac{s + 2\gamma\omega_n}{\left(s^2 + 2\gamma\omega_n s + \omega_n^2\right)}\right] \quad \text{(G)}$$

We now require the inverse transforms of the two terms in the square brackets. The first, $1/s$, is no problem, but we cannot find the inverse transform of the second term in Appendix A and this requires a little more algebra. Writing $\left(s^2 + 2\gamma\omega_n s + \omega_n^2\right)$ as $\left(s + \gamma\omega_n\right)^2 + \omega_d^2$ where $\omega_d^2 = \omega_n^2\left(1 - \gamma^2\right)$, the expression (G) can be written as:

$$\frac{1}{\omega_n^2}\left[\frac{1}{s} - \frac{s + \gamma\omega_n}{\left(s + \gamma\omega_n\right)^2 + \omega_d^2} - \frac{\left[\frac{\gamma}{\sqrt{1-\gamma^2}}\right]\omega_d}{\left(s + \gamma\omega_n\right)^2 + \omega_d^2}\right] \quad \text{(H)}$$

The inverse transforms of the three parts of expression (H) can now be found, using Appendix A, and the result (now re-inserting the constant, P/m) is

$$z = \frac{P}{m\omega_n^2}\left[1 - e^{-\gamma\omega_n t}\left(\cos \omega_d t + \frac{\gamma}{\sqrt{(1-\gamma^2)}}\sin \omega_d t\right)\right] \qquad (I)$$

agreeing with Eq. (G) of Example 2.3, since $m\omega_n^2 = k$.

3.1.2 The Convolution or Duhamel Integral

The 'convolution', 'Duhamel' or 'superposition' integral enables the exact response of linear systems to be found when the input function, usually an applied force, is known analytically. It is therefore a possible alternative to direct solution of the equation of motion by classical methods, as discussed in Chapter 2, or the Laplace transform method, discussed in Section 3.1.1. When the input function is known, but not analytically, perhaps because it was derived from test data, the convolution method can also form the basis of one of the 'semi-analytical' methods discussed in Section 3.2. It will be seen later that the impulse response of a system, usually called $h(t)$, is a fundamental way of describing the system, with several applications, for example in random vibration. It is first introduced here as an exact analytical method for finding the response of a single-DOF system. However, first we must define the *unit impulse function*.

The unit impulse function

If a force, F, acts on a mass, m, for time δt, we know from Newton's second law that the impulse, $\int_0^{\delta t} F \cdot dt$, is equal to the change in momentum, $m \cdot \delta v$, where δv is the change in the velocity of the mass, so

$$\int_0^{\delta t} F \cdot dt = m \cdot \delta v \qquad (3.14)$$

If the time δt is very small, the actual shape of the impulse is not important, and all that matters, as far as the change in velocity, δv, is concerned, is the value of the integral $\int_0^{\delta t} F \cdot dt$. A *unit impulse* is defined as one where,

$$\int_0^{\delta t} F \cdot dt = 1 \qquad (3.15)$$

Now if the mass, m, is part of a spring–mass–damper system, and a *unit* impulse of force is applied to it, then it follows, from Eqs (3.14) and (3.15), that if δv is the change in velocity, then $\delta v = 1/m$, but the change in displacement is negligible. If, for example, the equation of motion of the system is given by Eq. (3.2),

$$\ddot{z} + 2\gamma\omega_n\dot{z} + \omega_n^2 z = F/m$$

then the unit impulse response will be that of the free system $\ddot{z} + 2\gamma\omega_n\dot{z} + \omega_n^2 z = 0$ with initial conditions $z = 0$ and $\dot{z} = 1/m$, at $t = 0$. If the area of the impulse is P rather than 1, then the initial conditions become $z = 0$, and $\dot{z} = P/m$, at $t = 0$. So the impulse response can be found using the methods already discussed in Section 2.2 for dealing with free response.

It should be pointed out that in the above discussion, 'impulse' retains its original physical meaning, literally the time integral of a force. It can have other units, for example control engineers often define it as the time integral of a voltage rather than a force.

Example 3.2

Find the unit impulse response $h(t)$ of the single-DOF system represented by Eq. (3.2), assuming that the damping is (a) zero and (b) non-zero, but less than critical.

Solution

In both cases, the response to a unit impulse at $t = 0$ is equal to the free response of the system when the initial conditions are $z = 0$, $\dot{z} = (1/m)$, at $t = 0$. The method is then the same as that used in Section 2.2 to find the free response

Case (a): with zero damping:

The equation of motion is $\ddot{z} + \omega_n^2 z = 0$. The displacement response is, from Eq. (2.28), in Chapter 2, noting that $\gamma = 0$ and $\omega_d = \omega_n$ in this case:

$$z = A\cos\omega_n t + B\sin\omega_n t \tag{A}$$

where A and B are given by substituting the initial conditions. Differentiating Eq. (A):

$$\dot{z} = -A\,\omega_n\sin\omega_n t + B\,\omega_n\cos\omega_n t \tag{B}$$

Substituting $z = 0$ and $t = 0$ in Eq. (A), and $\dot{z} = (1/m)$ and $t = 0$ in Eq. (B), we find that $A = 0$ and $B = (1/m\omega_n)$, so from Eq. (A), the unit impulse response of the system is

$$z = \frac{1}{m\omega_n}\sin\omega_n t = h(t) \tag{C}$$

Case (b): with non-zero damping, but less than critical:

The equation of motion is

$$\ddot{z} + 2\gamma\omega_n\dot{z} + \omega_n^2 z = 0 \tag{D}$$

From Eqs (A) and (B) of Example 2.2, the free response is given by:

$$z = e^{-\gamma\omega_n t}(A\cos\omega_d t + B\sin\omega_d t) \tag{E}$$

and

$$\dot{z} = e^{-\gamma\omega_n t}[(B\omega_d - A\gamma\omega_n)\cos\omega_d t - (A\omega_d + B\gamma\omega_n)\sin\omega_d t] \tag{F}$$

Substituting the same initial conditions as for Case (a) into Eqs (E) and (F), we find that $A = 0$, and $B = (1/m\omega_d)$, so from Eq. (E), the required unit impulse response is

$$z = \frac{1}{m\omega_d}(e^{-\gamma\omega_n t}\sin\omega_d t) = h(t) \tag{G}$$

where the damped natural frequency, ω_d, is given by $\omega_d = \omega_n\sqrt{1 - \gamma^2}$, and ω_n is the undamped natural frequency.

Derivation of the convolution integral

In Fig. 3.1(a), the curve represents the input force history applied to a system, plotted against time τ. Time τ is measured in the same units as the general time, t, but is equal to zero when the force pulse starts, which is not necessarily at $t = 0$.

T is the time at which the response is required, regarded as fixed for the present. Let the time between $\tau = 0$ and $\tau = T$ be divided into n strips, numbered $1-n$, where n is very large, each of width $\delta\tau$. Each strip, of which strip i is typical, is regarded as an impulse of magnitude (area) P_i where,

$$P_i = F_i\,\delta\tau \tag{3.16}$$

where F_i is the value of the force at strip i.

The response of the system, $h(t)$, to a unit impulse, is assumed to be known. Fig. 3.1(b) shows the response, z_i, to the impulse P_i, which is

$$z_i = P_i\,h(T - \tau_i) = F_i\,\delta\tau\,h(T - \tau_i) \tag{3.17}$$

Its particular value at time T is $z_i(T)$ and the total response of the system, $z(T)$, at time T is the sum of all such responses to impulses 1 to n:

$$z(T) = z_1(T) + z_2(T) + z_3(T) + \cdots + z_n(T)$$

or

$$z(T) = F_1 \cdot h(T - \tau_1)\delta\tau + F_2 \cdot h(T - \tau_2)\delta\tau + \cdots + F_n \cdot h(T - \tau_n)\delta\tau \tag{3.18}$$

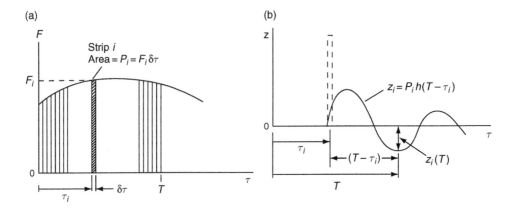

Fig. 3.1   Illustrating the derivation of the convolution integral.

or

$$z(T) = \sum_{i=1}^{n} F_i \cdot h(T - \tau_i)\delta\tau \qquad (3.19)$$

In the limit, as $\delta\tau \to 0$, this becomes a continuous integral:

$$z(T) = \int_0^T F(\tau) \cdot h(T - \tau)d\tau \qquad (3.20)$$

It should be noticed that the integration is with respect to τ, and that the integration limit, T, is regarded as a constant. However, T can take any value, and it is usually replaced by t, giving the following more usual expression for the convolution integral:

$$z(t) = \int_0^t F(\tau) \cdot h(t - \tau)d\tau \qquad (3.21)$$

Example 3.3

Use the convolution integral to find the response of the single-DOF system represented by Eq. (3.2), assuming that the damping is zero, when a step force of magnitude $\overline{F}$ is applied at $t = 0$, if the initial displacement, z, and the initial velocity, $\dot{z}$, are both zero at $t = 0$.

Solution

From Eq. (C) of Example 3.2, the unit impulse response of the system is

$$h(t) = \frac{1}{m\omega_n} \sin \omega_n t \qquad (A)$$

The applied force, $F(\tau)$, is

$$\begin{aligned} F(\tau) &= 0 & \tau < 0 \\ F(\tau) &= \overline{F} & 0 < \tau < \infty \end{aligned} \qquad (B)$$

From Eq. (3.21) and Eq. (A) (bearing in mind that $\sin(-\theta) = -\sin\theta$, and $\cos(-\theta) = \cos\theta$):

$$z(t) = \int_0^t F(\tau) \cdot h(t - \tau)d\tau = \frac{\overline{F}}{m\omega_n} \int_0^t \sin\,\omega_n(t - \tau)d\tau = \frac{\overline{F}}{m\omega_n} \int_0^t \sin(-\omega_n\tau + \omega_n t)d\tau$$

$$= \frac{\overline{F}}{m\omega_n} \int_0^t -\sin(\omega_n\tau - \omega_n t)d\tau = \frac{\overline{F}}{m\omega_n^2}\Big[\cos(\omega_n\tau - \omega_n t)\Big]_0^t$$

$$z(t) = \frac{\overline{F}}{m\omega_n^2}(1 - \cos\,\omega_n t) = \frac{\overline{F}}{k}(1 - \cos\,\omega_n t)$$

$$\left(\text{since } k = m\omega_n^2\right). \qquad (C)$$

3.1.3 *Listings of Standard Responses*

The three basic methods, discussed so far, for calculating the exact response of a single-DOF system were (1) direct solution of the equation of motion using classical

Table 3.1
Transient Responses of the System: $1/(s^2 + 2\gamma\omega_n s + \omega_n^2)$ $(\gamma < 1)$

Input function F					Response functions
Unit impulse $\delta(t)$	Unit step $H(t)$	Unit ramp t	$\sin \omega t$	$\cos \omega t$	
0	$\dfrac{1}{\omega_n^2}$	$\dfrac{-2\gamma}{\omega_n^3}$	0	0	Unit step $H(t)$
0	0	$\dfrac{1}{\omega_n^2}$	0	0	Unit ramp t
0	0	0	$\dfrac{\omega_n^2-\omega^2}{\left(\omega_n^2-\omega^2\right)^2+(2\gamma\omega_n\omega)^2}$	$\dfrac{2\gamma\omega_n\omega}{\left(\omega_n^2-\omega^2\right)^2+(2\gamma\omega_n\omega)^2}$	$\sin \omega t$
0	0	0	$\dfrac{-2\gamma\omega_n\omega}{\left(\omega_n^2-\omega^2\right)^2+(2\gamma\omega_n\omega)^2}$	$\dfrac{\omega_n^2-\omega^2}{\left(\omega_n^2-\omega^2\right)^2+(2\gamma\omega_n\omega)^2}$	$\cos \omega t$
$\dfrac{1}{\omega_d}$	$\dfrac{-\gamma}{\omega_n\omega_d}$	$\dfrac{2\gamma^2-1}{\omega_n^2\omega_d}$	$\dfrac{\omega\left[(\omega^2-\omega_n^2)+2\gamma^2\omega_n^2\right]}{\omega_d\left[\left(\omega_n^2-\omega^2\right)^2+(2\gamma\omega_n\omega)^2\right]}$	$\dfrac{-\gamma\omega_n\left(\omega_n^2+\omega^2\right)}{\omega_d\left[\left(\omega_n^2-\omega^2\right)^2+(2\gamma\omega_n\omega)^2\right]}$	$e^{-\gamma\omega_n t}\sin \omega_d t$
0	$\dfrac{-1}{\omega_n^2}$	$\dfrac{2\gamma}{\omega_n^3}$	$\dfrac{2\gamma\omega_n\omega}{\left(\omega_n^2-\omega^2\right)^2+(2\gamma\omega_n\omega)^2}$	$\dfrac{\omega^2-\omega_n^2}{\left(\omega_n^2-\omega^2\right)^2+(2\gamma\omega_n\omega)^2}$	$e^{-\gamma\omega_n t}\cos \omega_d t$

ω = forcing frequency (rad/s); ω_n = undamped natural frequency (rad/s), $\omega_n = \sqrt{k/m}$; ω_d = damped natural frequency (rad/s), $\omega_d = \omega_n\sqrt{1-\gamma^2}$; γ = non-dimensional viscous damping coefficient.

methods; (2) solution using Laplace transforms and (3), solution using the convolution integral. These are all tedious to carry out by hand, and in practice computer methods are used. For simpler problems, a table of standard responses, such as Table 3.1, is useful. This was based on a report by Huntley [3.1], intended for use in control engineering. Although it uses transfer functions in Laplace notation to characterize the equation of motion, the actual responses are obtained by superposition in the time domain. Table 3.1 lists the displacement response, z, of Eq. (3.2),

$$\ddot{z} + 2\gamma\omega_n\dot{z} + \omega_n^2 z = F/m$$

when the applied force, F, is any one of the *unit* functions listed at the top; the mass, m, is unity; the non-dimensional viscous damping coefficient is less than 1 and the initial conditions are $z = \dot{z} = 0$ at $t = 0$. The transfer function of the system in Laplace notation is therefore

$$\frac{1}{\left(s^2 + 2\gamma\omega_n s + \omega_n^2\right)}$$

Example 3.4

Use Table 3.1 to find the displacement response of the system shown in Fig. 2.1 and represented by Eq. (3.2), when,

(a) a step force of magnitude a force units is applied at $t = 0$ and the initial conditions are $z = 0$ and $\dot{z} = 0$ at $t = 0$.

(b) a ramp force, starting from zero and growing linearly at the rate of b force units/second, is applied at $t = 0$, with the same initial conditions, $z = 0$ and $\dot{z} = 0$ at $t = 0$.

Assume that the non-dimensional damping coefficient, γ, is less than unity.

Solution

Case (a), step input:

The response, z, to a *unit* force step, of a system with *unit* mass, is found by multiplying all the constants in the second column, 'unit step', by the corresponding response functions in the last column, and adding the products. For a step of magnitude a, and a system mass of m, the result must then be multiplied by a/m:

$$z = \frac{a}{m}\left[\left(\frac{1}{\omega_n^2}\right)H(t) + \left(\frac{-\gamma}{\omega_n\omega_d}\right)e^{-\gamma\omega_n t}\sin\omega_d t + \left(\frac{-1}{\omega_n^2}\right)e^{-\gamma\omega_n t}\cos\omega_d t\right] \qquad (A)$$

which can be simplified to the following, for $t > 0$ if the step function $H(t)$ starts at $t = 0$:

$$z = \frac{a}{m\omega_n^2}\left[1 - e^{-\gamma\omega_n t}\left(\cos\omega_d t + \frac{\gamma}{\sqrt{(1-\gamma^2)}}\sin\omega_d t\right)\right] \qquad (B)$$

Case (b), ramp input:

In this case, the constants in the third column, 'unit ramp', are individually multiplied by the response function in the last column, the products added, and the whole multiplied by b/m, as follows:

$$z = \frac{b}{m}\left[\left(\frac{-2\gamma}{\omega_n^3}\right)H(t) + \left(\frac{1}{\omega_n^2}\right)t + \left(\frac{2\gamma^2 - 1}{\omega_n^2\omega_d}\right)e^{-\gamma\omega_n t}\sin\omega_d t + \left(\frac{2\gamma}{\omega_n^3}\right)e^{-\gamma\omega_n t}\cos\omega_d t\right]$$
$$(C)$$

which can be simplified to

$$z = \frac{b}{m\omega_n^2}\left[t - \frac{2\gamma}{\omega_n}(1 - e^{-\gamma\omega_n t}\cos\omega_d t) + \left(\frac{2\gamma^2 - 1}{\omega_d}\right)e^{-\gamma\omega_n t}\sin\omega_d t\right] \qquad (D)$$

3.2 'Semi-analytical' methods

When the input function to a single-DOF system is not known analytically, for example it may have been obtained from test data, but the system is linear, it is not always necessary to use the step-by-step numerical methods (such as the Runge–Kutta method) described in Section 3.3. Instead, the input function, usually an applied force, can be broken down into sections, such as impulses, steps or ramps, and the problem is solved piecewise, using any of the exact methods discussed above. The solution is then partly analytical, making it, in theory, very accurate, and issues such as convergence do not arise.

3.2.1 Impulse Response Method

If the force input function is known only in graphical or tabular form, analytic solution using the convolution integral, Eq. (3.21), is not possible, but a summation based on Eq. (3.19) can still be used, i.e.:

$$z(t) \approx \sum_{i=1}^{n} F_i \cdot h(t - \tau_i)\delta t$$

where F_i represents the approximated force at intervals of δt, but $h(t - \tau_i)$ is the exact analytic unit impulse response evaluated at the same small intervals. Computer programs based on this method are available.

3.2.2 Straight-line Approximation to Input Function

An alternative to the impulse method described above is to approximate the input function, F, by straight-line ramp segments, as shown in Fig. 3.2.

In this approach, the force input, F, at t_0 consists of a constant (step) value, plus a constant slope, or ramp. The displacement and slope between t_0 and t_1 are calculated from the standard analytic responses of the single-DOF system to these inputs (see Example 3.4), and the values of these at t_1 become the initial conditions for a similar calculation between t_1 and t_2, and so on. Computer programs based on this method are available. Such programs are simplified if the time steps are made equal, since many of the intermediate calculations then only have to be done once, although this is not essential. The calculated response using this method is exact for the polygonal approximation to the force, but not, of course, for the underlying curve.

3.2.3 Superposition of Standard Responses

The response to simple force inputs, made up by the superposition of steps, ramps, sinusoids, etc., can easily be plotted using simple spreadsheet methods. The procedure is illustrated by the following example. It differs from the straight-line segment method, above, in that superposition is used, to avoid carrying initial conditions from section to section.

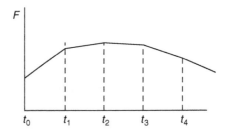

Fig. 3.2   Input function approximated by straight lines.

Example 3.5

Use an exact piecewise-linear method to find the displacement response, z, of the system represented by Eq. (3.2):

$$\ddot{z} + 2\gamma\omega_n\dot{z} + \omega_n^2 z = F/m \qquad (A)$$

from $t=0$ to $t=0.5$ s when F consists of the force pulse shown in Fig. 3.3(a), and the other numerical values are as follows: mass, $m = 100$ kg; damping coefficient $\gamma = 0.05$; the undamped natural frequency, $f_n = 10$ Hz ($\omega_n = 20\pi$ rad/s). Assume $z = \dot{z} = 0$ at $t=0$.

Solution

As shown in Fig. 3.3(b), the required input function can be formed by the super-position of:
(1) a positive ramp of slope 4000 N/s starting at $t = 0$;
(2) a similar negative ramp starting at $t = 0.05$ s; and
(3) a negative step of 200 N starting at $t = 0.2$ s
    Once started, these functions all continue to the end of the computation.
    From Example 3.4, the exact response to a ramp of slope b is

$$z = \frac{b}{m\omega_n^2}\left[t - \frac{2\gamma}{\omega_n}(1 - e^{-\gamma\omega_n t}\cos\omega_d t) + \left(\frac{2\gamma^2 - 1}{\omega_d}\right)e^{-\gamma\omega_n t}\sin\omega_d t\right] \qquad (B)$$

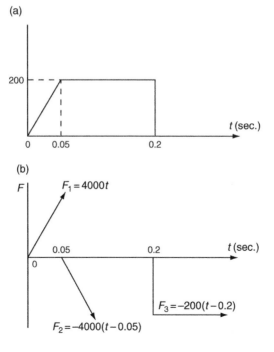

Fig. 3.3   Input pulse produced by superposition of standard functions.

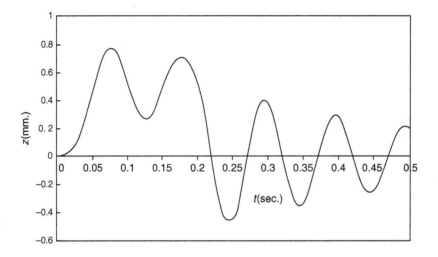

Fig. 3.4   Response of a single-DOF system to the input force pulse shown in Fig. 3.3.

and the exact response to a step of magnitude a is

$$z = \frac{a}{m\omega_n^2}\left[1 - e^{-\gamma\omega_n t}\left(\cos\omega_d t + \frac{\gamma}{\sqrt{(1-\gamma^2)}}\sin\omega_d t\right)\right] \tag{C}$$

where

$$\omega_d = \omega_n\sqrt{1-\gamma^2}$$

Inserting numerical values, the response to the first ramp, z_1, which is required for $0 < t < 0.5$, is

$$\begin{aligned} z_1 = 0.01013\{t &- 0.001591[1 - \exp(-3.1415t)\cos 62.753t] \\ &- 0.01585\exp(-3.1415t)\sin 62.753t\} \end{aligned} \tag{D}$$

The response, z_2, to the second ramp, which is required for $0.05 < t < 0.5$, is

$$\begin{aligned} z_2 = -0.01013\{(t-0.05) &- 0.001591[1 - \exp(-3.1415(t-0.05))\cos 62.753(t-0.05)] \\ &- 0.01585\exp(-3.1415(t-0.05))\sin 62.753(t-0.05)\} \end{aligned} \tag{E}$$

and the response, z_3, to the negative step, which is required for $0.2 < t < 0.5$, is

$$\begin{aligned} z_3 = -0.0005066\{1 &- \exp[-3.1415(t-0.2)] \\ &\times [\cos 62.753(t-0.2) + 0.05006\sin(t-0.2)]\} \end{aligned} \tag{F}$$

The response, plotted from a spreadsheet, is shown in Fig. 3.4.

3.3 Step-by-step numerical methods using approximate derivatives

The analytical or semi-analytical methods described above should always be used, when possible, to find time history solutions for any system, since there is no question about their accuracy. They cannot be used, however, when the system is non-linear, and step-by-step numerical methods must then be used. In these methods, the time, t, is broken down into a series of small steps, h, say, and the solution proceeds step by step, each small computation being based on the information available at the time.

In this section, the use of such methods to solve single-DOF problems only is considered; however, most methods can also be used for multi-DOF systems. The number of numerical methods available today is very large, and it will only be possible to introduce a few of the simpler methods here. However, these include the useful and popular Runge–Kutta fourth-order method. The application of the Runge–Kutta method to multi-DOF systems is discussed in Chapter 6.

In all methods, the general idea is to extend the solution, i.e. the function of t that we are trying to generate, say $z(t)$, by one time step, h, at a time. This will normally be from t_j to t_{j+1}, as shown in Fig. 3.5. At the very beginning of the computation, let $j = 0$, say, so we can refer to the starting time as t_0, the next point as t_1 and so on. The initial conditions at t_0 are assumed to be known.

In the case of the single-DOF system, considered here, the equation of motion to be solved is

$$m\ddot{z} + c\dot{z} + kz = F \qquad (3.22)$$

where the constants m, c and k are known numerically, as is the complete time history of the force F.

In some methods, Eq. (3.22) can be used as it stands. It is then simply rearranged as:

$$\ddot{z} = \frac{F}{m} - \frac{c}{m}\dot{z} - \frac{k}{m}z \qquad (3.23)$$

More usually, Eq. (3.22) is split into two simultaneous first-order differential equations, as follows. Let $u = \dot{z}$, then,

$$\dot{z} = u \qquad (3.24a)$$

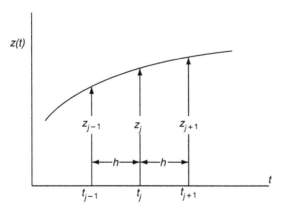

Fig. 3.5   Notation for step-by-step methods.

and, since $\ddot{z} = \dot{u}$,

$$\dot{u} = \frac{F}{m} - \frac{c}{m}u - \frac{k}{m}z \tag{3.24b}$$

Many methods use the Taylor series, and this should be reviewed before proceeding further. Applied to the present context, the Taylor series makes it possible to calculate the value of a function at a point in time different (i.e. either later or earlier) than another point where the value of the function, and all its derivatives, are known. In Fig. 3.5, for example, if z_j and its time derivatives $\dot{z}_j$, $\ddot{z}_j$, $\dddot{z}_j$, etc. are known at time t_j, then the value of z_{j+1} at t_{j+1} (an amount of time h later) can, in theory, be found. In practice, of course, it can only be found approximately, since only a few derivatives will be known. The Taylor series is used in the form:

$$z_{j+1} = z_j + h\dot{z}_j + \frac{h^2}{2!}\ddot{z}_j + \frac{h^3}{3!}\dddot{z}_j + \cdots \tag{3.25}$$

It also allows a past value, say z_{j-1}, to be found, since h can be negative, so,

$$z_{j-1} = z_j - h\dot{z}_j + \frac{h^2}{2!}\ddot{z}_j - \frac{h^3}{3!}\dddot{z}_j + \cdots \tag{3.26}$$

(where ! indicates a factorial value, e.g. $3! = 3 \times 2 \times 1 = 6$)

3.3.1 Euler Method

The Euler method uses the equation of motion, Eq. (3.22), in the form of the two first-order equations, Eqs (3.24a) and (3.24b). In this method, z and u are updated at each step in the following simple way:

$$z_{j+1} = z_j + hu_j \tag{3.27a}$$

and:

$$u_{j+1} = u_j + h\dot{u}_j \tag{3.27b}$$

The procedure will be clear from Table 3.2, which is the beginning of a simple spreadsheet to calculate the response when any input force history, defined by F_0, F_1, F_2, etc., is applied to the system of Eq. (3.22). The response may be required in terms of displacement, z, velocity $\dot{z}(= u)$ or acceleration $\ddot{z}(= \dot{u})$. The initial conditions $z_0 = 0$ and $u_0 = 0$ at t_0 have been inserted in the first row; these could be any other desired initial values. It will be seen that $\dot{u}_j$ is obtained from Eq. (3.24b), i.e.:

Table 3.2
Spreadsheet for the Euler Method

j	t_j	z_j	u_j	$\dot{u}_j$	F_j
0	$= 0$	$= 0$	$= 0$	$= \dfrac{F_0}{m} - \dfrac{c}{m}u_0 - \dfrac{k}{m}z_0$	$=F_0$
1	$= h$	$= z_0 + hu_0$	$= u_0 + h\dot{u}_0$	$= \dfrac{F_1}{m} - \dfrac{c}{m}u_1 - \dfrac{k}{m}z_1$	$=F_1$
2	$=2h$	$= z_1 + hu_1$	$= u_1 + h\dot{u}_1$	$\vdots$	$=F_3$
3	$=3h$	$\vdots$	$\vdots$		$\vdots$

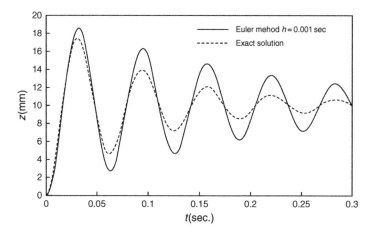

Fig. 3.6   Simple Euler method compared with exact result. Time step 0.001 s.

$$\dot{u}_j = \frac{F_j}{m} - \frac{c}{m}u_j - \frac{k}{m}z_j \qquad (3.28)$$

where F_j is the known value of the applied force, F.

In Fig. 3.6, the time history of z, as given by the spreadsheet, Table 3.2, is compared with the exact solution (from Example 2.3), using the following numerical values:

$F = \text{constant} = 100\,\text{N}$ (a step force of 100 N applied at $t = 0$).

$m = 1$ kg;

$k = 10\,000$ N/m;

$c = 20$ N/m/s;

$h = \text{the time step} = 0.001$ s

This is a disappointing result, and it appears to indicate that the wrong value of damping has been used. This is not the case, however, since when the time-step, h, is reduced to 0.0002 s., the improved result shown in Fig. 3.7 is obtained.

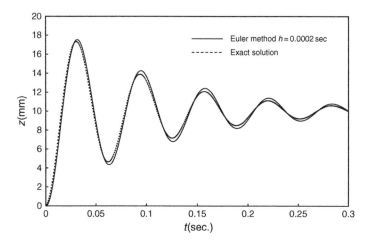

Fig. 3.7   Simple Euler method compared with exact result. Time step 0.0002 s.

These examples illustrate the generally poor performance of the simple Euler method. The *modified Euler method* is a considerable improvement, as discussed below.

3.3.2 Modified Euler Method

In the original Euler method, the change in z or u over one increment of time, h, was assumed to be given by the slope at the first point multiplied by h, for example $z_{j+1} = z_j + hu_j$ and $u_{j+1} = u_j + h\dot{u}_j$. A better approximation to the slope over the whole time step would be the average of the slopes at t_j and t_{j+1}. Thus, in Eq. (3.27a), u_j would be replaced by $\frac{1}{2}(u_j + u_{j+1})$, and in Eq. (3.27b), $\dot{u}_j$ would be replaced by $\frac{1}{2}(\dot{u}_j + \dot{u}_{j+1})$. This is the basis of the modified Euler method. Applying these changes to Eqs (3.27a) and (3.27b), they become

$$z_{j+1} = z_j + \frac{1}{2}h(u_j + u_{j+1}) \tag{3.29a}$$

and

$$u_{j+1} = u_j + \frac{1}{2}h(\dot{u}_j + \dot{u}_{j+1}) \tag{3.29b}$$

Since u_{j+1} and $\dot{u}_{j+1}$ have not been calculated at step j, they must be approximated from the information available at step j. In Eq. (3.29a) we can use Eq. (3.27b), i.e.

$$u_{j+1} = u_j + h\dot{u}_j$$

and in Eq. (3.29b) we can use Eq. (3.28), i.e.:

$$\dot{u}_j = \frac{F_j}{m} - \frac{c}{m}u_j - \frac{k}{m}z_j$$

However, in Eq. (3.29b) we also need an approximation for $\dot{u}_{j+1}$, and this can be $\dot{u}'_{j+1}$ where,

$$\dot{u}'_{j+1} = \frac{F_{j+1}}{m} - \frac{c}{m}u_{j+1} - \frac{k}{m}z_{j+1} = \frac{F_{j+1}}{m} - \frac{c}{m}(u_j + h\dot{u}_j) - \frac{k}{m}(z_j + hu_j) \tag{3.30}$$

noting that F_{j+1} *is* available at step j and Eqs (3.27a) and (3.27b) have been used.

Table 3.3 is a spreadsheet program based on these equations, again assuming that $z_0 = u_0 = 0$.

This was used to produce Fig. 3.8, using the same numerical values as before, with a time step of 0.0025 s. When compared with the exact answer (from Example 2.3), this result is better than that from the unmodified Euler method (Fig. 3.7), in spite of the step h being 12.5 times longer.

3.3.3 Central Difference Method

If Eq. (3.26) is subtracted from Eq. (3.25), and derivatives higher than $\dot{z}_j$ are ignored, we have

$$z_{j+1} - z_{j-1} = 2h\dot{z}_j \quad \text{or rearranging:} \quad \dot{z}_j = \frac{z_{j+1} - z_{j-1}}{2h} \tag{3.31}$$

Table 3.3
Spreadsheet for the Modified Euler Method

j	t_j	F_j	z_j	u_j	$\dot{u}_j$	$\dot{u}'_{j+1}$
0	$= hj$	$= F_0$	$= 0$	$= 0$	$= \dfrac{F_0}{m} - \dfrac{c}{m}u_0 - \dfrac{k}{m}z_0$	$= \dfrac{F_1}{m} - \dfrac{c}{m}(u_0 + h\dot{u}_0)$ $- \dfrac{k}{m}(z_0 + hu_0)$
1	$= hj$	$= F_1$	$= z_0 + hu_0 + \frac{1}{2}h^2\dot{u}_0$	$= u_0 + \frac{1}{2}h(\dot{u}_0 + \dot{u}'_0)$	$= \dfrac{F_1}{m} - \dfrac{c}{m}u_1 - \dfrac{k}{m}z_1$	$= \dfrac{F_2}{m} - \dfrac{c}{m}(u_1 + h\dot{u}_1)$ $- \dfrac{k}{m}(z_1 + hu_1)$
2	$= hj$	$= F_2$	$= z_1 + hu_1 + \frac{1}{2}h^2\dot{u}_1$	$= u_1 + \frac{1}{2}h(\dot{u}_1 + \dot{u}'_1)$	$\vdots$	$\vdots$

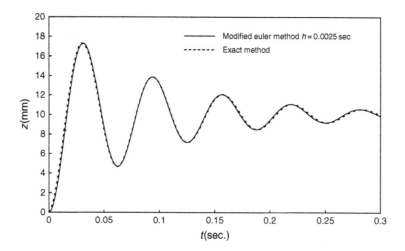

Fig. 3.8 Modified Euler method compared with exact result. Time step 0.0025 s.

If Eq. (3.26) is added to Eq. (3.25), ignoring derivatives higher than $\ddot{z}_j$, then we have

$$z_{j+1} + z_{j-1} = 2z_j + h^2\ddot{z}_j \quad \text{or rearranging:} \quad \ddot{z}_j = \frac{z_{j+1} - 2z_j + z_{j-1}}{h^2} \qquad (3.32)$$

It is not necessary, in this case, to split the second-order equation of motion, Eq. (3.22), into two first-order equations, and Eqs (3.31) and (3.32) can be substituted into it directly. However, it will be noticed that these equations both require a value for z_{j-1}, and this is not available at $t = 0$, so the method is not self-starting. An approximation for the required value of z_{-1} may, however, be found as follows. From Eq. (3.26) we have, ignoring derivatives above the second,

$$z_{j-1} = z_j - h\dot{z}_j + \frac{h^2}{2!}\ddot{z}_j$$

With $j = 0$ this is

$$z_{-1} = z_0 - h\dot{z}_0 + \frac{h^2}{2}\ddot{z}_0 \qquad (3.33)$$

Now z_0 and $\dot{z}_0$ are the known initial conditions, but we still need a value for $\ddot{z}_0$. This is given by the original equation of motion. If, for example, this was $m\ddot{z} + c\dot{z} + kz = F$, then at $t = 0$:

$$\ddot{z}_0 = \frac{F_0}{m} - \frac{c}{m}\dot{z}_0 - \frac{k}{m}z_0 \qquad (3.34)$$

Example 3.6

(a) Write a spreadsheet program to solve the equation $m\ddot{z} + c\dot{z} + kz = F$ by the central difference method.
(b) Plot the displacement time history, z, with $F = $ constant $= 100$ N; initial conditions $z_0 = \dot{z}_0 = 0$, and

$$m = 1 \text{ kg};$$
$$k = 10\,000 \text{ N/m};$$
$$c = 20 \text{ N/m/s}$$

Compare the plot with the exact answer from the expression derived in Example 2.3.

Solution

(a) Substituting Eqs (3.31) and (3.32) into the equation of motion, $m\ddot{z} + c\dot{z} + kz = F$, we have

$$m\left(\frac{z_{j+1} - 2z_j + z_{j-1}}{h^2}\right) + c\left(\frac{z_{j+1} - z_{j-1}}{2h}\right) + kz_j = F_j \qquad (A)$$

Rearranging,

$$z_{j+1} = a_1 z_j + a_2 z_{j-1} + a_3 F_j \qquad (B)$$

where

$$a_1 = \frac{2m - kh^2}{\frac{1}{2}ch + m}, \qquad a_2 = \frac{\frac{1}{2}ch - m}{\frac{1}{2}ch + m} \qquad a_3 = \frac{h^2}{\frac{1}{2}ch + m} \qquad (C)$$

The value of z_{-1} is taken from Eq. (3.26), i.e.

$$z_{-1} = z_0 - h\dot{z}_0 + \frac{h^2}{2}\ddot{z}_0$$

where

$$z_0 = \dot{z}_0 = 0 \quad \text{and} \quad \ddot{z}_0 = \frac{F_0}{m} - \frac{c}{m}\dot{z}_0 - \frac{k}{m}z_0$$

Table 3.4
Spreadsheet for Central Difference Method

j	t_j	F_j	z_j
-1	$= hj$		$= \dfrac{h^2 F_0}{2m}$
0	$= hj$	$= F_0$	$= 0$
1	$= hj$	$= F_1$	$= a_1 z_0 + a_2 z_{-1} + a_3 F_0$
2		$= F_2$	$= a_1 z_1 + a_2 z_{-1} + a_3 F_1$
3		$\vdots$	$\vdots$

from Eq. (3.34). Therefore,

$$z_{-1} = 0 - 0 + \frac{h^2 F_0}{2m} = \frac{h^2 F_0}{2m} \tag{D}$$

Table 3.4 is a spreadsheet to calculate z.

The constants a_1, a_2 and a_3 are given by Eqs (C).

(b) Inserting the values of m, k and c, and initial conditions $z_0 = \dot{z}_0 = 0$, Fig. 3.9 is a plot of z using a step length $h = 0.005\,\text{s}$, compared with the exact solution from the expression derived in Example 2.3.

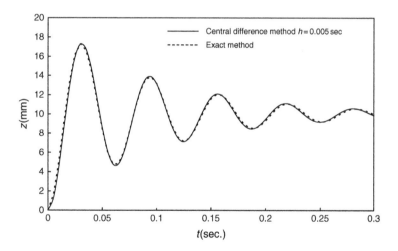

Fig. 3.9   Central difference method compared with exact result. Time step 0.005 s.

3.3.4 The Runge–Kutta Method

We have seen that the accuracy of the methods discussed so far depends critically upon the estimation of the average slope of the function from one point to the next. The simple Euler method uses only the slope at the first point to predict the change in value at the second. The modified Euler is an improvement, in that it also uses an estimate of the slope at the second point. The central difference method uses three values of the function, the past, present and future values. The Runge–Kutta method continues this series of improvements by using a more sophisticated method, based on Simpson's rule, to predict the average slope over the interval h. In the case of the fourth-order Runge–Kutta method, described next, it does this by introducing a temporary point half way between each pair of grid points. The mean slope of the function over each interval, h, is then based on a weighted average of *four* estimates of the slope, obtained from *three* different times, the beginning, the center and the end of each interval, the center point being used twice. The simple Euler method is used for the half-interval calculations within each main interval h.

The Runge–Kutta method is very accurate, self-starting and easily adapted to multi-DOF systems, as discussed in Chapter 6. It is therefore probably the most popular method in general use.

For the solution of a single-DOF system, as discussed here, the second-order equation, for example Eq. (3.22) or (3.23), is split into two first-order equations, Eqs (3.24a) and (3.24b), i.e.,

$$u = \dot{z}$$

and

$$\dot{u} = \ddot{z} = \frac{F}{m} - \frac{c}{m}u - \frac{k}{m}z$$

These are solved in the same way as for the simple Euler method, described above, except that the mean slopes $\langle u \rangle$ and $\langle \dot{u} \rangle$ are obtained by taking a weighted average over each interval h, for example the interval between times t_j and t_{j+1}, as follows:

$$\langle u \rangle = \frac{1}{6}\left(u_j^{(1)} + 2u_j^{(2)} + 2u_j^{(3)} + u_j^{(4)} \right) \tag{3.34}$$

and

$$\langle \dot{u} \rangle = \frac{1}{6}\left(\dot{u}_j^{(1)} + 2\dot{u}_j^{(2)} + 2\dot{u}_j^{(3)} + \dot{u}_j^{(4)} \right) \tag{3.35}$$

where the significance of the superscripts is explained below.

The solution of Eqs (3.24a) and (3.24b), for an arbitrary force input F, proceeds as follows. At each step in the computation, say step j, at t_j, we calculate four values each of z_j, u_j and $\dot{u}_j$ (designated $z_j^{(1)}$ to $z_j^{(4)}$, etc.), as shown in Table 3.5.

Table 3.5
Intermediate Calculations in the Runge–Kutta Method

$z_j^{(1)} = z_j$	$u_j^{(1)} = u_j$	$\dot{u}_j^{(1)} = \frac{F_j}{m} - \frac{c}{m}u_j^{(1)} - \frac{k}{m}z_j^{(1)}$
$z_j^{(2)} = z_j + \frac{h}{2}u_j^{(1)}$	$u_j^{(2)} = u_j + \frac{h}{2}\dot{u}_j^{(1)}$	$\dot{u}_j^{(2)} = \frac{F_j}{m} - \frac{c}{m}u_j^{(2)} - \frac{k}{m}z_j^{(2)}$
$z_j^{(3)} = z_j + \frac{h}{2}u_j^{(2)}$	$u_j^{(3)} = u_j + \frac{h}{2}\dot{u}_j^{(2)}$	$\dot{u}_j^{(3)} = \frac{F_j}{m} - \frac{c}{m}u_j^{(3)} - \frac{k}{m}z_j^{(3)}$
$z_j^{(4)} = z_j + hu_j^{(3)}$	$u_j^{(4)} = u_j + h\dot{u}_j^{(3)}$	$\dot{u}_j^{(4)} = \frac{F_j}{m} - \frac{c}{m}u_j^{(4)} - \frac{k}{m}z_j^{(4)}$

It is seen in Table 3.5 that the first quantity, denoted by $^{(1)}$, is at the beginning of the interval, the second *and* third are half way through, i.e. at $h/2$, and the fourth is at the end.

These results are then used to calculate z_{j+1} and u_{j+1}, as follows:

$$z_{j+1} = z_j + \frac{h}{6}\left(u_j^{(1)} + 2u_j^{(2)} + 2u_j^{(3)} + u_j^{(4)}\right) \tag{3.36}$$

$$u_{j+1} = u_j + \frac{h}{6}\left(\dot{u}_j^{(1)} + 2\dot{u}_j^{(2)} + 2\dot{u}_j^{(3)} + \dot{u}_j^{(4)}\right) \tag{3.37}$$

The whole procedure, using Table 3.5, and Eqs (3.36) and (3.37) again, is repeated to calculate z_{j+2} and u_{j+2} and so on.

Incidentally, the method for finding the average slope over the interval h can be seen to be related to Simpson's rule, which for *three* slopes, u_1, u_2 and u_3 spaced at intervals of h, would give

$$\langle u \rangle = \frac{1}{2h}\left[\frac{1}{3}h(u_1 + 4u_2 + u_3)\right] = \frac{1}{6}(u_1 + 4u_2 + u_3)$$

The similarity to Eqs (3.34) and (3.35) can be seen if the center term is split into two.

Example 3.7

(a) Write a spreadsheet program to solve the equation $m\ddot{z} + c\dot{z} + kz = F$ by the fourth-order Runge–Kutta method.
(b) Plot the displacement time history, z, with F = constant = 100 N; initial conditions $z_0 = u_0 = 0$ and

   m = 1 kg;
   k = 10 000 N/m and
   c = 20 N/m/s.

Compare the plot with the exact answer from the expression derived in Example 2.3.

Solution

(a) A suitable spreadsheet layout, based on Table 3.5 and Eqs (3.36) and (3.37), using one row per value of j, is shown in Table 3.6. The initial conditions z_0 and u_0 in the first row are set to zero in this case. F_0 is the value of F at t_0.
(b) Fig. 3.10 is a plot of the Runge–Kutta result compared with the exact solution from Example 2.3. The time-step, h, was 0.005 s in this case. An acceptable result was also obtained with h = 0.0075 s.

Table 3.6
Spreadsheet for the Fourth-order Runge–Kutta Method

j	t_j	$z_j^{(1)}$ and response z_j	$z_j^{(2)}$	$z_j^{(3)}$	$z_j^{(4)}$	$u_j^{(1)}$ and response z_j	$u_j^{(2)}$	$u_j^{(3)}$	$u_j^{(4)}$	$\ddot{u}_j^{(1)}$ and $\ddot{z}_j$	$\ddot{u}_j^{(2)}$	$\ddot{u}_j^{(3)}$	$\ddot{u}_j^{(4)}$
0	0	$=z_0$ (= initial value of $z=0$)	$=z_0$ $+\dfrac{h}{2}u_0^{(1)}$	$=z_0$ $+\dfrac{h}{2}u_0^{(2)}$	$=z_0$ $+hu_0^{(3)}$	$=u_0$ (= initial value of $u=0$)	$=u_0$ $+\dfrac{h}{2}\dot{u}_0^{(1)}$	$=u_0$ $+\dfrac{h}{2}\dot{u}_0^{(2)}$	$=u_0$ $+h\dot{u}_0^{(3)}$	$=\dfrac{F_0}{m}$ $-\dfrac{c}{m}u_0^{(1)}$ $-\dfrac{k}{m}z_0^{(1)}$	$=\dfrac{F_0}{m}$ $-\dfrac{c}{m}u_0^{(2)}$ $-\dfrac{k}{m}z_0^{(2)}$	$=\dfrac{F_0}{m}$ $-\dfrac{c}{m}u_0^{(3)}$ $-\dfrac{k}{m}z_0^{(3)}$	$=\dfrac{F_0}{m}$ $-\dfrac{c}{m}u_0^{(4)}$ $-\dfrac{k}{m}z_0^{(4)}$
1	h	$=z_0+$ $\dfrac{h}{6}\left(u_0^{(1)}+2u_0^{(2)}+2u_0^{(3)}+u_0^{(4)}\right)$	$=z_1$ $+\dfrac{h}{2}u_1^{(1)}$	$=z_1$ $+\dfrac{h}{2}u_1^{(2)}$	$=z_1$ $+hu_1^{(3)}$	$=u_0+$ $\dfrac{h}{6}\left(\dot{u}_0^{(1)}+2\dot{u}_0^{(2)}+2\dot{u}_0^{(3)}+\dot{u}_0^{(4)}\right)$	$=u_1$ $+\dfrac{h}{2}\dot{u}_1^{(1)}$	$=u_1$ $+\dfrac{h}{2}\dot{u}_1^{(2)}$	$=u_1$ $+h\dot{u}_1^{(3)}$	$=\dfrac{F_1}{m}$ $-\dfrac{c}{m}u_1^{(1)}$ $-\dfrac{k}{m}z_1^{(1)}$	$=\dfrac{F_1}{m}$ $-\dfrac{c}{m}u_1^{(2)}$ $-\dfrac{k}{m}z_1^{(2)}$	$=\dfrac{F_1}{m}$ $-\dfrac{c}{m}u_1^{(3)}$ $-\dfrac{k}{m}z_1^{(3)}$	$=\dfrac{F_1}{m}$ $-\dfrac{c}{m}u_1^{(4)}$ $-\dfrac{k}{m}z_1^{(4)}$
2	$2h$	$=z_1+$ $\dfrac{h}{6}\left(u_1^{(1)}+2u_1^{(2)}+2u_1^{(3)}+u_1^{(4)}\right)$	$=z_2$ $+\dfrac{h}{2}u_2^{(1)}$	$=z_2$ $+\dfrac{h}{2}u_2^{(2)}$	$=z_2$ $+hu_2^{(3)}$	$=u_1$ $+\dfrac{h}{6}\left(\dot{u}_1^{(1)}+2\dot{u}_1^{(2)}+2\dot{u}_1^{(3)}+\dot{u}_1^{(4)}\right)$	$=u_2$ $+\dfrac{h}{2}\dot{u}_1^{(1)}$	$=u_2$ $+\dfrac{h}{2}\dot{u}_1^{(2)}$	$=u_2$ $+h\dot{u}_1^{(3)}$	$\vdots$	$\vdots$	$\vdots$	$\vdots$
3	$\vdots$	$\vdots$	$\ldots$	$\ldots$	$\ldots$	$\ldots$	$\ldots$	$\ldots$					

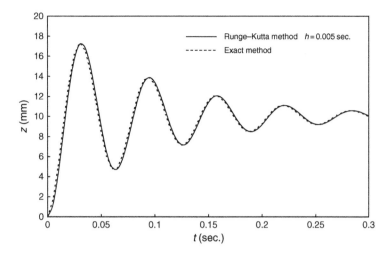

Fig. 3.10   Fourth-order Runge–Kutta method compared with exact result. Time step 0.005 s.

3.3.5 Discussion of the Simpler Finite Difference Methods

There are many more finite difference methods available, such as predictor–corrector methods, the Newmark-β method, and more. The four methods described above, however, have introduced the subject. The relative performance of these simple methods can now be assessed roughly, since the same single-DOF problem was used in each case. This was chosen to be the excitation of a linear system by a step force input, so that the exact solution could be found for comparison. The undamped natural frequency was 100 rad/s (15.9 Hz) and the natural period about 0.063 s.

The simple Euler method cannot be recommended, and required extremely short time steps, around 300 steps per period of the oscillation, to produce acceptable results. Worse, it sometimes gave believable but inaccurate results. The modified Euler method was a considerable improvement, giving a good result using a time step of 0.0025 s., or about 25 steps per period.

The central difference method gave a good result with a time step of 0.005 s, about 12 steps per period. The Runge–Kutta method gave a very good result using the same time step, and an acceptable result with a step as long as 0.0075 s, only about 8 steps per period of the oscillation. These results are in line with the usual advice given, when using these methods, that there should be at least 10 time steps per period of the highest expected response frequency.

The results from the central difference method and the Runge–Kutta method are actually quite similar, and better than the Euler and modified Euler methods. However, a disadvantage of the central difference method is that it is not self-starting, and of the methods discussed here, the Runge–Kutta method is recommended.

3.4 Dynamic factors

The *dynamic factor* is a useful engineering concept that, at its simplest, compares the maximum dynamic displacement response of a system with the static displacement that would be produced by a steady force with the same magnitude as the peak value of the actual force. The idea can be extended to responses other than displacement, and inputs other than force, such as base motion. Although the concept can be applied to systems with any number of degrees of freedom, the approach is over-simplistic for multi-DOF systems, and it is usually applied only to single-DOF systems. The dynamic factor is a fairly crude concept, and should not be used to replace a proper dynamic analysis. However, as an aid to engineering judgement, in the early stages of a design, it can be of considerable value.

3.4.1 Dynamic Factor for a Square Step Input

As a simple example of a dynamic factor, consider the undamped system shown in Fig. 3.11(a). The equation of motion is

$$m\ddot{z} + kz = F \tag{3.38}$$

and it was shown in Example 3.3 that the displacement response, z, when the applied force, F, consists of a step force of magnitude $\overline{F}$, as shown in Fig. 3.11(b), is

$$z = \frac{\overline{F}}{k}(1 - \cos\omega_n t) \tag{3.39}$$

where ω_n is the natural frequency, $\sqrt{k/m}$, in rad/s

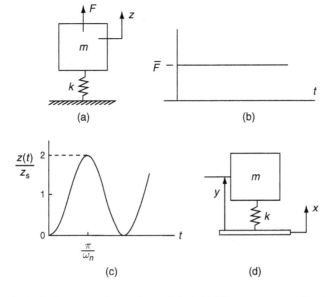

Fig. 3.11   Response of an undamped single-DOF system to a step input.

The static displacement, z_s, say, of the same system, if a *steady* force equal in magnitude to the step, $\bar{F}$, was applied (since then $\ddot{z} = 0$), is

$$z_s = \frac{\bar{F}}{k} \tag{3.40}$$

Then,

$$\frac{z}{z_s} = 1 - \cos \omega_n t \tag{3.41}$$

This is sketched in Fig. 3.11(c). The *dynamic factor (DF)* is defined in this case as the maximum displacement of the system, divided by the static displacement, when a static load equal to the peak value of the dynamic load is applied, i.e.,

$$DF = \frac{z_{max}}{z_s} \tag{3.42}$$

where z_{max} is the maximum value of z occurring anywhere in its time history. The dynamic factor is seen to be the maximum value of Eq. (3.41), which is 2 in this case, and it occurs at time $t = \pi/\omega_n$.

Dynamic factors can also be defined for systems excited via the base, such as that shown in Fig. 3.11(d). The equation of motion is in this case given by Eq. (2.4), but with the damping omitted, i.e.,

$$m\ddot{y} + ky = -m\ddot{x} \tag{3.43}$$

By comparing Eqs (3.38) and (3.43), we see that the spring displacement, y, relative to the base, when the base acceleration changes by a step, $\ddot{x}$, is

$$y = \frac{-m\ddot{x}}{k}(1 - \cos \omega_n t) \tag{3.44}$$

Also, from Eq. (3.43), it can be seen that the steady spring displacement, y_s, when a steady base acceleration, equal in magnitude to the acceleration step $\ddot{x}$, is applied, is:

$$y_s = \frac{-m\ddot{x}}{k} \tag{3.45}$$

then from Eqs (3.44) and (3.45),

$$\frac{y}{y_s} = (1 - \cos \omega_n t) \tag{3.46}$$

which is seen to be of the same form as Eq. (3.41), and the non-dimensional response y/y_s is the same as z/z_s. Clearly, the same result would be obtained whenever the applied force and base acceleration functions were of the same shape, and it follows that the same dynamic factor applies for both. It should be noted, however, that this applies only to the non-dimensional responses and dynamic factors, and that the displacements z and y are defined differently.

The approach used above is easily extended to damped single-DOF systems. In the case of a step force the non-dimensional response z/z_s was derived for $\gamma < 1$ in Example 2.3 as Eq. (H), i.e.,

$$\frac{z}{z_s} = 1 - e^{-\gamma\omega_n t}\left(\cos \omega_d t + \frac{\gamma}{\sqrt{1-\gamma^2}}\sin \omega_d t\right), \tag{3.47}$$

which can be solved for its maximum value for any value of γ. For small γ the sine term can be neglected and ω_d can be approximated by ω_n.

3.5 Response spectra

A response or shock spectrum, sometimes known as a *shock response spectrum, (SRS)*, describes the variation of the dynamic factor, as defined above, when a parameter that it depends upon is varied.

Response spectra will be derived for two important cases, a rectangular pulse and a sloping step. The input function will be assumed to be a force time history applied to the mass, i.e. the system shown in Fig. 3.11(a), but, as has been seen in the previous section, the results will, in fact, also apply to the system of Fig. 3.11(d), provided the base input function is acceleration. These are only two of the many ways that the input and response could be defined: the input could be base displacement or velocity, for example and the response could be velocity or acceleration, defined either as an absolute value at the mass or as a relative value between the mass and the base. Response spectra for many of these combinations, with many different input functions, are given by Ayre [3.2].

Response spectra are usually derived for an undamped system since this gives the worst case response and is not unduly pessimistic for light damping.

3.5.1 Response Spectrum for a Rectangular Pulse

The displacement response of an undamped single-DOF system to a rectangular force pulse can be found by adding the response to a positive step to that of a delayed negative step, as shown in Fig. 3.12. The positive step, of magnitude $\overline{F}$, and its response are shown in Fig. 3.12(a). This response is, from Eq. (3.39),

$$z = \frac{\overline{F}}{k}(1 - \cos \omega_n t)$$

or, non-dimensionally, from Eq. (3.41),

$$\frac{z}{z_s} = (1 - \cos \omega_n t) \tag{3.41}$$

Defining the duration of the pulse as T_0, the similar but negative step, $-\overline{F}$, and its response are shown in Fig. 3.12(b). This response is delayed by T_0. It does not exist for $t < T_0$, but for $t > T_0$ it is given by:

$$z = -\frac{\overline{F}}{k}[1 - \cos \omega_n(t - T_0)] \tag{3.48}$$

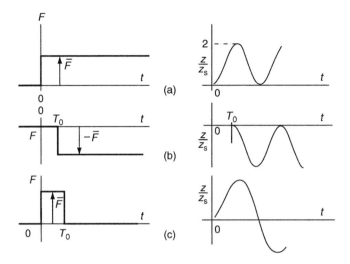

Fig. 3.12   Force pulse generated by superposition of two steps.

or, non-dimensionally:

$$\frac{z}{z_s} = -[1 - \cos \omega_n(t - T_0)] \tag{3.49}$$

In Fig. 3.12(c), the two steps are combined to form the pulse, and the response, z/z_s is as follows:

For $t < T_0$, from Eq. (3.41),

$$\frac{z}{z_s} = (1 - \cos \omega_n t)$$

For $t > T_0$, by adding the responses from Eq. (3.41) and Eq. (3.49),

$$\frac{z}{z_s} = \{(1 - \cos \omega_n t) - [1 - \cos \omega_n(t - T_0)]\}$$

or

$$\frac{z}{z_s} = [\cos \omega_n(t - T_0) - \cos \omega_n t] \tag{3.50}$$

Defining $T = 2\pi/\omega_n$ as the natural period of the system, if $T_0/T > 1/2$, the first peak will always occur before the end of the pulse, and the maximum value of z/z_s, equal to z_{max}/z_s, will be given by Eq. (3.41), and is equal to 2. However, if $T_0/T < 1/2$, z_{max}/z_s will be given by Eq. (3.50), and will be some value between 0 and 2 depending on T_0/T. This can be found by differentiating Eq. (3.50) and equating to zero, to find the time at which z_{max}/z_s occurs. Substituting this time into Eq. (3.50) then gives z_{max}/z_s as a function of T_0/T. This can be shown to be

$$\frac{z_{max}}{z_s} = 2 \sin\left(\pi \frac{T_0}{T}\right) \tag{3.51}$$

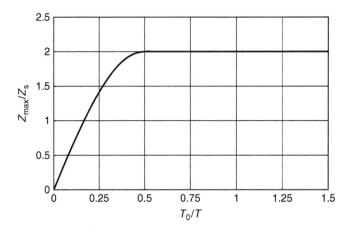

Fig. 3.13   Response spectrum for a rectangular pulse.

The complete response spectrum for the rectangular pulse is shown in Fig. 3.13.

This shows only the maximum value reached at any time, sometimes known as the *maximax*, which is usually of most interest. Some response spectra found in the literature also show the magnitude of subsequent peaks, known as the *residual spectrum*.

3.5.2 Response Spectrum for a Sloping Step

The displacement response of the same single-DOF system, as shown in Fig. 3.11(a), to a step with a sloping leading edge, can be found by combining the response to a positive ramp function, Fig. 3.14(a), with that of a negative and delayed ramp, Fig. 3.14(c). Adding these inputs produces the desired sloped step, as shown in Fig. 3.14(e).

The responses to the two ramps, shown in Figs 3.14(b) and (d), are similarly added, to produce the complete response, as shown in Fig. 3.14(f).

In this case, T_0 is defined as the rise-time of the sloping step, and T is the natural period of the system, as before. The response spectrum in this case is the peak value of the non-dimensional response shown in Fig. 3.14(f), z_{max}/z_s, plotted against T_0/T.

Using Example 3.4(b), but setting $\gamma = 0$, $m\omega_n^2 = k$, and $\omega_d = \omega_n$, the displacement response of an undamped single-DOF system to a ramp input force of slope $\overline{F}/T_0$ is

$$z = \frac{\overline{F}}{kT_0}\left(t - \frac{1}{\omega_n}\sin\omega_n t\right) \tag{3.52}$$

Since the static displacement, z_s, due to a steady force $\overline{F}$ is $z_s = \overline{F}/k$, Eq. (3.52) can be written as:

$$\frac{z}{z_s} = \frac{1}{T_0}\left(t - \frac{1}{\omega_n}\sin\omega_n t\right) \tag{3.53}$$

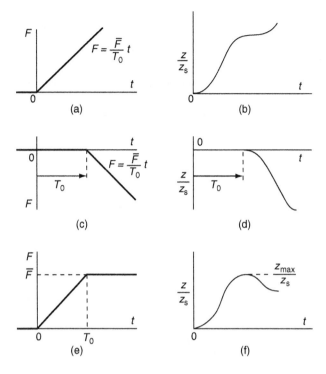

Fig. 3.14   Sloping step generated by superposition of two ramps.

and this is the response to the first, positive ramp. For $t < T_0$, this is the only response, and there are no response peaks during this time, all peaks occurring when $t > T_0$. The response to the second, negative ramp exists only for $t > T_0$, and is given by:

$$\frac{z}{z_s} = -\frac{1}{T_0}\left[(t - T_0) - \frac{\sin \omega_n(t - T_0)}{\omega_n}\right] \tag{3.54}$$

The response to the complete sloping step for $t > T_0$ is given by the sum of the responses from Eqs (3.53) and (3.54), and is

$$\frac{z}{z_s} = 1 + \frac{\sin \omega_n(t - T_0)}{T_0 \omega_n} - \frac{\sin \omega_n t}{T_0 \omega_n} \tag{3.55}$$

By differentiating Eq. (3.55) and equating to zero, the values of t at which maxima occur can be found, and substituting these into Eq. (3.55) gives

$$\frac{z_{max}}{z_s} = 1 + \frac{1}{T_0 \omega_n}\sqrt{2(1 - \cos T_0 \omega_n)} \tag{3.56}$$

or, since $T = 2\pi/\omega_n$,

$$\frac{z_{max}}{z_s} = 1 + \frac{1}{2\pi(T_0/T)}\sqrt{2[1 - \cos 2\pi(T_0/T)]} \tag{3.57}$$

Figure 3.15 is the response spectrum for the sloping step, plotted from Eq. (3.57) as a function of T_0/T:

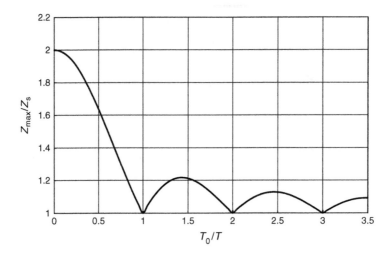

Fig. 3.15   Response spectrum for a sloping step.

References

3.1 Huntley, E (1966). *A Matrix Formulation for the Time Response of Time-Invariant Linear Systems to Discrete Inputs.* (British) Royal Aircraft Establishment, Technical Report No. 66038, February 1966.
3.2 Harris, CM and Piersol, AG, eds (2002). *Shock and Vibration Handbook.* Fifth Edition, Chapter 8 by Ayre, R.S. McGraw-Hill.

4 The Linear Single Degree of Freedom System: Response in the Frequency Domain

Contents

In this chapter, the response of single-DOF systems to sinusoidal or simple harmonic excitation is discussed. The two single-DOF configurations discussed in Chapter 3 are included, i.e. excitation by a directly applied force and excitation by base, or support, motion. However, there is now a third case to consider, excitation by a rotating unbalanced mass. Figure 4.1 shows the three standard configurations.

The following notation is used throughout:

For all configurations:

m = total moving mass;

k = spring stiffness, $k = m\omega_n$ for single-DOF system;

c = dimensional viscous damping coefficient;

ω = forcing frequency, rad/s;

f = forcing frequency in Hz, where $f = \omega/2\pi$;

ω_n = undamped natural frequency, rad/s;

f_n = undamped natural frequency, Hz, where $f_n = \omega_n/2\pi$;

ω_d = damped natural frequency, rad/s, $\omega_d = \omega_n\sqrt{1 - \gamma^2}$;

γ = non-dimensional viscous damping coefficient. $\gamma = c/2m\omega_n$;

Ω = frequency ratio $\omega/\omega_n = f/f_n$.

For Fig. 4.1(c) only:

$\overline{m}$ = rotating unbalanced mass.

r = radius arm for unbalanced mass.

4.1 Response of a single degree of freedom system with applied force

4.1.1 Response Expressed as Amplitude and Phase

The first configuration to consider is that shown in Fig. 4.1(a). The equation of motion is

$$m\ddot{z} + c\dot{z} + kz = F \tag{4.1}$$

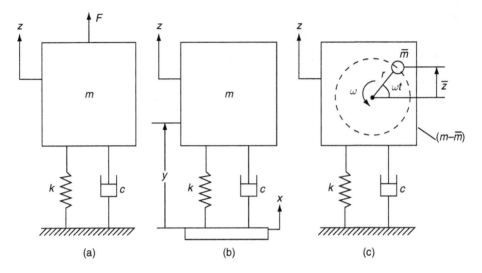

Fig. 4.1   Common single-DOF configurations in schematic form: (a) direct force excitation, (b): base motion excitation, (c) rotating unbalance excitation.

or, written in 'standard form', by substituting $k = m\omega_n^2$ and $c = 2m\omega_n$:

$$\ddot{z} + 2\gamma\omega_n\dot{z} + \omega_n^2 z = F/m \tag{4.2}$$

When the force, F, is, for example, defined as $F = P \sin \omega t$, the response consists of two parts, the complementary function and the particular integral. It was shown in Example 2.4 that the total response is

$$z = e^{-\gamma\omega_n t}(A \cos \omega_d t + B \sin \omega_d t) + C \cos \omega t + D \sin \omega t \tag{4.3}$$

where the first two terms, those including $e^{-\gamma\omega_n t}$, are the 'complementary function' or transient terms. If damping is present, these components soon die away, leaving only the last two terms, due to the 'particular integral'. The *steady-state* displacement response to a sinusoidal force $F = P \sin \omega t$ is therefore simply:

$$z = C \cos \omega t + D \sin \omega t \tag{4.4}$$

The constants C and D were given in Example 2.4, as Eqs (J) and (K) and are

$$C = \frac{P(-2\gamma\omega_n\omega)}{m\left[(\omega_n^2 - \omega^2)^2 + (2\gamma\omega_n\omega)^2\right]} \tag{4.5}$$

$$D = \frac{P(\omega_n^2 - \omega^2)}{m\left[(\omega_n^2 - \omega^2)^2 + (2\gamma\omega_n\omega)^2\right]} \tag{4.6}$$

It is neater to express Eqs (4.5) and (4.6) in terms of the frequency ratio, Ω, where $\Omega = \omega/\omega_n = f/f_n$. With this substitution, and also putting $k = m\omega_n^2$:

$$C = \frac{P}{k} \cdot \frac{-2\gamma\Omega}{(1 - \Omega^2)^2 + (2\gamma\Omega)^2} \tag{4.7}$$

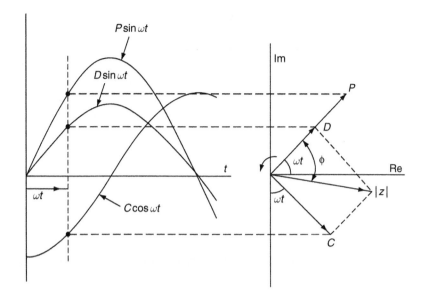

Fig. 4.2

$$D = \frac{P}{k} \cdot \frac{1 - \Omega^2}{\left(1 - \Omega^2\right)^2 + (2\gamma\Omega)^2} \tag{4.8}$$

So the steady-state relationship between the force, F, and the displacement, z, is such that when $F = P \sin \omega t$, then $z = C \cos \omega t + D \sin \omega t$, where C and D are as given above.

It can be seen from Eq. (4.7) that C is negative for any value of Ω, and from Eq. (4.8) that D is positive for $\Omega < 1$, and negative for $\Omega > 1$. So considering a case when $\Omega < 1$, the waveforms $P \sin \omega t$, $C \cos \omega t$, and $D \sin \omega t$ must be as sketched on the left in Fig. 4.2.

It is possible to find the amplitude and phase of the displacement response, z, relative to the force, F, by trigonometry alone, but as is often the case, using complex rotating vectors makes it easier. The waveforms on the left of Fig. 4.2 can be related to the vectors on the right, using the relationships introduced in Chapter 1. It can be seen from the right side of Fig. 4.2 that the magnitude, or modulus, of z, say $|z|$, is given by:

$$|z| = \sqrt{C^2 + D^2}$$

Substituting for C and D from Eqs (4.7) and (4.8) leads to:

$$|z| = \frac{P}{k} \cdot \frac{1}{\sqrt{\left(1 - \Omega^2\right)^2 + (2\gamma\Omega)^2}} \tag{4.9}$$

The phase angle, ϕ, by which the displacement vector, $|z|$, lags the force, P, is given by:

$$\tan \phi = \frac{-C}{D} = \frac{2\gamma\Omega}{1 - \Omega^2} \tag{4.10}$$

Equation (4.9) can also be expressed in non-dimensional form. From Eq. (4.1), the static displacement z_s of the system, when a static force equal to the peak value, P, of the applied force, $F = P \sin \omega t$, is applied (since $\dot{z} = \ddot{z} = 0$) is

$$z_s = \frac{P}{k} \qquad\qquad (4.11)$$

Using Eq. (4.11), Eq. (4.9) can be written in the form:

$$\frac{|z|}{z_s} = \frac{1}{\sqrt{\left(1 - \Omega^2\right)^2 + (2\gamma\Omega)^2}} \qquad\qquad (4.12)$$

The phase lag, ϕ, is, of course, unaffected, and is still given by Eq. (4.10).

Example 4.1

Plot graphs showing $|z|/z_s$ and ϕ versus Ω for the damping ratios, $\gamma = 0.5, 0.2, 0.1, 0.05, 0.03$ and 0.02, for the system shown in Fig. 4.1(a), and comment on the results.

Solution

The magnitude of the steady-state displacement amplitude, as a multiple of the static displacement for a static load equal to the peak value of the exciting force, is given by Eq. (4.12), and the phase angle, ϕ, by which the displacement lags the force is given by Eq. (4.10). These are plotted as Figs 4.3(a) and (b), respectively.

All the magnitude plots show a clear resonant peak near $\Omega = 1$, i.e. where the excitation frequency, ω, equals the undamped natural frequency, ω_n, except for the case with $\gamma = 0.5$, i.e. half of critical damping. It can be shown that the true peak response occurs at the frequency ratio, $\Omega\sqrt{1 - 2\gamma^2}$, and its value is $1/(2\gamma\sqrt{1 - \gamma^2})$. For small damping, however, these values are so close to Ω and $1/2\gamma$, respectively, that most plots do not show the difference.

An amplitude response, such as Fig. 4.3(a), can be used to obtain a rough estimate of damping from a measured response. It can be shown that at frequencies f_1 and f_2 which are, respectively, γf_n below and γf_n above, f_n, the amplitude is $1/\sqrt{2}$ or 0.707 times the maximum value. Then the damping coefficient, γ, is given, to a good approximation, by: $\gamma = (f_2 - f_1)/2f_n$. The frequencies f_1 and f_2 are known as the half-power points.

When the excitation frequency is much smaller than the natural frequency, the system behaves like a spring, and the dynamic response is close to the static displacement, so $|z|/z_s \approx 1$. At the other extreme, when the excitation frequency is much higher than the natural frequency, the system tends to behave like a mass.

From the phase plot, Fig. 4.3(b), it can be seen that the displacement lags the force by an angle that is near zero at very low excitation frequency, and approaches π, or 180°, at high excitation frequency. However, for any value of damping, the phase lag is always $\pi/2$, or 90°, when the excitation is at the undamped natural frequency, a fact useful in test work.

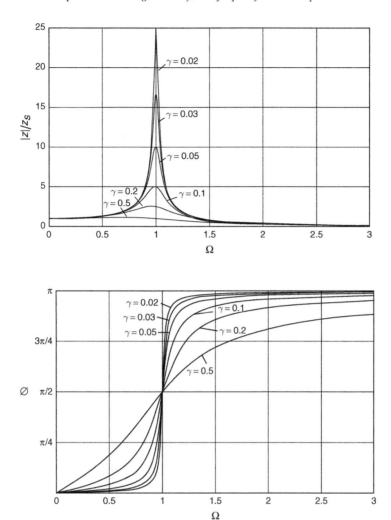

Fig. 4.3   (a) Non-dimensional magnitude response of a forced dingle-DOF system, (b) phase angle corresponding to (a).

4.1.2 Complex Response Functions

In Fig. 4.2, it was seen that the functions of time on the left are given by the projections, on the vertical axis, of the rotating vectors on the right. We know that at any frequency, ω, the displacement vector, $\underline{z}$, has a fixed complex relationship, $(a + ib)$, say, to the force vector, $\underline{F}$, i.e.,

$$\underline{z} = (a + ib)\underline{F} \qquad (4.13)$$

Thus if we define the force vector by:

$$F = \underline{F}e^{i\omega t} \qquad (4.14)$$

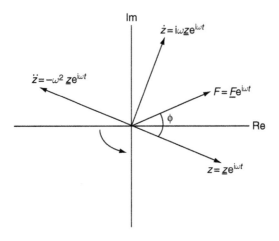

Fig. 4.4

then the displacement vector, $\underline{z}$, is defined by:

$$z = \underline{z}e^{i\omega t} = (a + ib)\underline{F}e^{i\omega t}, \tag{4.15}$$

and the velocity and acceleration vectors are defined by:

$$\dot{z} = \underline{\dot{z}}e^{i\omega t} = i\omega(a + ib)\underline{F}e^{i\omega t} \tag{4.16}$$

and

$$\ddot{z} = \underline{\ddot{z}}e^{i\omega t} = -\omega^2(a + ib)\underline{F}e^{i\omega t} \tag{4.17}$$

These quantities are shown in Fig. 4.4.

Substituting Eqs (4.14)–(4.17) into the equation of motion, Eq. (4.2), $\ddot{z} + 2\gamma\omega_n\dot{z} + \omega_n^2 z = F/m$, and dividing through by $\underline{F}e^{i\omega t}$ gives

$$(a + ib)\left(-\omega^2 + i\omega \cdot 2\gamma\omega_n + \omega_n^2\right) = \frac{1}{m} \tag{4.18}$$

But from Eq. (4.13), $(a + ib) = \underline{z}/\underline{F}$, so:

$$\frac{\underline{z}}{\underline{F}} = \frac{1}{m\left(\omega_n^2 - \omega^2 + i\omega \cdot 2\gamma\omega_n\right)} \tag{4.19}$$

or

$$\frac{\underline{z}}{\underline{F}} = \frac{1}{m\omega_n^2(1 - \Omega^2 + i \cdot 2\gamma\Omega)} \tag{4.20}$$

where

$$\Omega = \omega/\omega_n$$

Rationalizing Eq. (4.20) by multiplying numerator and denominator by $(1 - \Omega^2 - i \cdot 2\gamma\Omega)$:

$$\frac{z}{F} = \frac{\left(1 - \Omega^2\right) - i(2\gamma\Omega)}{m\omega_n^2\left[\left(1 - \Omega^2\right)^2 + (2\gamma\Omega)^2\right]} \tag{4.21}$$

Since $m\omega_n^2 = k$, it is sometimes convenient to write Eq. (4.21) in the form:

$$\frac{z}{F} = \frac{\left(1 - \Omega^2\right) - i(2\gamma\Omega)}{k\left[\left(1 - \Omega^2\right)^2 + (2\gamma\Omega)^2\right]} \tag{4.22}$$

The quantity z/F is an example of a *frequency response function,* FRF, in this case expressed in complex form. FRFs in general are discussed in more detail in the next section.

4.1.3 Frequency Response Functions

The response of a single-DOF system to a directly applied force has been expressed above as a displacement, either in the form of amplitude and phase or as a complex number. The response can equally well be expressed in other ways, such as velocity or acceleration. Once we have an expression for one of these, it is easy to derive the others; for example, if the response is given in the form of complex displacement, versus the excitation frequency, ω, in radians/second, it is simply multiplied by $i\omega$ to produce complex velocity or by $(i\omega)^2 = -\omega^2$ to produce complex acceleration. In the case of displacement expressed as amplitude and phase, the same effect is achieved by multiplying the displacement magnitude by ω to produce velocity, and increasing the phase angle, ϕ, by $\pi/2$ radians or 90°. To convert from displacement to acceleration, the amplitude is multiplied by ω^2 and the phase angle is increased by π radians or 180°.

The response in the form of displacement, velocity or acceleration, divided by the force, is known as a *standard frequency response function,* irrespective of whether it is plotted as a complex quantity or as amplitude and phase. The inverse quantities, the applied force divided by displacement, velocity or acceleration, are known as *inverse FRFs*. The six resulting FRFs have names, in some cases several alternate names, as shown in Table 4.1. Although given here for reference, the use of names for FRFs can cause confusion; for example, the name *impedance* could be thought to mean 'force per unit displacement', but actually means 'force per unit velocity'. It is therefore recommended that if the names are used, the meaning should also be given, for example: ... 'mobility (velocity per unit force)'.

Table 4.1
Names of Frequency Response Functions

Standard FRFs		Inverse FRFs	
z/F	Receptance, Admittance, Dynamic Compliance, or Dynamic Flexibility.	F/z	Dynamic stiffness
$\dot{z}/F$	Mobility	$F/\dot{z}$	Impedance
$\ddot{z}/F$	Inertance or Accelerance	$F/\ddot{z}$	Apparent mass

Example 4.2

(a) Write expressions for:
   (i) the complex receptance, $\underline{z}/\underline{F}$; (ii) the complex mobility, $\underline{\dot{z}}/\underline{F}$; (iii) the complex inertance, $\underline{\ddot{z}}/\underline{F}$, of a single-DOF system with direct force excitation, having the following parameters:
   mass, $m = 1$ kg;
   undamped natural frequency, $f_n = 10$ Hz;
   viscous damping coefficient, $\gamma = 0.05$.
(b) Plot the results for the excitation frequency range $f = 0$–30 Hz in each case, as 'Nyquist' or 'circle' plots. These are plots where the real part is plotted versus the imaginary part on an Argand diagram. Show spot frequencies on the plots in Hz.
(c) Comment on the results.

Solution

Part (a)(i):
The complex receptance, $\underline{z}/\underline{F}$, is given by Eq. (4.21):

$$\frac{\underline{z}}{\underline{F}} = \frac{\left(1 - \Omega^2\right) - i(2\gamma\Omega)}{m\omega_n^2\left[\left(1 - \Omega^2\right)^2 + (2\gamma\Omega)^2\right]} \qquad (A)$$

where

$$\Omega = \frac{f}{f_n} = \frac{\omega}{\omega_n}$$

f = forcing frequency in Hz. $f = 0$–30 Hz
f_n = undamped natural frequency in Hz = 10 Hz
ω = forcing frequency in rad/s
ω_n = undamped natural frequency in rad/s = 20π
m = mass = 1 kg
γ = non-dimensional viscous damping coefficient = 0.05.

Part (a)(ii):
The complex mobility, $\underline{\dot{z}}/\underline{F}$, is given by multiplying the complex receptance $\underline{z}/\underline{F}$ by $i\omega$:

$$\frac{\underline{\dot{z}}}{\underline{F}} = \frac{i\omega\left[\left(1 - \Omega^2\right) - i(2\gamma\Omega)\right]}{m\omega_n^2\left[\left(1 - \Omega^2\right)^2 + (2\gamma\Omega)^2\right]} = \frac{2\pi f\left[(2\gamma\Omega) + i\left(1 - \Omega^2\right)\right]}{m\omega_n^2\left[\left(1 - \Omega^2\right)^2 + (2\gamma\Omega)^2\right]} \qquad (B)$$

Part (a)(iii):
The complex inertance, $\underline{\ddot{z}}/\underline{F}$, is given by multiplying the mobility by $i\omega$ or the receptance $\underline{z}/\underline{F}$ by $(i\omega)^2 = -\omega^2$:

$$\frac{\underline{\ddot{z}}}{\underline{F}} = \frac{-\omega^2\left[\left(1 - \Omega^2\right) - i(2\gamma\Omega)\right]}{m\omega_n^2\left[\left(1 - \Omega^2\right)^2 + (2\gamma\Omega)^2\right]} = \frac{-\Omega^2\left[\left(1 - \Omega^2\right) - i(2\gamma\Omega)\right]}{m\left[\left(1 - \Omega^2\right)^2 + (2\gamma\Omega)^2\right]} \qquad (C)$$

Part (b):

The three plots are shown as Figs 4.5(a), (b) and (c). The following notes refer to the units used.

(i)   The unit of the receptance, as given by Eq. (A), is m/N (meters/newton). This has been changed to mm/N, by multiplying by 1000, in the plot, Fig. 4.5(a).

(ii) The unit of mobility, as given by Eq. (B), is m/s/N, and are plotted in this form in Fig. 4.5(b).

(iii) The unit of inertance, as given by Eq. (C), is $m/s^2/N$. This has been changed to the more usual g/N, by dividing by 9.81, in the plot, Fig. 4.5(c)

Part (c):

'Circle' or 'Nyquist' plots were introduced into vibration analysis by Kennedy and Pancu [4.1] in 1947, and still provide a simple and fairly accurate method for analysing vibration test data. Plots from vibration tests can be used to estimate a natural frequency and damping coefficient of a system, as follows:

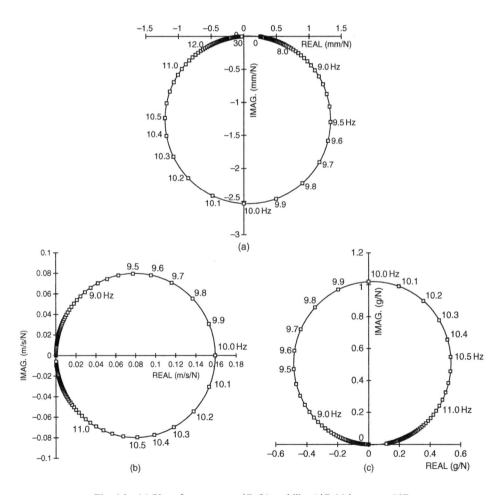

Fig. 4.5   (a) Plot of receptance $\underline{z}/\underline{F}$, (b) mobility $\underline{\dot{z}}/\underline{F}$, (c) inertance $\underline{\ddot{z}}/\underline{F}$.

(1) The natural frequency can be identified as the point at which the spacing, on the plot, for equal increments of the excitation frequency, is widest. So from any of the three figures shown, the natural frequency can be seen to be 10 Hz.

(2) The method of estimating the damping coefficient of a system from the half-power points of the amplitude response, mentioned in Example 4.1, can usually be applied, somewhat more accurately, in the case of Nyquist plots such as Figs 4.5(a)–(c). The 'half-power' frequencies f_1 and f_2 can now be identified as:

f_1 = excitation frequency at a phase angle 45° below the phase of the resonant frequency, f_n;

f_2 = excitation frequency at a phase angle 45° above the phase of the resonant frequency, f_n.

Then, as before, an estimate of γ is $\gamma = ((f_2 - f_1)/2f_n)$. From any of the three plots, this gives $\gamma = ((10.5 - 9.5)/(2 \times 10.0)) = 0.05$, which was, of course, the value of γ used in plotting the curves.

Although all three plots, Figs 4.5(a)–(c), look circular, it can be shown that with viscous damping, as used here, only the mobility plot, Fig. 4.5(b), is a true circle. With hysteretic damping, as discussed in Chapter 5, only the receptance plot is a true circle.

Frequency response functions can be plotted in a number of ways, depending upon the application. The use of the 'Nyquist' or 'circle' plot has been illustrated in Example 4.2. This is sometimes the best way to plot FRFs, but it does not give a good overall view of the response versus frequency. Amplitude and phase plots, of which Figs 4.3(a) and (b), together, form an example, are better in this respect, and in vibration tests it is usual to plot these first, using Nyquist plots to define the areas near resonance more precisely.

When log scales are used for both magnitude and frequency, amplitude plots are sometimes known as 'Bode plots'.

A third possible way to plot FRFs is as graphs of the real and imaginary parts versus excitation frequency.

For a useful discussion of different ways to plot FRFs, Chapter 2 of 'Modal Testing' by Ewins [4.2] is recommended.

4.2 Single-DOF system excited by base motion

In the second standard configuration, shown in Fig. 4.1(b), the excitation is provided by motion of the base instead of by a force applied to the mass. The schematic diagram is repeated as Fig. 4.6.

There are now more possible input/output combinations to consider. The input base motion may be defined, for example, in any of the three ways: as a displacement, x, a velocity, $\dot{x}$, or an acceleration, $\ddot{x}$. The response can also be expressed in any of these ways, and there are two different responses to consider: the absolute motion of the mass, z, $\dot{z}$ or $\ddot{z}$, or its motion y, $\dot{y}$ or $\ddot{y}$ relative to the base.

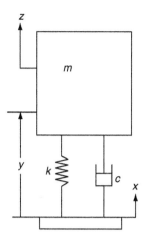

Fig. 4.6   Schematic diagram for base excitation of a single-DOF system.

4.2.1 Base Excitation, Relative Response

The equation of motion of a single-DOF system with base excitation was derived in Section 2.1.2. From Eq. (2.4), we have

$$m\ddot{y} + c\dot{y} + ky = -m\ddot{x}$$

Expressing this equation in complex vector form, so that, for example, y and x, which are functions of time, t, become $\underline{y}$ and $\underline{x}$, to indicate that they are functions of $i\omega$, and also making the usual substitutions, $c = 2\gamma\omega_n m$ and $k = m\omega_n^2$, this becomes.

$$\ddot{\underline{y}} + 2\gamma\omega_n\dot{\underline{y}} + \omega_n^2\underline{y} = -\ddot{\underline{x}} \tag{4.23}$$

If we require, say, the complex frequency response ratio $\underline{y}/\underline{x}$, we can use the method given in Section 4.1.2, as follows. The following substitutions are made into Eq. (4.23):

$$\dot{\underline{y}} = i\omega\underline{y} \qquad \ddot{\underline{y}} = -\omega^2\underline{y} \qquad \ddot{\underline{x}} = -\omega^2\underline{x} \tag{4.24}$$

With these substitutions, Eq. (4.23) can be written as:

$$\frac{\underline{y}}{\underline{x}} = \frac{\omega^2}{\left(\omega_n^2 - \omega^2 + i\omega.2\gamma\omega_n\right)} = \frac{\Omega^2\left[\left(1 - \Omega^2\right) - i(2\gamma\Omega)\right]}{\left[\left(1 - \Omega^2\right)^2 + (2\gamma\Omega)^2\right]} \tag{4.25}$$

where

$$\Omega = \omega/\omega_n$$

Equation (4.25) can also be expressed in amplitude and phase form. The amplitude ratio, $|y|/|x|$, is

$$\frac{|y|}{|x|} = \frac{\Omega^2}{\sqrt{\left(1 - \Omega^2\right)^2 + (2\gamma\Omega)^2}} \tag{4.26}$$

and the phase angle, ϕ, by which the relative displacement, y, lags the base displacement, x, is given by:

$$\tan \phi = \frac{2\gamma\Omega}{1 - \Omega^2} \tag{4.27}$$

Equations (4.26) and (4.27) are shown graphically in Figs 4.7(a) and (b).

The relative motion between the mass and the base, and the input base motion, can easily be expressed in terms of velocity or acceleration, as well as displacement.

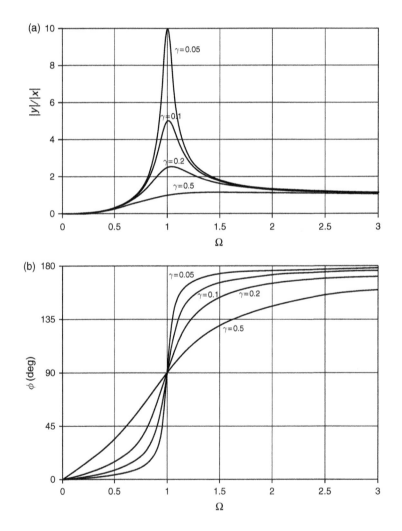

Fig. 4.7   (a) Relative displacement between mass and base per unit base displacement, (b) phase angle corresponding to (a).

If, as in Eq. (4.25), complex notation is being used, then the following relationships apply:

$$\underline{\dot{x}} = i\omega\underline{x}, \quad \underline{\ddot{x}} = -\omega^2\underline{x}, \quad \underline{\dot{y}} = i\omega\underline{y}, \quad \underline{\ddot{y}} = -\omega^2\underline{y} \tag{4.28}$$

Then it can be seen that

$$\frac{\underline{\dot{y}}}{\underline{\dot{x}}} = \frac{i\omega\underline{y}}{i\omega\underline{x}} = \frac{\underline{y}}{\underline{x}} \quad \text{and} \quad \frac{\underline{\ddot{y}}}{\underline{\ddot{x}}} = \frac{-\omega^2\underline{y}}{-\omega^2\underline{x}} = \frac{\underline{y}}{\underline{x}} \tag{4.29}$$

so Eq. (4.25) applies unaltered for the complex ratios $\underline{\dot{y}}/\underline{\dot{x}}$ and $\underline{\ddot{y}}/\underline{\ddot{x}}$.

Other complex ratios that could be required are easily found using Eqs (4.28). Some examples are

$$\frac{\underline{y}}{\underline{\dot{x}}} = \frac{1}{i\omega}\left(\frac{\underline{y}}{\underline{x}}\right), \quad \frac{\underline{y}}{\underline{\ddot{x}}} = \frac{-1}{\omega^2}\left(\frac{\underline{y}}{\underline{x}}\right), \quad \frac{\underline{\dot{y}}}{\underline{x}} = i\omega\left(\frac{\underline{y}}{\underline{x}}\right), \quad \frac{\underline{\ddot{y}}}{\underline{x}} = -\omega^2\left(\frac{\underline{y}}{\underline{x}}\right) \tag{4.30}$$

where $(\underline{y}/\underline{x})$ is given by Eq. (4.25) in all cases.

If an amplitude and phase representation, rather than a complex ratio, is required, Eq. (4.28) can be expressed in the form:

$$|\dot{x}| = \omega|x|, \quad |\ddot{x}| = \omega^2|x|, \quad |\dot{y}| = \omega|y|, \quad |\ddot{y}| = \omega^2|y| \tag{4.31}$$

where $|x|$, $|\dot{x}|$, $|\ddot{y}|$, etc. are single-peak or modulus amplitudes. Corresponding versions of Eq. (4.30) can also be formed, i.e,

$$\frac{|y|}{|\dot{x}|} = \frac{1}{\omega} \cdot \left(\frac{|y|}{|x|}\right), \quad \frac{|y|}{|\ddot{x}|} = \frac{1}{\omega^2} \cdot \left(\frac{|y|}{|x|}\right), \quad \frac{|\dot{y}|}{|x|} = \omega \cdot \left(\frac{|y|}{|x|}\right), \quad \frac{|\ddot{y}|}{|x|} = \omega^2 \cdot \left(\frac{|y|}{|x|}\right) \tag{4.32}$$

where $|y|/|x|$ is given by Eq. (4.26) in all cases.

Remembering that the phase angle, ϕ, given by Eq. (4.27) is defined as the angle by which the relative displacement, y, lags the base displacement, x, then ϕ is reduced by $\pi/2$ or 90° for each multiplication by i in the complex plane.

Example 4.3

Most accelerometers consist of a small single-DOF mass–spring–damper system, as represented by Fig. 4.6. The displacement, y, of the mass relative to the base is measured, typically by semi-conductor strain gages on the spring, or by the piezo-electric effect. For the frequency range from zero up to some fraction of the natural frequency, the output from the gages is nearly proportional to the acceleration applied at the base.

(a) Find expressions for the frequency response of such an accelerometer:
    (i) in complex form;
    (ii) in the form of ampltude and phase.
(b) Comment on the results.

Solution

Part (a)(i):
The complex ratio giving the displacement, y, of the mass relative to the base, per unit base displacement, x, is given by Eq. (4.25):

$$\frac{y}{x} = \frac{\Omega^2\left[(1-\Omega^2) - i(2\gamma\Omega)\right]}{\left[(1-\Omega^2)^2 + (2\gamma\Omega)^2\right]} \quad\quad\quad (A)$$

However, in the case of the accelerometer, we require the complex ratio $y/\ddot{x}$. From the second part of Eq. (4.30):

$$\frac{y}{\ddot{x}} = \frac{-1}{\omega^2} \cdot \left(\frac{y}{x}\right), \quad\quad\quad (B)$$

so the required complex response is given by substituting Eq. (A) into Eq. (B), giving:

$$\frac{y}{\ddot{x}} = \frac{-\Omega^2\left[(1-\Omega^2) - i(2\gamma\Omega)\right]}{\omega^2\left[(1-\Omega^2)^2 + (2\gamma\Omega)^2\right]} \quad\quad\quad (C)$$

or since $\Omega = \omega/\omega_n$:

$$\frac{y}{\ddot{x}} = -\frac{1}{\omega_n^2}\frac{\left[(1-\Omega^2) - i(2\gamma\Omega)\right]}{\left[(1-\Omega^2)^2 + (2\gamma\Omega)^2\right]} \qu\quad\quad (D)$$

This is the complex relationship between the base acceleration and the output of the accelerometer, which is proportional to the spring displacement.

Part (a)(ii):
Equation (D) can also be expressed in amplitude and phase form. The spring displacement magnitude $|y|$ per unit of base acceleration magnitude, $|\ddot{x}|$, is

$$\frac{|y|}{|\ddot{x}|} = \frac{1}{\omega_n^2} \cdot \frac{1}{\sqrt{(1-\Omega^2)^2 + (2\gamma\Omega)^2}} \quad\quad\quad (E)$$

By considering the complex response $y/\ddot{x}$ given by Eq. (D), y apparently lags $\ddot{x}$ by 180° at $\Omega = 0$, by 270° at $\Omega = 1$, and approaches 360° lag at high values of Ω. This is essentially due to the way we have defined x and y, which is arbitrary, and we can reverse the phase of the output, so that y is in phase with $\ddot{x}$ at $\Omega = 0$, lags by 90° at $\Omega = 1$, and so on. The phase lag ϕ is then given by:

$$\tan\phi = \frac{2\gamma\Omega}{1 - \Omega^2} \quad\quad\quad (F)$$

which is the same as both Eqs (4.10) and (4.27).

Part (b):

It is interesting to note that, ignoring overall constants, the response of an accelerometer, expressed as relative base to mass displacement for base acceleration input, as shown by Eqs (D)–(F), is the same as the displacement response of a single-DOF system for direct force excitation. The latter was given by Eqs (4.9), (4.10), (4.21) and (4.22). Therefore Figs 4.3(a) and (b) can also be interpreted as the amplitude and phase characteristics of an accelerometer, ignoring constant factors.

It can be seen from Fig. 4.3(a) that high damping is required to obtain a 'flat' response for frequencies up to about half the natural frequency. However, this also introduces phase error. With low damping, the phase error is small for a wider frequency band, but the magnitude error becomes large at frequencies approaching the natural frequency, and accelerometers with low damping, which are the majority, are used only for frequencies up to about 20% of the natural frequency, limiting the theoretical error to about 6% compared with the flat portion of the curve.

4.2.2 Base Excitation: Absolute Response

For some applications, the absolute motion of the mass, shown as z in Fig. 4.6, is of more interest than its motion relative to the base, y, as discussed above. The equation of motion relating base motion, x, to z was derived in Chapter 2 as Eq. (2.5):

$$m\ddot{z} + c\dot{z} + kz = c\dot{x} + kx$$

Using the complex method, as before, and introducing the usual substitutions, $c = 2\gamma\omega_n m$ and $k = m\omega_n^2$, this becomes

$$\left[-\omega^2 + i\omega(2\gamma\omega_n) + \omega_n^2\right]\underline{z} = \left[i\omega(2\gamma\omega_n) + \omega_n^2\right]\underline{x} \tag{4.33}$$

or

$$\frac{\underline{z}}{\underline{x}} = \frac{1 + i(2\gamma\Omega)}{(1 - \Omega^2) + i(2\gamma\Omega)} \tag{4.34}$$

where

$$\Omega = \omega/\omega_n$$

From Eq. (4.34) the amplitude ratio is

$$\frac{|z|}{|x|} = \sqrt{\frac{1 + (2\gamma\Omega)^2}{(1 - \Omega^2)^2 + (2\gamma\Omega)^2}} \tag{4.35}$$

and the corresponding phase lag is

$$\tan\varphi = \frac{2\gamma\Omega^3}{1 - \Omega^2 + (2\gamma\Omega)^2} \tag{4.36}$$

Equation (4.35) gives the absolute displacement of the mass per unit displacement of the base, known as the *displacement transmissibility*. This is the ratio by which the motion of a mass supported by springs and dampers is reduced (or increased) compared with the motion of the base or supports. This should not be confused with *force transmissibility*, which is the ratio of the *force* transmitted to the supports, assumed fixed, when a sinusoidal force is applied to the mass, as discussed in Section 4.3. Surprisingly, it will be seen that force transmissibility, although a completely different concept from displacement transmissibility, is given by the same expression.

Example 4.4

(a) Plot the vertical displacement transmissiblity, $|z|/|x|$, of a mass–spring–damper system, as shown in Fig. 4.6, versus Ω, the excitation frequency divided by the undamped natural frequency, for the following viscous damping ratios: $\gamma = 0.05$, 0.1, 0.2, 0.5 and 1.0.
(b) Discuss the application of the curves to vibration isolation systems.

Solution

(a) The displacement transmissibility is shown in Fig. 4.8, plotted from Eq. (4.35).
(b) Electronics equipment, and other delicate objects, can be isolated from the vibrating structure to which they are attached, by interposing some form of spring and damper between them. Considering sinusoidal motion of the base as the input, and the absolute response of the mass as the output, it is easily shown that this configuration magnifies components of vibration for $\Omega < \sqrt{2}$, and attenuates them for $\Omega > \sqrt{2}$. So vibration reduction only occurs for frequencies greater than $\sqrt{2}$ times the natural frequency of the system, and it is important to recognize

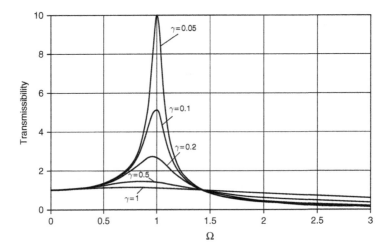

Fig. 4.8

that for input frequencies lower than this, the vibration level will be increased, possibly by a very large factor if the excitation frequency is close to the natural frequency.

The effect of damping is to reduce the response close to the resonant frequency, which is desirable if the excitation frequency is ever in that band. On the other hand, increased damping actually reduces the vibration attenuation at higher frequencies.

Vibration isolation is a specialized subject, further discussed in Chapter 11.

4.3 Force transmissibility

Figure 4.9 repeats the mass–spring–damper model shown in Fig. 2.1, but this time, instead of considering only the free body diagram for the mass, we also consider the forces applied to the base, or support structure, which in this case is assumed fixed.

From Eq. (2.1a), we have

$$m\ddot{z} + c\dot{z} + kz = F \tag{4.37}$$

which was, of course, obtained from the free body diagram. However, Fig. 4.9 also shows the reaction forces applied to the base. If the sum of these forces is F_T, then,

$$F_T = kz + c\dot{z} \tag{4.38}$$

The *force transmissibility* is defined as the ratio $|F_T|/|F|$, the magnitude of the total reaction force divided by the magnitude of the applied force.

Expressing Eqs (4.37) and (4.38) in complex form, and making the usual substitutions, $c = 2\gamma\omega_n m$ and $k = m\omega_n^2$, they become

$$\underline{F} = \left(-\omega^2 m + i\omega \cdot 2\gamma m\omega_n + m\omega_n^2\right)\underline{z} \tag{4.39}$$

and

$$\underline{F}_T = \left(i\omega \cdot 2\gamma m\omega_n + m\omega_n^2\right)\underline{z} \tag{4.40}$$

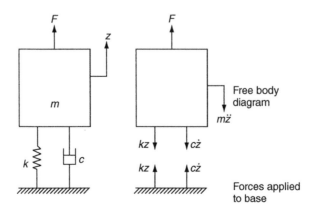

Fig. 4.9   Single-DOF model for force transmissibilty.

It is easily shown that

$$\frac{F_T}{F} = \frac{1 + i \cdot 2\gamma\Omega}{(1 - \Omega^2) + i \cdot 2\gamma\Omega} \tag{4.41}$$

where $\Omega = \omega/\omega_n$, and the magnitude ratio is

$$\frac{|F_T|}{|F|} = \sqrt{\frac{1 + (2\gamma\Omega)^2}{(1 - \Omega^2)^2 + (2\gamma\Omega)^2}} \tag{4.42}$$

The right side of Eq. (4.42) is seen to be precisely the same as that of Eq. (4.35), the expression for displacement transmissibility, $|z|/|x|$, when a similar system is excited via the base. However, the base is fixed in the case of Eq. (4.42), so although displacement transmissibility and force transmissibility are given by the same expression, the system configurations are not quite the same.

It follows, of course, that Fig. 4.8 is a plot of force transmissibility as well as of displacement transmissibility.

4.4 Excitation by a rotating unbalance

A case of considerable practical interest is when the exciting force applied to the mass is an inertia force due to an unbalanced rotating mass, $\overline{m}$, at a radius, r, from a shaft fixed with respect to the main mass. The analysis is simplified if the main mass, m, is considered to include the rotating mass, $\overline{m}$. The non-rotating mass is therefore $(m - \overline{m})$. The configuration of the system is shown in Fig. 4.1(c), repeated here as Fig. 4.10.

The notation is given by the figure. The unbalanced mass $\overline{m}$ at radius r rotates at angular velocity ω rad/s, and its vertical displacement *relative to the main mass* at any instant is $\overline{z}$. The main mass is assumed to have only vertical motion, and lateral forces are ignored.

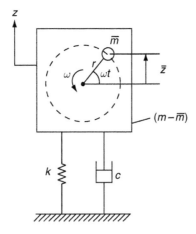

Fig. 4.10   Model for excitation by rotating unbalance.

4.4.1 Displacement Response

The equation of motion is easily found by considering the equilibrium of the main mass:

$$(m - \overline{m})\ddot{z} + \overline{m}(\ddot{z} + \ddot{\overline{z}}) + kz + c\dot{z} = 0 \tag{4.43}$$

Equation (4.43) can also be written as:

$$m\ddot{z} + c\dot{z} + kz = -\overline{m}\ddot{\overline{z}} \tag{4.44}$$

But,

$$-\overline{m}\ddot{\overline{z}} = \overline{m}r\omega^2 \sin \omega t \tag{4.45}$$

and so

$$m\ddot{z} + c\dot{z} + kz = \overline{m}\,r\,\omega^2 \sin \omega t \tag{4.46}$$

It should be noted that in Eq. (4.44), the quantity $-\overline{m}\ddot{\overline{z}}$ (or $\overline{m}r\omega^2 \sin \omega t$) is not the inertia force applied to the system by the mass $\overline{m}$, which is $\overline{m}(\ddot{z} + \ddot{\overline{z}})$, and that $\ddot{\overline{z}}$ is a relative, not an absolute, acceleration.

If P is substituted for $\overline{m}\,r\,\omega^2$, it is seen that Eq. (4.46) is the same as an equation already considered, i.e.,

$$m\ddot{z} + c\dot{z} + kz = P \sin \omega t \tag{4.47}$$

The magnitude of the response, $|z|$, was given by Eq. (4.9) as:

$$|z| = \frac{P}{k} \cdot \frac{1}{\sqrt{(1 - \Omega^2)^2 + (2\gamma\Omega)^2}}$$

so it follows that, with P replaced by $\overline{m}r\omega^2$:

$$|z| = \frac{\overline{m}r\omega^2}{k} \cdot \frac{1}{\sqrt{(1 - \Omega^2)^2 + (2\gamma\Omega)^2}} \tag{4.48}$$

Since $k = m\omega_n^2$, Eq. (4.48) can be rearranged in the non-dimensional form:

$$\left(\frac{m}{\overline{m}r}\right)|z| = \frac{\Omega^2}{\sqrt{(1 - \Omega^2)^2 + (2\gamma\Omega)^2}} \tag{4.49}$$

The right side of Eq. (4.49) will be recognized as being the same as that of Eq. (4.26), which gives the relative displacement, $|y|$, between mass and base, per unit base displacement $|x|$, for the system shown in Fig. 4.6. Thus, Fig. 4.7(a) will also serve as a plot of $(m/\overline{m}r)|z|$ versus Ω.

It must be remembered, when using Eq. (4.49), that m includes the out of balance mass $\overline{m}$.

4.4.2 Force Transmitted to Supports

The magnitude of the force transmitted to the ground or support structure, $|F_T|$, can be found from the force transmissibility equation, Eq. (4.42):

$$\frac{|F_T|}{|F|} = \sqrt{\frac{1 + (2\gamma\Omega)^2}{(1 - \Omega^2)^2 + (2\gamma\Omega)^2}}$$

where $|F|$ is given by:

$$|F| = \overline{m}r\omega^2 \tag{4.50}$$

Example 4.5

The washing machine shown in Fig. 4.11 consists of a rotating drum, containing the washing, inside a non-rotating cylindrical 'tub'. The whole assembly, complete with ballast masses (to increase the mass), and wet washing, has a mass of 40 kg. It is supported within the casing on an arrangement of springs and dampers, so that its undamped natural frequency, for vertical motion, is 5 Hz, and the linear viscous damping coefficient is 0.2. The design spin speed is 1200 rpm. It is estimated that the maximum likely unbalance due to the wet washing is 0.2 kg m.

At the design spin speed of 1200 rpm, find the following:
(a) the vertical tub vibration displacement relative to the casing, which is assumed fixed;
(b) the total vertical force transmitted to the casing;
(c) the vertical force that would have been transmitted to the casing if the tub and drum assembly had been rigidly attached to it.

Forces other than vertical forces may be ignored.

Solution

Part (a):
The design spin speed of 1200 rpm corresponds to a forcing frequency, f, of 20 Hz. The undamped natural frequency, f_n, is 5 Hz, so at this speed, $\Omega = f/f_n = 4$.

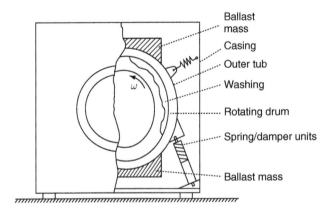

Fig. 4.11   Washing machine discussed in Example 4.5.

The single-peak displacement amplitude of the tub, $|z|$, is given by Eq. (4.49):

$$\left(\frac{m}{\overline{mr}}\right)|z| = \frac{\Omega^2}{\sqrt{(1-\Omega^2)^2+(2\gamma\Omega)^2}}$$

or

$$|z| = \left(\frac{\overline{mr}}{m}\right)\frac{\Omega^2}{\sqrt{(1-\Omega^2)^2+(2\gamma\Omega)^2}} \tag{A}$$

Inserting the numerical values:

   m = total mass of non-rotating and rotating parts = 40 kg;
   $\overline{mr}$ = 0.2 kg m;
   γ = 0.2;
   Ω = 4.
Then from Eq. (A),
   $|z|$ = 0.00530 m = 5.30 mm

Part (b):
The peak force transmitted to the casing, $|F_T|$, is given by Eq. (4.42):

$$\frac{|F_T|}{|F|} = \sqrt{\frac{1+(2\gamma\Omega)^2}{(1-\Omega^2)^2+(2\gamma\Omega)^2}} \tag{B}$$

where from Eq. (4.50),

$$|F| = \overline{mr}\omega^2 \tag{C}$$

From Eqs (B) and (C),

$$|F_T| = \overline{mr}\omega^2\sqrt{\frac{1+(2\gamma\Omega)^2}{(1-\Omega^2)^2+(2\gamma\Omega)^2}}$$

Using the numerical values above, with $\omega^2 = 2\pi f^2 = (40\pi)^2$,

   $|F_T|$ = 395 N

Part (c):
With the tub and drum rigidly fixed to the casing in a vertical sense, the peak out of balance force transmitted to the casing would be $\overline{mr}\omega^2 = 0.2 \times (40\pi)^2 = 3158$ N.

So the effect of the suspension system is to reduce the force transmitted to the casing by a factor of nearly 8.

References

4.1 Kennedy, CC and Pancu CDP. (1947). Use of vectors in vibration measurement and analysis. *J. Aero. Sci.* 14, 603–25.

4.2 Ewins, DJ (2000). *Modal Testing: Theory, Practice, and Application.* Research Studies Press Ltd.

5 Damping

Contents

Damping was introduced in Chapter 1 as one of the 'building blocks' making up the systems we deal with in structural dynamics, and some of the ways in which it can be generated were discussed. The effects of linear, viscous, damping, on the response of a single-DOF system, were discussed in Chapters 2, 3 and 4. In this chapter, we also look at another form of linear damping, *hysteretic damping*, and compare it with viscous damping. We then discuss how damping is related to energy loss; the different ways in which damping is quantified; two common forms of non-linear damping; and methods of approximating them with linear models. Damping in multi-DOF systems is discussed, with those systems, in Chapter 6.

5.1 Viscous and hysteretic damping models

These are the two linear damping models, and so far we have only discussed the viscous model. Hysteretic damping is sometimes called *structural* or *solid* damping, so unfortunately, the phrase 'structural damping' can mean either the damping (of any kind) in a structure or hysteretic damping. Therefore, only the latter term will be used here.

Comparing viscous and hysteretic damping is only possible by comparing frequency responses, since systems with hysteretic damping do not have an analytic transient response.

We first rearrange the equation of motion of a single-DOF system with viscous damping, into a slightly unusual form more easily compared with that for a system with hysteretic damping, as follows.

The familiar system shown in Fig. 5.1 has the equation of motion:

$$m\ddot{z} + c\dot{z} + kz = F \qquad (5.1)$$

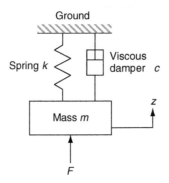

Fig. 5.1   Spring, mass and damper system represented by Eq. (5.1).

From Chapter 2, the non-dimensional, viscous, damping coefficient, γ, of this system is defined as:

$$\gamma = \frac{c}{c_c} \tag{5.2}$$

where c_c is the critical value of damping, given by:

$$c_c = 2m\omega_n \tag{5.3}$$

and ω_n is the undamped natural frequency, in rad/s, where

$$\omega_n = \sqrt{\frac{k}{m}} \tag{5.4}$$

Using Eqs (5.2)–(5.4), the dimensional damping coefficient, c, can be replaced in Eq. (5.1) by $2\gamma(k/\omega_n)$, giving:

$$m\ddot{z} + \left(\frac{2\gamma k}{\omega_n}\right)\dot{z} + kz = F \tag{5.5}$$

Equation 5.5 can also be expressed in complex frequency response form.

$$\left[-m\omega^2 + i\omega\left(\frac{2\gamma k}{\omega_n}\right) + k\right]\underline{z} = \underline{F} \tag{5.6}$$

where ω is the excitation frequency in rad/s. The displacement, z and the force, F, both functions of t, have been replaced by $\underline{z}$ and $\underline{F}$, respectively, rotating vectors in the complex plane.

Finally, making the substitution $\Omega = \omega/\omega_n$, Eq. (5.6) can be written as:

$$[-m\omega^2 + i(2\gamma\Omega)k + k]\underline{z} = \underline{F} \tag{5.7}$$

In the case of hysteretic damping, the damping force is proportional to the displacement, z, but in phase with the velocity $\dot{z}$. Thus a single-DOF equation with hysteretic damping can be formed simply by making the stiffness complex:

$$m\ddot{z} + k(1 + ig)z = F \tag{5.8}$$

where g is the *hysteretic damping coefficient*.

Equation (5.8), however, is very questionable mathematically. Since z is a real function of time and g and k are real, the damping force $igkz$ is an imaginary function of time, which makes no sense physically. Therefore it is only permissible to write it in that form if it will later be converted to frequency form.

If Eq. (5.8) is written in complex vector form, i.e.,

$$[-m\omega^2 + igk + k]\underline{z} = \underline{F} \tag{5.9}$$

the result, Eq. (5.9), can be compared directly with Eq. (5.7), and we see that they are the same, except that in the case of hysteretic damping, g replaces $2\gamma\Omega$. If the excitation frequency, ω, is equal to the undamped natural frequency, ω_n, then $\Omega = 1$, and

$$g = 2\gamma \tag{5.10}$$

At higher excitation frequencies the viscous damping becomes larger than the value given by Eq. (5.10), and at lower frequencies it becomes smaller. This would be expected to make a considerable difference to the response at values of Ω not near to 1, but the following example shows how little the frequency response of a single-DOF system with viscous damping differs from that of a system with hysteretic damping, provided Eq. (5.10) is applied.

Example 5.1

For sinusoidal forcing, at a range of frequencies from zero to 5 times the resonant frequency, plot the non-dimensional magnitude of the displacement response of the single-DOF system shown in Fig. 5.1, with:
(a) Viscous damping, with $\gamma = 0.02$, and 'equivalent' hysteretic damping with $g = 0.04$, typical of a metal or composite structure.
(b) Viscous damping, with $\gamma = 0.20$, and 'equivalent' hysteretic damping with $g = 0.40$, typical of a vibration isolation system using elastomeric supports.

Solution

We require expressions giving the magnitudes of the non-dimensional frequency responses of the two systems; that for viscous damping was derived in Chapter 4 (Section 4.1.1), as Eq. (4.12):

$$\frac{|z|}{z_s} = \frac{1}{\sqrt{(1 - \Omega^2)^2 + (2\gamma\Omega)^2}} \tag{A}$$

where z_s is the displacement of the system when a static force equal in magnitude to $\underline{F}$ is applied.

An expression for the same system, but with hysteretic damping, can be obtained from Eq. (5.9):

$$[-m\omega^2 + igk + k]\underline{z} = \underline{F} \tag{B}$$

Using Eq. (5.4), and making the substitution $\Omega = (\omega/\omega_n)$, as before:

$$m\omega^2 = m\omega_n^2\Omega^2 = k\Omega^2$$

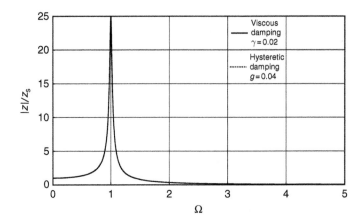

Fig. 5.2   Comparison of the frequency response of a single-DOF system with: (1) a viscous damping
coefficient of 0.02 and (2) a hysteretic damping coefficient of 0.04.

and

$$\frac{z}{F} = \frac{1}{k[(1 - \Omega^2) + ig]}$$ (C)

The modulus or magnitude ratio is

$$\frac{|z|}{|F|} = \frac{1}{k\sqrt{(1 - \Omega^2)^2 + g^2}}$$ (D)

Again introducing the static displacement, z_s, when a static load equal in magnitude
to $|F|$ is applied to the system, then $z_s = |F|/k$ and Eq. (D) becomes

$$\frac{|z|}{z_s} = \frac{1}{\sqrt{(1 - \Omega^2)^2 + g^2}}$$ (E)

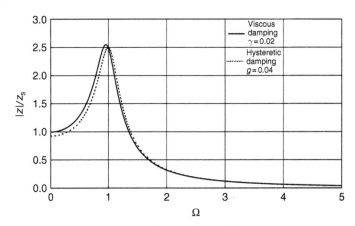

Fig. 5.3   Comparison of the frequency response of a single-DOF system with: (1) a viscous damping
coefficient of 0.20 and (2), a hysteretic damping coefficient of 0.40.

Figure 5.2 is a plot of the non-dimensional magnification $|z|/z_s$ versus Ω for viscous damping, using Eq. (A), with $\gamma = 0.02$, and hysteretic damping, using Eq. (E), with $g = 0.04$. The curves are, in fact, so similar over the whole frequency range that they cannot be distinguished at the scale used.

Figure 5.3 is a similar comparison with the viscous damping set at $\gamma = 0.2$ and the hysteretic damping at $g = 0.4$. In this case it is possible to distinguish the two curves, but they are very similar.

Example 5.1 illustrates the fact that, for most purposes, it makes very little difference which of the two linear damping models is used, in spite of the apparently large difference in damping force at frequencies well above, and well below, resonance. This is, of course, because, in most cases, nearly all the response is confined to a narrow band of frequencies close to the natural frequency, where the damping force has been matched. Also, in any case, well below the natural frequency, the response is largely controlled by stiffness, and well above the natural frequency, by mass. In practice, therefore, the choice of damping model is usually made on the basis of convenience. So, if we require a time solution, only viscous damping will work. On the other hand, provided only solutions in the frequency domain are required, hysteretic damping is sometimes more convenient.

If the hysteretic damping model produces frequency responses practically indistinguishable from the viscous model, and it cannot be used for solutions in time, we may wonder why it was introduced in the first place. The first reason appears to be mainly historical, and is related to the solution of aircraft flutter equations. In the early days of flutter analysis, it was realized that this strictly requires a complex eigenvalue solution, which was, at that time, virtually impossible. However, by introducing fictitious hysteretic damping into the equations, in such a sense, *positive or negative*, that it just allowed a steady-state flutter condition, a much easier *real* eigenvalue solution could be used. The second reason is that it provides a very simple way to add damping to a multi-DOF system. This topic is discussed further in Section 6.4.

5.2 Damping as an energy loss

Damping is an energy loss, and relating it to a hysteresis diagram gives some insight into the mechanism of damping, and helps to introduce some fundamental measures of damping such as *specific damping capacity*.

5.2.1 Energy Loss per Cycle – Viscous Model

In the system shown in Fig. 5.4, consisting of a spring and viscous damper in parallel, with no mass, let the displacement x be prescribed as:

$$x = X \sin \omega t \tag{5.11}$$

then the velocity

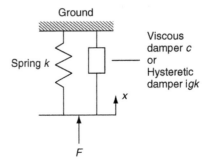

Fig. 5.4   Model for energy loss.

$$\dot{x} = X\omega \cos \omega t \tag{5.12}$$

The force input F required to produce this displacement must also be sinusoidal, and given by:

$$F = kx + c\dot{x} = kX \sin \omega t + cX\omega \cos \omega t \tag{5.13}$$

The power dissipated at any instant is $F\dot{x}$, and the energy lost over one complete cycle, e_d, is equal to the integral of $F\dot{x}$ over that period of time, $2\pi/\omega$ seconds in this case. Therefore

$$e_d = \int_0^{2\pi/\omega} [(kX \sin \omega t + cX\omega \cos \omega t)(X\omega \cos \omega t)]\mathrm{d}t \tag{5.14}$$

which simplifies to

$$e_d = cX^2\omega^2 \int_0^{2\pi/\omega} \cos^2 \omega t \cdot \mathrm{d}t = \pi cX^2\omega \tag{5.15}$$

If the damper is part of a complete spring–mass–damper system, with stiffness, k, and mass, m, then, using Eqs (5.2)–(5.4), the dimensional damping c can be written in terms of the non-dimensional damping coefficient, γ, as follows:

$$c = 2\gamma m\omega_n = \frac{2\gamma k}{\omega_n} \tag{5.16}$$

so the energy loss per cycle, e_d, can also be expressed as:

$$e_d = 2\pi kX^2\gamma\Omega \tag{5.17}$$

where $\Omega = (\omega/\omega_n)$, the excitation frequency divided by the undamped natural frequency.

5.2.2 Energy Loss per Cycle – Hysteretic Model

By definition, the hysteretic damping force is proportional to the displacement, x, but in phase with the velocity, $\dot{x}$. The displacement is again assumed to be prescribed by Eq. (5.11), and Eq. (5.12) again gives the velocity. However, the force F becomes

$$F = kX \sin \omega t + kXg \cos \omega t \tag{5.18}$$

The energy loss per cycle, e_d, is, as always, $\int_0^{2\pi/\omega} F\dot{x} \cdot dt$, which in this case is

$$e_d = \int_0^{2\pi/\omega} [kX(\sin \omega t + g\cos \omega t)(\omega X \cos \omega t)]dt \qquad (5.19)$$

which reduces to

$$e_d = \pi k X^2 g \qquad (5.20)$$

This is the same as Eq. (5.17) except that g has replaced $2\gamma\Omega$.

The maximum energy stored in the spring, e_s, will be required later, and is

$$e_s = \frac{1}{2}kX^2 \qquad (5.21)$$

5.2.3 Graphical Representation of Energy Loss

The expressions derived above for the energy loss due to damping can also be obtained graphically. For the viscous damping model, if the force, F, from Eq. (5.13), is plotted against the displacement, x, from Eq. (5.11), as in Fig. 5.5, the elliptical area traced out on the graph represents the energy dissipated in one complete cycle. The area of the ellipse, equal to the energy dissipated per cycle, is seen to be $\pi c X^2\omega$, agreeing with Eq. (5.15).

In the case of hysteretic damping, Fig. 5.6, the force, F, and the displacement, x, are given by Eqs (5.18) and (5.11), respectively. The area of the ellipse, representing the energy dissipated per cycle, is in this case seen to be $e_d = \pi k g X^2$, agreeing with Eq. (5.20).

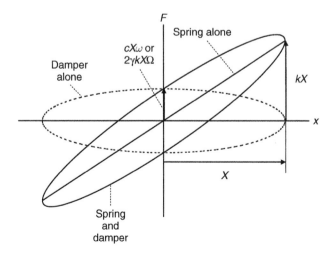

Fig. 5.5 Spring and viscous damper in parallel.

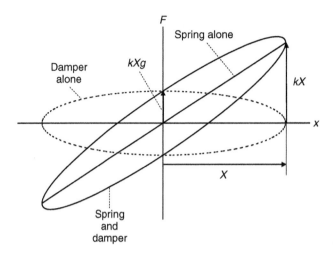

Fig. 5.6   Spring and hysteretic damper in parallel.

5.2.4 Specific Damping Capacity

Young's modulus, E, and the corresponding shear modulus, G, both defined as (stress/strain), provide the information required about a material to derive its stiffness properties, but tell us nothing about its ability to dissipate energy. Most structures dissipate very little energy, but some materials, including elastomers, essentially natural and synthetic rubbers, can be formulated to have quite high internal damping, as well as stiffness. This makes them very suitable for the manufacture of low cost vibration isolators, engine mountings, etc., that combine stiffness and damping in one molding. In order to model such materials, Young's modulus, E (or G) can be used to describe the stiffness at a fundamental level, but we also need a simple measure of internal damping. The *specific damping capacity*, usually given the symbol Ψ, is just what is required. This is defined as the ratio:

$$\Psi = \frac{e_d}{e_s} \tag{5.22}$$

where e_d is the energy dissipated in one complete cycle of an oscillation, assumed to be simple harmonic, and e_s is the *maximum* energy stored during the same cycle. In the following derivation of Ψ, tension–compression loading is assumed, but shear loading could be assumed equally well.

A simple block of material with length L and cross-section area A is assumed to behave like a linear spring and a hysteretic damper in parallel, as in Fig. 5.6. Let the displacement, x, be prescribed, as before, by Eq. (5.11), i.e., $x = X \sin \omega t$, then the applied force is, as before, from Eq. (5.18):

$$F = kx \sin \omega t + kXg \cos \omega t$$

However, since the cross-sectional area is A

$$F = \sigma A \tag{5.23}$$

where σ is the stress, and

$$x = \varepsilon L \tag{5.24}$$

and also

$$X = \varepsilon_m L \tag{5.25}$$

where L is the original length of the block, ε the strain and ε_m its magnitude or 'peak value'. Then,

$$k = F/x = \sigma A/\varepsilon L = EA/L \tag{5.26}$$

where E is Young's Modulus (stress/strain) or σ/ε.

Substituting Eqs (5.24) and (5.25) into Eq. (5.11) gives the strain:

$$\varepsilon = \varepsilon_m \sin \omega t \tag{5.27}$$

and substituting Eqs (5.23), (5.25) and (5.26) into Eq. (5.18) gives the stress:

$$\sigma = E \varepsilon_m (\sin \omega t + g \cos \omega t) \tag{5.28}$$

We can now plot stress against strain, as shown in Fig. 5.7. The energy dissipated per cycle is the area of the ellipse, $e_d = \pi E \varepsilon_m^2 g$, and the max energy stored is $e_s = \frac{1}{2} E \varepsilon_m^2$.

From Eq. (5.22) the specific damping capacity (SDC), Ψ, is

$$\Psi = e_d/e_s = (\pi E \varepsilon_m^2 g)/(\tfrac{1}{2} E \varepsilon_m^2) = 2\pi g \tag{5.29}$$

The *loss coefficient*, sometimes given the symbol, η, is the SDC expressed per radian rather than per cycle, that is, divided by 2π, so $\eta = g$, the hysteretic damping coefficient!

Finally, it can be seen that although E and G are real quantities, hysteretic damping can be introduced into a system at the same time as the stiffness by regarding them as complex, i.e.,

$$\underline{E} = E(1 + \mathrm{i}g) \tag{5.30}$$

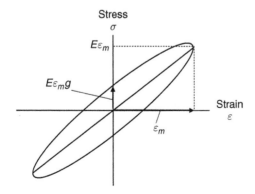

Fig. 5.7   Energy diagram expressed as stress versus strain.

and

$$\underline{G} = G(1 + \mathrm{i}g) \tag{5.31}$$

where $\underline{E}$ and $\underline{G}$ replace E and G, respectively. It must, of course, be remembered that this is only valid in the frequency domain.

5.3 Tests on damping materials

The energy diagrams discussed in the foregoing sections suggest a simple way of assessing the stiffness and energy dissipation properties of materials. If the test is arranged so that stress is plotted against strain, or what amounts to the same thing, except for scaling factors, load is plotted against displacement, the resulting figure gives some insight into the nature of the stiffness and damping properties of the material. First, assuming sinusoidal forcing, an elliptical plot indicates linearity at that particular frequency and excitation level. Secondly, hysteretic damping is indicated by the ellipse remaining of constant aspect ratio as the frequency is changed. On the other hand, a tendency for the ellipse to become proportionally fatter with increasing frequency points to viscous damping.

Tests of this nature are commonly used to measure the two essential properties, Young's modulus, E and specific damping capacity, Ψ (or one of its equivalents). Young's modulus, E, is, of course, given by the slope of a mean line through the hysteresis diagram, and Ψ can be found from the areas representing energy lost and energy stored.

5.4 Quantifying linear damping

We have already discussed several ways of quantifying damping:
(a) The dimensional viscous damping coefficient, c;
(b) The non-dimensional viscous damping coefficient, γ;
(c) The hysteretic damping coefficient, g;
(d) The specific damping capacity (SDC), Ψ;
(e) The loss coefficient, η.
A few methods not listed above are now introduced.

5.4.1 Quality Factor, Q

A term borrowed from the electronics field is the 'quality factor', Q. This is the magnification, at resonance, of a circuit containing an inductor and a capacitor, and the name comes from the fact that it is a measure of the 'quality' of the inductor, which is reduced by its internal resistance. It is analogous to the magnification of a spring–mass–damper system at resonance, which is $1/2\gamma$ for viscous damping or $1/g$ for hysteretic damping. These are therefore equivalents, so *roughly*

$$Q = 1/2\gamma = 1/g \tag{5.32}$$

Another definition of Q is that it is equal to the center frequency of a resonant response divided by the 'half-power bandwidth', which is the width of the frequency band over which the magnification is more than $1/\sqrt{2}$ (about 0.707) of the response at the resonant frequency.

5.4.2 Logarithmic Decrement

The decay rate of a single-DOF spring–mass–damper system can, of course, be used to estimate the damping, and the 'logarithmic decrement', δ, has sometimes been used to quantify the system damping also, as an alternative to γ or g.

The log dec. is defined as the natural log (base e) of the ratio of the heights of two successive peaks, for example the ratio (X_1/X_2) in Fig. 5.8.

Any freely decaying single-DOF response (see Chapter 2) can be represented by:

$$x = e^{-\gamma \omega_n t}(A \cos \omega_d t + B \sin \omega_d t) \tag{5.33}$$

where ω_n is the undamped natural frequency and ω_d is the damped natural frequency, related to ω_n by:

$$\omega_d = \omega_n \sqrt{1 - \gamma^2} \tag{5.34}$$

A and B are constants depending on initial conditions.

In the decaying oscillation shown in Fig. 5.8, if the height of the first peak, X_1, is

$$X_1 = Ce^{-\gamma \omega_n t_1} \tag{5.35}$$

where C is a constant depending on A and B, then the height of the second peak, X_2 which occurs $2\pi/\omega_d$ seconds later, is

$$X_2 = Ce^{[-\gamma \omega_n (t_1 + 2\pi/\omega_d)]} \tag{5.36}$$

From Eqs (5.34), (5.35) and (5.36), the ratio (X_1/X_2) is given by:

$$\frac{X_1}{X_2} = e^{\left(\frac{2\pi\gamma}{\sqrt{1-\gamma^2}}\right)} \tag{5.37}$$

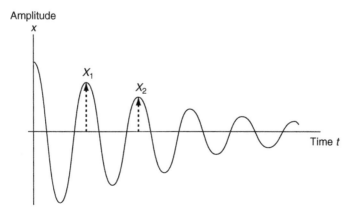

Fig. 5.8

The log dec., δ, is defined as:

$$\delta = \log_e \frac{X_1}{X_2} = \frac{2\pi\gamma}{\sqrt{1-\gamma^2}} \tag{5.38}$$

or for practical purposes, if γ is small,

$$\delta = \log \text{dec.} = 2\pi\gamma \tag{5.39}$$

Example 5.2

Find the approximate value of the non-dimensional viscous damping coefficient, γ, for the decay record shown in Fig. 5.8, given that $X_1 = 0.7304$ and $X_2 = 0.5335$.
  From Eq. (5.38), $\delta = \log_e (0.7304/0.5335) = 0.314$
  Then from Eq. (5.39), $\gamma = 0.314/2\pi = 0.05$.

5.4.3 Number of Cycles to Half Amplitude

The number of cycles of a decaying oscillation required for the amplitude to fall to a half, or some other fraction, of a given peak height is sometimes used to estimate, or as a measure of, damping. If we generalize Eq. 5.36 to give the height of the nth peak, X_{1+n}, n complete cycles after X_1, i.e.:

$$X_{1+n} = Ce^{\left[-\gamma\omega_n\left(\frac{t_1+2\pi n}{\omega_d}\right)\right]} \tag{5.40}$$

Then from Eqs 5.34, 5.35 and 5.40:

$$\frac{X_{1+n}}{X_1} = e^{\left(-2\pi n\gamma\frac{\omega_n}{\omega_d}\right)} = e^{-\frac{2\pi n\gamma}{\sqrt{1-\gamma^2}}} \tag{5.41}$$

or approximately for small γ :

$$\frac{X_{1+n}}{X_1} = \exp(-2\pi n\gamma)$$

or

$$\log_e \frac{X_{1+n}}{X_1} = (-2\pi n\gamma) \tag{5.42}$$

where X_{1+n}/X_1 is $\frac{1}{2}$, or whatever other fraction of the initial amplitude is required after n complete cycles.

Example 5.3

Find the approximate number of cycles to half amplitude corresponding to $\gamma = 0.01$, 0.02 and 0.05.

Solution

From Eq. (5.42), for small γ,

$$\log_e\left(\frac{X_{1+n}}{X_1}\right) = (-2\pi n\gamma)$$

where in this case

$$\left(\frac{X_{1+n}}{X_1}\right) = 1/2$$

Now $\log_e \frac{1}{2} = -0.6931$, so the number of cycles to half amplitude, $n_{1/2}$, is

$$n_{1/2} = (-\log_e \frac{1}{2}/2\pi\gamma) = (0.6931/2\pi\gamma) = 0.110/\gamma$$

Table 5.1 gives the value of $n_{1/2}$ corresponding to each value of γ.

Table 5.1

γ	Number of cycles to half amplitude, $n_{1/2}$
0.01	11.0
0.02	5.5
0.05	2.2

5.4.4 Summary Table for Linear Damping

Table 5.2 below lists eight ways of expressing linear damping, summarizing the methods covered in this section. All are non-dimensional except c.

Table 5.2
Different Ways of Expressing Linear Damping.

Name	Usual symbol	Validity and equivalents
Viscous damping: dimensional	c	Valid for any system, including a discrete damper. Dimensions are force (or moment) per unit linear (or angular) velocity.
Viscous damping: non-dimensional coefficient	γ	Valid only in a system containing mass, m, stiffness, k, and damping, c. $\gamma = c/c_c$ where $c_c = 2m\omega_n = 2\sqrt{km}$
Hysteretic damping: coefficient	g	The system must also contain stiffness. $g = 2\gamma(\omega/\omega_n)$. $g \approx 2\gamma$ for most purposes.
Quality factor	Q	$Q = 1/2\gamma$ or $1/g$
Log decrement	δ	$\delta = 2\pi\gamma$
Cycles to half amplitude	$n_{\frac{1}{2}}$	$n_{\frac{1}{2}} = \left(-\log_e \frac{1}{2}/(2\pi\gamma)\right)$ and similarly for other fractions.
Specific damping capacity	Ψ	The system must also contain stiffness. $\Psi = 2\pi g$
Loss coefficient	η	The system must also contain stiffness. $\eta = \Psi/2\pi = g$

The 'validity' in the last column is a reminder that non-dimensional ways of expressing viscous damping only make sense in the context of a system that also contains mass and stiffness. Non-dimensional ways of expressing hysteretic damping require stiffness but not necessarily mass.

5.5 Heat dissipated by damping

The heat generated by damping should not be forgotten. This point is illustrated by the following example.

Example 5.4

A vibration isolation system, consisting of a rigid equipment tray supported on four vibration isolators, is being tested. The isolators are manufactured from an elastomer having a specific damping capacity, Ψ, equal to 2.5 at all frequencies. The tray, with payload, has a mass, m, of 20 kg, and its natural frequency, f_n, is 20 Hz. A proposed accelerated endurance test on the complete assembly requires it to be placed on a large shaker, and excited at its resonant frequency, so that the elastomeric mounts develop a relative displacement amplitude of ± 5 mm. Calculate the total heat dissipation of the four mounts.

Solution

The combined stiffness of the four isolators, assumed constant, is given by $k = m \omega_n^2 = m (2\pi f_n)^2$, where $m = 20$ kg and $f_n = 20$ Hz. Then $k = 20(2\pi \times 20)^2 = 315\,800$ N/m.

The hysteretic damping coefficient, g, from Table 5.2 is $(\Psi/2\pi) = (2.5/2\pi) = 0.40$ approximately. From Eq. (5.20), the energy dissipated per cycle (in J) is given by

$$e_d = \pi k g X^2$$

where in this case $X = 5$ mm or 0.005 m, so

$$e_d = (\pi \times 315800 \times 0.4 \times 0.005^2) = 9.92 \text{ J per cycle or } (9.92 \times 20) = 198 \text{ J/s or W}$$

The power dissipation of each of the four isolators is therefore about 50 W, and since they would be quite small, probably about 50 mm in diameter, and 25 mm in depth, high temperatures, and possible failure, are to be expected.

5.6 Non-linear damping

While the damping inherent in structures is often slightly non-linear, linear models usually work quite well, provided that the damping constants, which usually have to be measured, or estimated from similar structures, are well chosen: for example, they should allow for the possibility that the damping may change with amplitude, time or temperature. Some discrete dampers, such as friction or hydraulic units, however, are inherently non-linear. Two examples, dealt with here, are *Coulomb damping*, associated with sliding friction, where a friction force opposes the motion, and *square-law damping*, which is obtained naturally when fluid is forced through a simple orifice.

5.6.1 Coulomb Damping

This is produced when the movement of the system generates a friction force opposing the motion. The magnitude of this force is the product of the normal force acting between the rubbing surfaces and the coefficient of friction of the material. In the simple analysis below it is assumed constant. Such devices were used on early road vehicles, and are still sometimes used, in conjunction with metal springs, in vibration isolation systems, especially on washing machines. They suffer from the disadvantage that the system is locked, and ineffective as an isolator, below some level of applied force determined by the static friction. Another problem is that the friction force actually produced is somewhat variable in practice.

5.6.2 Square Law Damping

This is usually produced, in practice, by forcing hydraulic fluid through a small orifice by some form of piston. The resulting turbulent flow means that the pressure acting on the piston increases as the square of the velocity. If the flow could be made laminar, linear damping would result, but this is rarely done, for practical reasons. In practice, therefore, a hydraulic damper usually produces a force proportional to the square of the piston velocity. Such a device can only work effectively when the fluid is in compression, since a fluid will 'cavitate' below some value of absolute pressure, so in order to obtain damping in both directions, either the device must be double-acting or two units, acting in opposite senses, must be used. The force, F, produced by a damper when the fluid is being compressed is of the form:

$$F = C\dot{x}^2 \tag{5.43}$$

where C is the square-law damping coefficient, analogous to, but not to be confused with, the linear viscous coefficient c introduced above, and $\dot{x}$ is the velocity in the direction causing fluid compression. In the case of a single-acting unit, the damping produced in the opposite direction will be close to zero. If a double-acting unit is used, the damping produced on the return stroke will also be as given by Eq. (5.43), but the value of the constant C will not necessarily be the same in the two directions. However, if the unit is designed so that C has the same value in both directions, the force F will be given by:

$$F = C\dot{x}^2 \mathrm{sgn}(\dot{x}) \tag{5.44}$$

where $\mathrm{sgn}(\dot{x})$ is taken to mean 'the sign of $\dot{x}$'. In computer applications, Eq. (5.44) can conveniently be written as:

$$F = C\dot{x}|\dot{x}| \tag{5.45}$$

which has the same effect as Eq. (5.44).

Square-law devices such as this are used in shock-absorbing applications, but can also be used to control vibration. It should be mentioned that an 'automotive damper' as fitted in road vehicle suspensions, although operating on the same principle, usually has a special valve which changes the orifice area with flow rate, so that the device approximates to a linear viscous damper, not a square-law damper, as might be expected.

The shock-absorber struts used in aircraft landing gear are usually of the square-law type, since this provides efficient energy absorption during impact with the ground. To minimize bounce, the value of C is usually made higher for rebound than for the initial compression. However, an aircraft landing gear has a secondary role as a vehicle suspension unit during taxy. Here the square-law characteristic is less desirable, and the value of C may be made dependent on strut closure, closure velocity, etc. to improve the performance during this phase of operation.

Also in the aerospace field, a hydraulically operated power control unit may be used, part of the time, as a hydraulic damper. This requires only minor modifications to the device, such as switching in a bypass orifice. The damper so formed will then usually be of the square-law type, and can be dealt with as such. Examples of this practice are (1) the use of a nose gear steering actuator as a part-time shimmy damper, and (2) the use of aileron, elevator or rudder power units as flutter dampers in certain failure cases.

5.7 Equivalent linear dampers

It will be seen from the above discussion that, although, it is unusual for an engineer in industry to have to deal with a truly non-linear structure, non-linear *damping*, particularly the square-law type, does crop up from time to time. When only a few degrees of freedom are involved, time-stepping methods will probably be used to investigate these non-linear systems, since few analytical methods exist. However, this may not always be practical for, say, a multi-mode aircraft flutter investigation. When only the damping is non-linear, there is an approximate method of solution that can give good results in many cases. This uses the concept of an *equivalent linear damper*, a hypothetical damper, usually assumed viscous, giving the same energy dissipation per cycle as the actual non-linear damper. Using this concept, ordinary linear methods can still give useful results. Another application for the method is when a time-stepping method of solution is proposed, but we perhaps do not know what non-linear damping coefficient to use. The equivalent linear damper method can then be used to find a good approximation to the best value before running the time-stepping solution.

Expressions giving the equivalent viscous damping for systems with Coulomb and square-law dampers are derived below. These assume that the displacement time history, x, say, and therefore the velocity and acceleration time histories, $\dot{x}$ and $\ddot{x}$, remain truly sinusoidal throughout. This is not strictly true, but the method can give surprisingly good results.

5.7.1 *Viscous Equivalent for Coulomb Damping*

The equivalent viscous damper can be determined, for sinusoidal motion, as follows. In Fig. 5.9, let the displacement, x, be prescribed as:

$$x = -X \cos \omega t \tag{5.46}$$

then

$$\dot{x} = \omega X \sin \omega t \tag{5.47}$$

$\underline{F}$ is the friction force, and is assumed constant, but can act in either direction, depending on the sign of the velocity, $\dot{x}$. In the half-cycle $t=0$ to $t=\pi/\omega$, when $\dot{x}$ is upwards, $\underline{F}$ is as shown in Fig. 5.9, and

$$F = kx + \underline{F} \tag{5.48}$$

but in the half-cycle $t=\pi/\omega$ to $t=2\pi/\omega$, the velocity is reversed, and

$$F = kx - \underline{F} \tag{5.49}$$

The energy, e_{d}, dissipated per complete cycle is $\int_0^{2\pi/\omega} F\dot{x} \cdot \mathrm{d}t$, equal, in this case, to $2 \int_0^{\pi/\omega} F\dot{x} \cdot \mathrm{d}t$. Substituting for F and $\dot{x}$:

$$e_{\mathrm{d}} = 2 \int_0^{\pi/\omega} (-kX \cos \omega t + \underline{F})(\omega X \sin \omega t) \cdot \mathrm{d}t = 4\underline{F}X \tag{5.50}$$

From Eq. (5.15) the energy dissipation per cycle, e_{d}, for a spring and viscous damper, is $\pi c X^2 \omega$, so for the system with Coulomb damping to have the same dissipation per cycle as the system with viscous damping:

$$e_{\mathrm{d}} = \pi c_{\mathrm{e}} X^2 \omega = 4\underline{F}X \tag{5.51}$$

so

$$c_{\mathrm{e}} = \frac{4\underline{F}}{\pi \omega X} \tag{5.52}$$

where c_{e} is the required equivalent viscous damper.

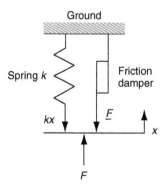

Fig. 5.9

It can be seen that in the usual case, when $\underline{F}$ and ω remain constant, the equivalent viscous damping, c_e, falls inversely with increasing amplitude, X. This is usually an undesirable characteristic; the damping effectively reduces just when it is most needed.

5.7.2 Viscous Equivalent for Square Law Damping

In Fig. 5.10, the square law damper is assumed double-acting, so that the damping force, F_D, is given by Eq. (5.44): $F_D = C\dot{x}^2\mathrm{sgn}(\dot{x})$

To derive an expression for a viscous damper, c_e, 'equivalent' to a square-law damper, we assume, say, that

$$x = -X\cos\omega t \tag{5.53}$$

then

$$\dot{x} = \omega X\sin\omega t \tag{5.54}$$

For the half-cycle $t=0$ to $t=\pi/\omega$,

$$F = kx + C\dot{x}^2$$

and for $t=\pi/\omega$ to $t = 2\pi/\omega$

$$F = kx - C\dot{x}^2$$

So the energy dissipated per complete cycle is

$$e_d = 2\int_0^{\pi/\omega}(-kX\cos\omega t + C\omega^2 X^2\sin^2\omega t)(\omega X\sin\omega t)\cdot dt$$

$$e_d = 2\int_0^{\pi/\omega}C\omega^3 X^3\sin^3\omega t\cdot dt = \frac{8}{3}C\omega^2 X^3 \tag{5.55}$$

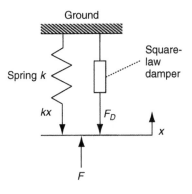

Fig. 5.10

From Eq. (5.15), the energy dissipation per cycle, e_d, for a linear spring and viscous damper, is $\pi c X^2 \omega$, so for the system with square-law damping to have the same dissipation per cycle as the system with viscous damping:

$$e_d = \pi c_e X^2 \omega = \frac{8}{3} C \omega^2 X^3$$

and so

$$c_e = \frac{8 C \omega X}{3\pi} \tag{5.56}$$

where c_e is the equivalent viscous damper defined above.

5.7.3 Limit Cycle Oscillations with Square-Law Damping

It can be seen from Eq. (5.56) that for constant frequency, ω, the equivalent viscous damper, c_e, tends to increase proportionately with amplitude, which means that when the amplitude is very small, so is the fraction of critical damping. This characteristic is important when attempting to use a square-law damper to stabilize a dynamically unstable system, that is, one that would otherwise flutter or shimmy. Usually, some particular value of linear viscous damping coefficient is required to stop the oscillation altogether, and at very small amplitudes the square-law equivalent of this may not be developed. The system vibration amplitude therefore increases until sufficient damping to stabilize it *is* produced, and a steady-state limit cycle oscillation (LCO) at that amplitude results. This may nevertheless be acceptable if it is both non-destructive, and occurs only rarely.

An equation like Eq. (5.56) can be used to find the amplitude at which the LCO will stabilize. The equivalent viscous damper, c_e, is set to the amount of damping, known from a calculation, or test, to be necessary just to stop the instability, and ω is the frequency of the instability in rad/s. Then X is the LCO displacement corresponding to any given value of C.

5.8 Variation of damping and natural frequency in structures with amplitude and time

Variations of the measured values of damping coefficients and natural frequencies in structures must be considered at the same time, because the two tend to go together. It is tempting to believe that the natural frequencies and damping coefficients of real structures have accurately reproducible, constant, values, like the mathematical models used to represent them. This is actually far from true, and the following variations are quite normal in metal aerospace structures, for example:
(a) Variation of up to about 1% in measured natural frequency with different forcing levels;
(b) Variation of up to about 50% in measured damping coefficients with forcing level, and also with time.

Generally, plotting the *variation* of natural frequency against the *variation* of damping tends to produce a consistent trend; that is, those structures showing the largest frequency variation also show the largest damping variation. This is not too surprising when it is realized that stiffness and damping are two aspects of the same hysteresis diagram.

These effects are not generally sufficiently important to make us abandon the idea that structures can safely be treated as linear systems. The typical variation in measured natural frequency, of up to about 1%, for example, corresponds to a stiffness variation of about 2%, and most structural dynamics engineers would be happy to regard such a system as linear. However, there is an area where relatively small variations of natural frequency and damping can have effects disproportionate to their importance. This is in the field of modal testing. Some methods for adjusting the relative force levels in multi-point excitation, for example, rely on trial phase measurements. It can be seen from the 'circle plots' in Example 4.2 that in the region of resonance, phase is extremely sensitive to changes of frequency, and a change of natural frequency is just as effective as a change of excitation frequency. So the typical 1% change of natural frequency mentioned above is worth 45° of phase change in a typical structure, with, say, $\gamma = 0.02$. It can also be seen that a typical 50% change in damping, other things being equal, from say $\gamma = 0.02\text{–}0.03$, can also produce a considerable phase change.

6 Introduction to Multi-degree-of-freedom Systems

Contents

Multi-degree-of-freedom (multi-DOF) systems are defined as those requiring two or more coordinates to describe their motion. This excludes continuous systems, which theoretically have an infinite number of freedoms. However, with the almost universal application of the finite element (FE) method, systems that would, in the past, have been treated as continuous have now become multi-DOF systems, so these, with single-DOF systems, now cover most practical tasks in structural dynamics.

Multi-DOF systems fall into two groups:

(1) Simple lumped-parameter systems, consisting of only a few degrees of freedom, not generally requiring the use of the finite element method to set them up.
(2) Structures, which are now almost always set up by using the finite element method. The displacement coordinates will be those produced by the FE package, and will be Cartesian (x, y, z) linear displacements at 'nodes' or 'grid points', usually with rotations about them, known as global coordinates.

In this chapter, the basic principles are first introduced by discussing very simple systems, which fall into the first group above. While Lagrange's equations are regarded as the basic method for setting up the equations of motion, other methods are also discussed. Matrix methods, essential for dealing with larger systems, are then introduced, leading on to transformation into modal coordinates, discussing assumed and normal modes, with an introduction to eigenvalues and eigenvectors, and their properties. After a discussion of damping in multi-DOF systems, the two main methods for calculating the response of multi-DOF systems, the normal mode summation method, and the alternative, direct solution of the equations in the time domain, are introduced.

6.1 Setting up the equations of motion for simple, undamped, multi-DOF systems

The undamped 2-DOF system shown in Fig. 6.1(a) will be used to illustrate three different methods for setting up the equations of motion of simple multi-DOF systems. These are (1) using Newton's second law, and D'Alembert's principle, as for

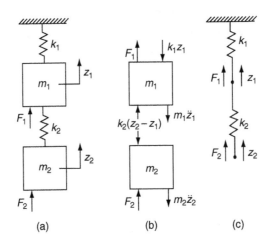

Fig. 6.1   Undamped 2-DOF system in schematic form.

the single-DOF systems discussed in earlier chapters (2) generating the static stiffness (or flexibility) matrix and adding the inertia forces; and (3) using Lagrange's equations.

6.1.1 Equations of Motion from Newton's Second Law and d'Alembert's Principle

This method, introduced in Chapter 2 for single-DOF systems, can also be used for simple 2- or 3-DOF systems.

Figure 6.1(b) shows all the forces acting on the two free bodies, m_1 and m_2. Taking forces as positive when upwards, and equating all the forces acting on each mass to zero, we have

for m_1:

$$F_1 - m_1\ddot{z}_1 - k_1 z_1 + k_2(z_2 - z_1) = 0 \tag{6.1a}$$

and for m_2:

$$F_2 - m_2\ddot{z}_2 - k_2(z_2 - z_1) = 0 \tag{6.1b}$$

Writing Eqs (6.1a) and (6.1b) in matrix form gives

$$\begin{bmatrix} m_1 & 0 \\ 0 & m_2 \end{bmatrix} \begin{Bmatrix} \ddot{z}_1 \\ \ddot{z}_2 \end{Bmatrix} + \begin{bmatrix} (k_1 + k_2) & -k_2 \\ -k_2 & k_2 \end{bmatrix} \begin{Bmatrix} z_1 \\ z_2 \end{Bmatrix} = \begin{Bmatrix} F_1 \\ F_2 \end{Bmatrix} \tag{6.2}$$

This is one of the several possible forms of the equations of motion for this system.

6.1.2 Equations of Motion from the Stiffness Matrix

If we remove the masses from Fig. 6.1(a), we are left with the two springs shown in Fig. 6.1(c). We can find the stiffness matrix for this simple system using the same procedure as in Example 1.4. This gives

$$\begin{Bmatrix} F_1 \\ F_2 \end{Bmatrix} = \begin{bmatrix} (k_1 + k_2) & -k_2 \\ -k_2 & k_2 \end{bmatrix} \begin{Bmatrix} z_1 \\ z_2 \end{Bmatrix} \tag{6.3}$$

Now replacing the masses m_1 and m_2, the forces F_1 and F_1 in Eq. (6.3) become $(F_1 - m_1\ddot{z}_1)$ and $(F_2 - m_2\ddot{z}_2)$, respectively:

$$\left\{ \begin{array}{c} F_1 - m_1\ddot{z}_1 \\ F_2 - m_2\ddot{z}_2 \end{array} \right\} = \left[\begin{array}{cc} (k_1 + k_2) & -k_2 \\ -k_2 & k_2 \end{array} \right] \left\{ \begin{array}{c} z_1 \\ z_2 \end{array} \right\} \tag{6.4}$$

which can be written as:

$$\left[\begin{array}{cc} m_1 & 0 \\ 0 & m_2 \end{array} \right] \left\{ \begin{array}{c} \ddot{z}_1 \\ \ddot{z}_2 \end{array} \right\} + \left[\begin{array}{cc} (k_1 + k_2) & -k_2 \\ -k_2 & k_2 \end{array} \right] \left\{ \begin{array}{c} z_1 \\ z_2 \end{array} \right\} = \left\{ \begin{array}{c} F_1 \\ F_2 \end{array} \right\}, \tag{6.5}$$

which is the same as Eq. (6.2). Thus, the equations of motion can also be obtained by regarding the inertia forces as additional external forces acting on the stiffness matrix. These inertia forces are, of course, negative (downwards in this case) for positive (upwards) acceleration, by D'Alembert's principle. It will often be found easier to first derive the flexibility matrix, and invert it to form the stiffness matrix. It will be seen later that if the equations are to be used to find eigenvalues and eigenvectors, the flexibility matrix can be used without inversion.

6.1.3 Equations of Motion from Lagrange's Equations

The use of Lagrange's equations is the standard method for setting up the equations of motion of multi-DOF systems. If the kinetic energy does not depend upon the displacements of the masses (which is usually true except in rotating systems, when centrifugal and Coriolis forces play a part), and damping is ignored, then Lagrange's equations can be written as:

$$\frac{d}{dt}\left(\frac{\partial T}{\partial \dot{q}_i} \right) + \frac{\partial U}{\partial q_i} = Q_i \quad (i = 1, 2) \tag{6.6}$$

where T is the total kinetic energy of the system, and U the total potential energy. The q_i (q_1 and q_2 in this case) are generalized coordinates, and the Q_i are generalized forces.

Certain restrictions apply to the choice of these generalized quantities:
(a) The coordinates must be independent of each other;
(b) The generalized forces, Q_i, must do the same work as the actual forces, F_1 and F_2.
These requirements are satisfied if we let $q_1 = z_1$, $q_2 = z_2$, $Q_1 = F_1$ and $Q_2 = F_2$.
Writing down the kinetic energy, T, and the potential energy, U:

$$T = \tfrac{1}{2}m_1\dot{z}_1^2 + \tfrac{1}{2}m_2\dot{z}_2^2 \tag{6.7}$$

$$U = \tfrac{1}{2}k_1z_1^2 + \tfrac{1}{2}k_2(z_2 - z_1)^2 = \tfrac{1}{2}k_1z_1^2 + \tfrac{1}{2}k_2z_2^2 - k_2z_1z_2 + \tfrac{1}{2}k_2z_1^2 \tag{6.8}$$

Then,

$$\frac{d}{dt}\left(\frac{\partial T}{\partial \dot{q}_1} \right) = \frac{d}{dt}\left(\frac{\partial T}{\partial \dot{z}_1} \right) = m_1\ddot{z}_1 \qquad \frac{d}{dt}\left(\frac{\partial T}{\partial \dot{q}_2} \right) = \frac{d}{dt}\left(\frac{\partial T}{\partial \dot{z}_2} \right) = m_2\ddot{z}_2$$

$$\frac{\partial U}{\partial q_1} = \frac{\partial U}{\partial z_1} = k_1z_1 - k_2z_2 + k_2z_1 \qquad \frac{\partial U}{\partial q_2} = \frac{\partial U}{\partial z_2} = k_2z_2 - k_2z_1$$

Substituting these into Eq. (6.6), together with $F_1 = Q_1$ and $F_2 = Q_2$:

$$m_1\ddot{z}_1 + k_1 z_1 - k_2 z_2 + k_2 z_1 = F_1 \tag{6.9a}$$

$$m_2\ddot{z}_2 + k_2 z_2 - k_2 z_1 = F_2 \tag{6.9b}$$

or expressed in matrix form:

$$\begin{bmatrix} m_1 & 0 \\ 0 & m_2 \end{bmatrix} \begin{Bmatrix} \ddot{z}_1 \\ \ddot{z}_2 \end{Bmatrix} + \begin{bmatrix} (k_1 + k_2) & -k_2 \\ -k_2 & k_2 \end{bmatrix} \begin{Bmatrix} z_1 \\ z_2 \end{Bmatrix} = \begin{Bmatrix} F_1 \\ F_2 \end{Bmatrix} \tag{6.10}$$

which is, of course, the same as both Eqs (6.2) and (6.5).

6.2 Matrix methods for multi-DOF systems

The analysis of larger multi-DOF systems will be seen to involve several coordinate transformations. The first of these is usually from basic mass and stiffness elements to equations of motion in global coordinates. This first stage is often carried out automatically by a finite element program, as discussed in Chapter 8. Secondly, the coordinates are often transformed from global coordinates to modal coordinates. These transformations are carried out by matrix methods to be developed in this and the next section.

6.2.1 Mass and Stiffness Matrices: Global Coordinates

We first consider the transformation of the mass data, from basic elements to the mass matrix in global coordinates. Displacements of the global coordinates will imply some related displacements of the individual mass elements. This is often based on the static displacements within a finite element when a unit displacement is applied at each global coordinate. Alternatively, the mass elements closest to a given node, or grid point, may be assumed to move with the global coordinates, in rigid-body fashion. Whatever the exact relationship, which we do not need to define now, let us assume only that it is linear, and is given by:

$$\{r\} = [X_m]\{z\} \tag{6.11}$$

where $\{r\}$ is a vector of displacements of the individual mass particles, and $\{z\}$ represents global coordinate displacements.

Equation (6.11) can be differentiated with respect to time, giving

$$\{\dot{r}\} = [X_m]\{\dot{z}\} \tag{6.12}$$

Let the mass particles be m_1, m_2, m_3, etc. Since the kinetic energy of a single mass particle, m_i, say, is $\frac{1}{2}m_i\dot{r}_i^2$, where $\dot{r}_i$ is the velocity of that particle, the kinetic energy, T, in the whole system is

$$T = \tfrac{1}{2}m_1\dot{r}_1{}^2 + \tfrac{1}{2}m_2\dot{r}_2{}^2 + \tfrac{1}{2}m_3\dot{r}_3{}^2 + \cdots \tag{6.13}$$

This can be written in matrix form as:

$$T = \tfrac{1}{2}\{\dot{r}\}^{T}[\overline{m}]\{\dot{r}\} \tag{6.14}$$

where $[\overline{m}]$ is a diagonal matrix of mass particles, $\{\dot{r}\}$ a column vector of the velocities of the individual mass particles and $\{\dot{r}\}^T$ its transpose.

Substituting Eq. (6.12) into Eq. (6.14), noting that transposing both sides of Eq. (6.12) gives $\{\dot{r}\}^T = \{\dot{z}\}^T [X_m]^T$, we have

$$T = \tfrac{1}{2}\{\dot{z}\}^T [X_{\mathrm{m}}]^T [\overline{m}][X_{\mathrm{m}}]\{\dot{z}\} \tag{6.15}$$

or

$$T = \tfrac{1}{2}\{\dot{z}\}^T [M]\{\dot{z}\} \tag{6.16}$$

where

$$[M] = [X_{\mathrm{m}}]^T [\overline{m}][X_{\mathrm{m}}]. \tag{6.17}$$

It can be shown that $[M]$ is the mass matrix of the system expressed in terms of global coordinates. Equation (6.16) is an important result giving the kinetic energy of the system in terms of the equations of motion in global coordinates.

Just as the mass properties of a structure can be traced back to individual particles of mass, the stiffness properties can be traced back to individual stiffness elements or 'springs', k_1, k_2, k_3, etc. Then, since the potential energy stored in a single spring of stiffness k_i and compression or extension, s_i, is $\tfrac{1}{2}k_i s_i{}^2$, the potential energy, U, in the whole system is

$$U = \tfrac{1}{2}k_1 s_1^2 + \tfrac{1}{2}k_2 s_2^2 + \cdots \tag{6.18}$$

where s_1, $s_2, \ldots$, etc. are the compressions or extensions of the individual stiffness elements. Writing Eq. (6.18) in matrix form:

$$U = \tfrac{1}{2}\{s\}^T [\overline{k}]\{s\} \tag{6.19}$$

where $\{s\}$ is a column vector of individual spring element compressions or extensions, $\{s\}^T$ its transpose and $[\overline{k}]$ a diagonal matrix of spring element stiffnesses.

Now suppose that the spring compressions or extensions, $\{s\}$, are related to the global displacement coordinates, $\{z\}$, by the transformation:

$$\{s\} = [X_{\mathrm{s}}]\{z\} \tag{6.20}$$

Substituting Eq. (6.20) into Eq. (6.19), noting that $\{s\}^T = \{z\}^T [X_{\mathrm{s}}]^T$, gives

$$U = \tfrac{1}{2}\{z\}^T [X_{\mathrm{s}}]^T [\overline{k}][X_{\mathrm{s}}]\{z\} \tag{6.21}$$

or

$$U = \tfrac{1}{2}\{z\}^T [K]\{z\} \tag{6.22}$$

where

$$[K] = [X_{\mathrm{s}}]^T [\overline{k}][X_{\mathrm{s}}] \tag{6.23}$$

It can be shown that $[K]$ is the stiffness matrix of the system, expressed in global coordinates, in the same way that $[M]$ is the mass matrix.

Equation (6.22) is an important result, giving the potential energy of the system in terms of the equations of motion in global coordinates, and is seen to be analogous to Eq. (6.16) which gives the kinetic energy.

Example 6.1

The equipment box shown in Fig. 6.2, to be fitted into an aircraft, is vibration isolated by two pairs of springs, of combined stiffness k_1 and k_2. When empty, the mass of the box is m and its mass center is at G. Its mass moment of inertia about G, when empty, is I. Three heavy items, which can be treated as point masses, m_1, m_2 and m_3, are fixed in the box as shown. The motion of the box is defined by two global coordinates, z and θ, the translation and rotation, respectively, of the reference center, point C.

Use matrix methods to derive the equations of motion of the box and contents in terms of the coordinates z and θ. Assume that the system is undamped.

Solution

Since there are no external forces or damping, the equations of motion will be of the form:

$$[M]\{\ddot{z}\} + [K]\{z\} = 0 \tag{A}$$

where $\{z\} = \left\{ {z \atop \theta} \right\}$. It is now required to find the 2×2 matrices $[M]$ and $[K]$.

The mass matrix, $[M]$, is given by Eq. (6.17):

$$[M] = [X_m]^T [\overline{m}][X_m] \tag{B}$$

where

$$[\overline{m}] = \begin{bmatrix} m & 0 & 0 & 0 & 0 \\ 0 & I & 0 & 0 & 0 \\ 0 & 0 & m_1 & 0 & 0 \\ 0 & 0 & 0 & m_2 & 0 \\ 0 & 0 & 0 & 0 & m_3 \end{bmatrix} \tag{C}$$

i.e. a diagonal matrix of the individual mass or inertia elements. Note that there are no cross terms between the mass of the empty box, m, and its moment of inertia, I, because the latter is defined about the mass center, G.

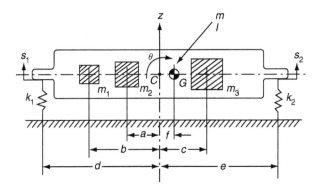

Fig. 6.2   Vibration-isolated equipment box discussed in Example 6.1.

The matrix $[X_{\mathrm{m}}]$ is defined by Eq. (6.11), $\{r\} = [X_{\mathrm{m}}]\{\underline{z}\}$, and is the transformation matrix, for small angles, between the five displacements, $r_m, r_I, r_{m_1}, r_{m_2}, r_{m_3}$ of the five 'mass' elements m, I, m_1, m_2, m_3 and the two global coordinates z and θ, as shown in Fig. 6.2. It is therefore of size 5×2, i.e. five rows and two columns. To derive the matrix $[X_{\mathrm{m}}]$, the first column consists of the displacements $r_m, r_I, r_{m_1}, r_{m_2}, r_{m_3}$, when the global coordinate z is increased from 0 to 1 unit, i.e. displaces one unit upwards. The second column of $[X_{\mathrm{m}}]$ consists of the displacements $r_m, r_I, r_{m_1}, r_{m_2}, r_{m_3}$ when the global coordinate θ is increased by one unit (one radian) clockwise.

The two columns of $[X_{\mathrm{m}}]$, and the complete matrix $[X_{\mathrm{m}}]$, are therefore as follows:

$$\begin{Bmatrix} r_m \\ r_I \\ r_{m_1} \\ r_{m_2} \\ r_{m_3} \end{Bmatrix}_1 = \begin{Bmatrix} 1 \\ 0 \\ 1 \\ 1 \\ 1 \end{Bmatrix} \tag{D}$$

$$\begin{Bmatrix} r_m \\ r_I \\ r_{m_1} \\ r_{m_2} \\ r_{m_3} \end{Bmatrix}_2 = \begin{Bmatrix} -f \\ 1 \\ a \\ b \\ -c \end{Bmatrix} \tag{E}$$

and

$$[X_{\mathrm{m}}] = \begin{bmatrix} 1 & -f \\ 0 & 1 \\ 1 & a \\ 1 & b \\ 1 & -c \end{bmatrix} \tag{F}$$

The displacement r_I in the first column is zero, because it is defined as a rotation, and displacement has no effect.

The small angle approximations are justified here because the expected vibration displacements are all very small compared with the dimensions of the box. The assumed unit displacements of one meter and one radian (say) should not, of course, be taken literally.

The mass matrix, in global coordinates, is now given by Eq. (B):

$$[M] = [X_{\mathrm{m}}]^{\mathrm{T}}[\overline{m}][X_{\mathrm{m}}] = \begin{bmatrix} 1 & -f \\ 0 & 1 \\ 1 & a \\ 1 & b \\ 1 & -c \end{bmatrix}^{\mathrm{T}} \begin{bmatrix} m & 0 & 0 & 0 & 0 \\ 0 & I & 0 & 0 & 0 \\ 0 & 0 & m_1 & 0 & 0 \\ 0 & 0 & 0 & m_2 & 0 \\ 0 & 0 & 0 & 0 & m_3 \end{bmatrix} \begin{bmatrix} 1 & -f \\ 0 & 1 \\ 1 & a \\ 1 & b \\ 1 & -c \end{bmatrix} \tag{G}$$

which multiplies out to:

$$[M] = \begin{bmatrix} (m + m_1 + m_2 + m_3) & (am_1 + bm_2 - cm_3 - fm) \\ (am_1 + bm_2 - cm_3 - fm) & (I + f^2 m + a^2 m_1 + b^2 m_2 + c^2 m_3) \end{bmatrix} \tag{H}$$

The stiffness matrix in global coordinates is given by Eq. (6.23):

$$[K] = [X_{\mathrm{s}}]^{\mathrm{T}}[\overline{k}][X_{\mathrm{s}}] \tag{I}$$

where $[\bar{k}]$ is a diagonal matrix of spring elements, in this case $\begin{bmatrix} k_1 & 0 \\ 0 & k_2 \end{bmatrix}$, and $[X_s]$ a matrix of individual spring displacements, when unit displacements of the global coordinates z and θ are applied. Either compression or extension may be taken as positive, but the same convention must be used throughout. Taking the spring extensions s_1 and s_2, as shown in Fig. 6.2, as positive, Eq. (6.20), $\{s\} = [X_s]\{z\}$, becomes:

$$\begin{Bmatrix} s_1 \\ s_2 \end{Bmatrix} = \begin{bmatrix} 1 & d \\ 1 & -e \end{bmatrix} \begin{Bmatrix} z \\ \theta \end{Bmatrix} \tag{J}$$

Then from Eq. (I):

$$[K] = \begin{bmatrix} 1 & d \\ 1 & -e \end{bmatrix}^T \begin{bmatrix} k_1 & \\ & k_2 \end{bmatrix} \begin{bmatrix} 1 & d \\ 1 & -e \end{bmatrix} = \begin{bmatrix} (k_1 + k_2) & (dk_1 - ek_2) \\ (dk_1 - ek_2) & (d^2 k_1 + e^2 k_2) \end{bmatrix} \tag{K}$$

Now that we have the mass matrix $[M]$ and the stiffness matrix $[K]$ in global coordinates, the equations of motion, in the form of Eq. (A):

$$[M]\{\ddot{z}\} + [K]\{z\} = 0$$

are simply:

$$\begin{bmatrix} (m + m_1 + m_2 + m_3) & (am_1 + bm_2 - cm_3 - fm) \\ (am_1 + bm_2 - cm_3 - fm) & (I + f^2 m + a^2 m_1 + b^2 m_2 + c^2 m_3) \end{bmatrix} \begin{Bmatrix} \ddot{z} \\ \ddot{\theta} \end{Bmatrix}$$
$$+ \begin{bmatrix} (k_1 + k_2) & (dk_1 - ek_2) \\ (dk_1 - ek_2) & (d^2 k_1 + e^2 k_2) \end{bmatrix} \begin{Bmatrix} z \\ \theta \end{Bmatrix} = 0 \tag{L}$$

6.2.2 Modal Coordinates

Up to this point, we have considered coordinates that are easily recognizable as such; for example, in the system shown in Fig. 6.1(a), the two coordinates are simply the actual displacements, z_1 and z_2, defined as positive upwards, of the masses m_1 and m_2, respectively. In general, with a large structure, the global coordinates will be identifiable translations and rotations at nodes (or 'grid points'), and indicate the displacements only at the point concerned.

In a multi-DOF system, however, it is possible to define modal coordinates instead, which define the amplitude of a deflection shape known as a mode. Thus, changing a single modal coordinate can affect the displacement over all or part of a structure.

There are two important classes of modes.

(1) The first type are simply assumed, or arbitrary, shapes. Assumed modes are used in component mode methods, often combined with normal modes of the components, and in the finite element method, the 'displacement functions' are essentially assumed modes defined between grid points.

(2) The other class of modes is that of normal modes. These have the important property of making the system matrices diagonal, i.e. separating the freedoms so that they can be treated as a series of single degrees of freedom, and therefore

much easier to deal with than a coupled system. Calculation of normal modes, however, requires the use of eigenvalues and eigenvectors.

6.2.3 Transformation from Global to Modal Coordinates

Suppose that we wish to transform the equations in global coordinates, say,

$$[M]\{\ddot{z}\} + [K]\{z\} = \{F\} \tag{6.24}$$

into an alternate form in modal coordinates, say,

$$[\underline{M}]\{\ddot{q}\} + [\underline{K}]\{q\} = \{Q\} \tag{6.25}$$

The first step is to define the transformation giving the relationship between the global displacements, $\{z\}$, and the modal displacements, $\{q\}$:

$$\{z\} = [X]\{q\} \tag{6.26}$$

from which

$$\{\dot{z}\} = [X]\{\dot{q}\} \tag{6.27}$$

$$\{z\}^{\mathrm{T}} = \{q\}^{\mathrm{T}}[X]^{\mathrm{T}} \tag{6.28}$$

and

$$\{\dot{z}\}^{\mathrm{T}} = \{\dot{q}\}^{\mathrm{T}}[X]^{\mathrm{T}} \tag{6.29}$$

If the number of displacements in $\{q\}$ is the same as in $\{z\}$, no information is lost, and $[X]$ is a square matrix. This is often the case for smaller systems, and means that the transformation is reversible. More usually, the number of modal coordinates in $\{q\}$ is far smaller than the number of global coordinates in $\{z\}$. There is then a theoretical loss of accuracy, and such a transformation must be justified on physical grounds, for example that the freedoms discarded represent motion at very high frequencies, beyond any possible excitation frequencies.

Transformation of the mass matrix
From Eq. (6.16), the kinetic energy in the global system is

$$T = \tfrac{1}{2}\{\dot{z}\}^{\mathrm{T}}[M]\{\dot{z}\}$$

From Eqs (6.16), (6.27) and (6.29):

$$T = \tfrac{1}{2}\{\dot{z}\}^{\mathrm{T}}[M]\{\dot{z}\} = \tfrac{1}{2}\{\dot{q}\}^{\mathrm{T}}[X]^{\mathrm{T}}[M][X]\{\dot{q}\} = \tfrac{1}{2}\{\dot{q}\}^{\mathrm{T}}[\underline{M}]\{\dot{q}\} \tag{6.30}$$

where the mass matrix in the new, modal system is $[\underline{M}]$, which from Eq. (6.30) can be seen to be given by:

$$[\underline{M}] = [X]^{\mathrm{T}}[M][X] \tag{6.31}$$

Transformation of the stiffness matrix

From Eq. (6.22), the potential energy in the global system is

$$U = \tfrac{1}{2}\{z\}^{\mathrm{T}}[K]\{z\}$$

From Eqs (6.22), (6.26) and (6.28):

$$U = \tfrac{1}{2}\{z\}^{\mathrm{T}}[K]\{z\} = \tfrac{1}{2}\{q\}^{\mathrm{T}}[X]^{\mathrm{T}}[K][X]\{q\} = \tfrac{1}{2}\{q\}^{\mathrm{T}}[\underline{K}]\{q\} \qquad (6.32)$$

where the new stiffness matrix, in the modal system, is $[\underline{K}]$, given by:

$$[\underline{K}] = [X]^{\mathrm{T}}[K][X] \qquad (6.33)$$

Transformation of the external forces

The expression for the transformation of the external forces, $\{F\}$, to the modal external forces, $\{Q\}$, can be found using the method of virtual work, as follows.

Suppose virtual displacements, $\{\hat{z}\}$, are applied to the global system. The work, W, done by the external forces, $\{F\}$, is then the sum of each virtual displacement multiplied by the appropriate external force, which is simply:

$$W = \{\hat{z}\}^{\mathrm{T}}\{F\} \qquad (6.34)$$

If $\{Q\}$ is the set of external modal forces giving the same loading as $\{F\}$, and $\{\hat{q}\}$ is the corresponding set of virtual displacements, the same work is done, and

$$W = \{\hat{z}\}^{\mathrm{T}}\{F\} = \{\hat{q}\}^{\mathrm{T}}\{Q\} \qquad (6.35)$$

The relationship between $\{z\}$ and $\{q\}$ has already been specified in Eq. (6.26) as:

$$\{z\} = [X]\{q\}$$

This transformation must also apply to the virtual displacements, so we can write:

$$\{\hat{z}\} = [X]\{\hat{q}\} \qquad (6.36)$$

and, of course,

$$\{\hat{z}\}^{\mathrm{T}} = \{\hat{q}\}^{\mathrm{T}}[X]^{\mathrm{T}} \qquad (6.37)$$

Substituting Eq. (6.37) into Eq. (6.35):

$$W = \{\hat{q}\}^{\mathrm{T}}[X]^{\mathrm{T}}\{F\} = \{\hat{q}\}^{\mathrm{T}}\{Q\} \qquad (6.38)$$

Therefore

$$\{Q\} = [X]^{\mathrm{T}}\{F\} \qquad (6.39)$$

which is the required expression relating the actual external forces, $\{F\}$, to the generalized or modal forces, $\{Q\}$.

A summary of the essential relationships

The matrix relationships derived in this section are of fundamental importance in the analysis of multi-DOF systems. In view of this, it is perhaps a good idea to run through them once more.

We started with the undamped equations of motion in global coordinates:

$$[M]\{\ddot{z}\} + [K]\{z\} = \{F\} \tag{6.24}$$

and showed that the equivalent equations in modal coordinates are

$$[\underline{M}]\{\ddot{q}\} + [\underline{K}]\{q\} = \{Q\} \tag{6.25}$$

provided that

$$[\underline{M}] = [X]^{\mathrm{T}}[M][X] \tag{6.31}$$

and that

$$[\underline{K}] = [X]^{\mathrm{T}}[K][X] \tag{6.33}$$

where $[X]$ defines a relationship between coordinates $\{z\}$ and $\{q\}$ given by:

$$\{z\} = [X]\{q\}. \tag{6.26}$$

It was then shown that the relationship between the actual external forces, $\{F\}$, and the external forces expressed as generalized forces, $\{Q\}$, is

$$\{Q\} = [X]^{\mathrm{T}}\{F\} \tag{6.39}$$

These relationships apply to any transformation from global into modal coordinates, *whether the modes are assumed or normal modes*. They can also be used, with appropriate change of notation, to transform any set of equations from one coordinate system to another.

Example 6.2

(a) Use assumed modes to transform the equations already derived for the system shown in Fig. 6.1(a), repeated here as Fig. 6.3, into a form having no stiffness coupling.

(b) Show how actual input forces, F_1 and F_2, can be applied to the transformed equations as generalized modal forces, and how generalized modal displacements can be converted back to actual local displacements.

Solution

Part (a):

The equations for this system were derived in Section 6.1 as Eqs (6.2), (6.5) and (6.10), which are all the same. They are

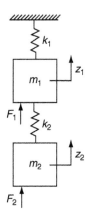

Fig. 6.3   Schematic 2-DOF system discussed in Example 6.2.

$$\begin{bmatrix} m_1 & 0 \\ 0 & m_2 \end{bmatrix} \begin{Bmatrix} \ddot{z}_1 \\ \ddot{z}_2 \end{Bmatrix} + \begin{bmatrix} (k_1 + k_2) & -k_2 \\ -k_2 & k_2 \end{bmatrix} \begin{Bmatrix} z_1 \\ z_2 \end{Bmatrix} = \begin{Bmatrix} F_1 \\ F_2 \end{Bmatrix} \tag{A}$$

These equations were formed by taking the displacements, z_1 and z_2, of the individual masses, m_1 and m_2, as the generalized coordinates. This resulted in a coupled stiffness matrix, but an uncoupled mass matrix.

We can now transform the coordinates, using the transformation defined by Eq. (6.26):

$$\{z\} = [X]\{q\} \tag{B}$$

Let us choose to represent the motion of the system by two coordinates, q_1 and q_2, where the actual displacements, z_1 and z_2, of the masses m_1 and m_2 are given by the following transformation, corresponding to Eq. (B):

$$\begin{Bmatrix} z_1 \\ z_2 \end{Bmatrix} = \begin{bmatrix} 1 & 0 \\ 1 & 1 \end{bmatrix} \begin{Bmatrix} q_1 \\ q_2 \end{Bmatrix} \tag{C}$$

The columns of the 2×2 matrix are simple examples of assumed modes; we have arbitrarily decided that the first mode is $\begin{Bmatrix} 1 \\ 1 \end{Bmatrix}$, i.e. the two masses move one unit upwards together, and the second mode is $\begin{Bmatrix} 0 \\ 1 \end{Bmatrix}$, i.e. m_1 does not move at all, and m_2 moves one unit upwards. From Fig. 6.3, we see that the first mode compresses k_1, but not k_2, and that the second mode compresses only k_2, not k_1. We would therefore expect that using these modes would remove the stiffness coupling.

We can now use the expressions developed in Section 6.2.3 to transform the equations from the form of Eq. (6.24):

$$[M]\{\ddot{z}\} + [K]\{z\} = \{F\} \tag{D}$$

to the form of Eq. (6.25):

$$[\underline{M}]\{\ddot{q}\} + [\underline{K}]\{q\} = \{Q\} \tag{E}$$

The new mass and stiffness matrices are given by Eqs (6.31) and (6.33), respectively, i.e.,

$$[\underline{M}] = [X]^{\mathrm{T}}[M][X] \tag{F}$$

and

$$[\underline{K}] = [X]^{\mathrm{T}}[K][X] \tag{G}$$

where

$$[X] = \begin{bmatrix} 1 & 0 \\ 1 & 1 \end{bmatrix} \tag{H}$$

Substituting Eqs (F) and (G) into Eq. (E) gives the new equations:

$$[X]^{\mathrm{T}}[M][X]\{\ddot{q}\} + [X]^{\mathrm{T}}[K][X]\{q\} = \{Q\} \tag{I}$$

or written out in full, using Eq. (H):

$$\begin{bmatrix} 1 & 0 \\ 1 & 1 \end{bmatrix}^{\mathrm{T}} \begin{bmatrix} m_1 & 0 \\ 0 & m_2 \end{bmatrix} \begin{bmatrix} 1 & 0 \\ 1 & 1 \end{bmatrix} \begin{Bmatrix} \ddot{q}_1 \\ \ddot{q}_2 \end{Bmatrix} + \begin{bmatrix} 1 & 0 \\ 1 & 1 \end{bmatrix}^{\mathrm{T}} \begin{bmatrix} (k_1 + k_2) & -k_2 \\ -k_2 & k_2 \end{bmatrix} \begin{bmatrix} 1 & 0 \\ 1 & 1 \end{bmatrix} \begin{Bmatrix} q_1 \\ q_2 \end{Bmatrix}$$
$$= \begin{Bmatrix} Q_1 \\ Q_2 \end{Bmatrix} \tag{J}$$

Multiplied out, these become

$$\begin{bmatrix} (m_1 + m_2) & m_2 \\ m_2 & m_2 \end{bmatrix} \begin{Bmatrix} \ddot{q}_1 \\ \ddot{q}_2 \end{Bmatrix} + \begin{bmatrix} k_1 & 0 \\ 0 & k_2 \end{bmatrix} \begin{Bmatrix} q_1 \\ q_2 \end{Bmatrix} = \begin{Bmatrix} Q_1 \\ Q_2 \end{Bmatrix} \tag{K}$$

This has had the desired effect of uncoupling the stiffness matrix, and the mass matrix is now coupled instead.

Part (b):
It is now easy to input actual external forces F_1 and F_2 by using Eq. (6.39),

$$\{Q\} = [X]^{\mathrm{T}}\{F\} \tag{L}$$

to convert them to modal external forces $\{Q\}$. In this example:

$$\{Q\} = \begin{Bmatrix} Q_1 \\ Q_2 \end{Bmatrix} = \begin{bmatrix} 1 & 0 \\ 1 & 1 \end{bmatrix}^{\mathrm{T}} \begin{Bmatrix} F_1 \\ F_2 \end{Bmatrix} = \begin{bmatrix} 1 & 1 \\ 0 & 1 \end{bmatrix} \begin{Bmatrix} F_1 \\ F_2 \end{Bmatrix} \tag{M}$$

If the equations were then solved to find the time histories of q_1 and q_2, Eq. (C) could then be used to find the local displacement time histories, z_1 and z_2

$$\{z\} = \begin{Bmatrix} z_1 \\ z_2 \end{Bmatrix} = \begin{bmatrix} 1 & 0 \\ 1 & 1 \end{bmatrix} \begin{Bmatrix} q_1 \\ q_2 \end{Bmatrix}$$

6.3 Undamped normal modes

In Section 6.2.3, we established rules for transforming a set of equations of motion in one coordinate system, say the global system, $\{z\}$:

$$[M]\{\ddot{z}\} + [K]\{z\} = \{F\} \tag{6.24}$$

into an alternate form in modal coordinates, $\{q\}$:

$$[\underline{M}]\{\ddot{q}\} + [\underline{K}]\{q\} = \{Q\} \tag{6.25}$$

In general, there will be coupling between the modes of a system after a coordinate transformation, and one or both of the matrices $[\underline{M}]$ and $[\underline{K}]$ will tend to be fully populated. However, there is a form of coordinate transformation that will remove all the coupling between the modal coordinates. This occurs when we choose to form the matrix $[X]$ in Eq. (6.26), $\{z\} = [X]\{q\}$, by making its columns consist of the eigen-vectors of the original global system. This transformation into *normal* or *principal modes* is of fundamental importance in structural dynamics. Before discussing it in detail, however, we should pause briefly to consider eigenvalues and eigenvectors.

6.3.1 Introducing Eigenvalues and Eigenvectors

Eigenvalues and eigenvectors are mathematical concepts, and their use is not confined to vibration theory. The words incidentally are derived from the German word *eigen*, meaning 'own', so the eigenvalues of a set of equations are its own values, and the eigenvectors are its own vectors. They tend to arise, for example, when the deflection shape of a system depends upon its loading, but the loading, in turn, depends upon the deflection shape. This is obviously true of a vibrating system with more than one degree of freedom, and it will be seen that such a system possesses as many eigenvalues, which are related to its natural frequencies, as there are degrees of freedom, and that each is associated with a shape function known as an eigenvector.

It is possible to find the eigenvalues and eigenvectors of a system explicitly only up to two degrees of freedom. Above that number, the process must be iterative, as will be discussed in Chapter 7.

Eigenvalues and eigenvectors can be real or complex, and this is determined entirely by the damping in the system. Systems with no damping have real eigenvalues and eigenvectors. If the damping is of the *proportional* kind (see Section 6.4.4), the eigenva-lues are complex, but the eigenvectors remain real. Other than in these special cases, both the eigenvalues and the eigenvectors become complex, and significantly more difficult to deal with. Fortunately, when the damping terms are small, the eigenvalues and the eigenvectors are 'nearly real' (i.e. the imaginary parts are small compared with the real parts), and in practice real eigenvectors can be used as a good approximation for practically all structures, and most systems generally. There are, however, some important exceptions, such as systems with significant gyroscopic effects, such as helicopter rotors, and aircraft flutter or response calculations using equations based on the p method, where aerodynamic damping terms can couple the modes.

Suppose we wish to find the eigenvalues and eigenvectors of the general undamped system represented by Eq. (6.24):

$$[M]\{\ddot{z}\} + [K]\{z\} = \{F\}$$

Since the eigenvalues and eigenvectors are unaffected by the applied forces, they can be obtained from the 'homogeneous' equations given by omitting the external forces:

$$[M]\{\ddot{z}\} + [K]\{z\} = 0 \qquad (6.40)$$

or, say,

$$\begin{bmatrix} m_{11} & m_{12} & \cdots \\ m_{21} & m_{22} & \cdots \\ \vdots & \vdots & \vdots \end{bmatrix} \begin{Bmatrix} \ddot{z}_1 \\ \ddot{z}_2 \\ \vdots \end{Bmatrix} + \begin{bmatrix} k_{11} & k_{12} & \cdots \\ k_{21} & k_{22} & \cdots \\ \vdots & \vdots & \vdots \end{bmatrix} \begin{Bmatrix} z_1 \\ z_2 \\ \vdots \end{Bmatrix} = 0 \qquad (6.41)$$

Since there is no damping, if the system is assumed to have been set in motion at frequency ω, all the elements of the vector $\{z\}$ will vibrate in phase, or exact antiphase, with each other. The responses at all the points z_1, z_2, etc. can be considered as horizontal or vertical projections of a rotating, complex unit vector, $e^{i\omega t}$, multiplied by the *real* constants $\bar{z}_1$, $\bar{z}_2$, etc. Then the responses of all the points, z_1, z_2, etc., are given by:

$$\{z\} = \begin{Bmatrix} z_1 \\ z_2 \\ \vdots \end{Bmatrix} = \begin{Bmatrix} \bar{z}_1 \\ \bar{z}_2 \\ \vdots \end{Bmatrix} e^{i\omega t} = \{\bar{z}\}e^{i\omega t} \qquad (6.42)$$

The vector of accelerations, $\{\ddot{z}\}$, is given by differentiating Eq. (6.42) twice with respect to time:

$$\{\ddot{z}\} = \begin{Bmatrix} \ddot{z}_1 \\ \ddot{z}_2 \\ \vdots \end{Bmatrix} = -\omega^2 \begin{Bmatrix} \bar{z}_1 \\ \bar{z}_2 \\ \vdots \end{Bmatrix} e^{i\omega t} = -\omega^2\{\bar{z}\}e^{i\omega t} \qquad (6.43)$$

Substituting Eqs (6.42) and (6.43) into Eq. (6.41) and dividing through by $e^{i\omega t}$ gives

$$-\omega^2 \begin{bmatrix} m_{11} & m_{12} & \cdots \\ m_{21} & m_{22} & \cdots \\ \vdots & \vdots & \vdots \end{bmatrix} \begin{Bmatrix} \bar{z}_1 \\ \bar{z}_2 \\ \vdots \end{Bmatrix} + \begin{bmatrix} k_{11} & k_{12} & \cdots \\ k_{21} & k_{22} & \cdots \\ \vdots & \vdots & \vdots \end{bmatrix} \begin{Bmatrix} \bar{z}_1 \\ \bar{z}_2 \\ \vdots \end{Bmatrix} = 0 \qquad (6.44)$$

or

$$\left(-\omega^2[M] + [K]\right)\{\bar{z}\} = 0$$

or writing $\lambda = \omega^2$

$$\left(-\lambda[M] + [K]\right)\{\bar{z}\} = 0 \qquad (6.45)$$

Equation (6.45) is an eigenvalue problem, although not in standard mathematical form. It can be satisfied by n different values of λ, say $\lambda_1, \lambda_2, \ldots, \lambda_n$, which are the *eigenvalues*, and n corresponding vectors, $\{\bar{z}\}$, say $\{\bar{z}\}_1, \{\bar{z}\}_2, \ldots, \{\bar{z}\}_n$, which are the *eigenvectors*, where n is the number of degrees of freedom.

There are several different ways of expressing Eq. (6.45), and many ways of extracting the eigenvalues and eigenvectors. Some of these will be discussed in Chapter 7, but for the present, we can use the following, very basic, method.

(1) Expansion of the determinant $|[K] - \lambda[M]|$ gives a polynomial of order n in λ, of which the n roots are the n eigenvalues, $\lambda_1, \lambda_2, \ldots, \lambda_n$.

(2) Substituting each eigenvalue in turn into Eq. (6.45) gives the eigenvectors. It can be seen from Eq. (6.45) that any eigenvector $\{\bar{z}\}$ could be multiplied by any overall factor and still satisfy the equations, so only the relative sizes of the elements can be found. There are two standard methods for normalizing the size of eigenvectors:

  (a) The element having the largest absolute value in each column (i.e. largest ignoring signs) is set equal to unity. Thus each element in the column is divided by that element. This method is useful when it is desired to plot the mode shapes.

  (b) The elements in each column are scaled so that the final mass matrix, formed by using the eigenvectors as modes, is a unit matrix. The modes are then sometimes described as *weighted normal* or *orthonormal modes*. This method is useful within programs, since once this convention is adopted, the generalized masses are always known to be unity.

However the eigenvectors are normalized, when the transformation matrix, $[X]$, in Eq. (6.26), instead of simply being based on assumed modes, as in the previous section, is formed by writing its columns as the eigenvectors of Eq. (6.24), something remarkable happens: both the mass and the stiffness matrices of the modal equations then become diagonal, and the system is transformed into a set of completely independent, single-DOF equations. This transformation into 'normal', or 'principal', coordinates is so useful that it is used in the analysis of almost all multi-DOF systems.

These processes will become clear after working through the following simple example.

Example 6.3

(a) Find the eigenvalues and eigenvectors of the undamped system shown in Fig. 6.3 with:

$m_1 = 1 \, \text{kg}; \quad m_2 = 2 \, \text{kg}; \quad k_1 = 10 \, \text{N/m} \quad k_2 = 10 \, \text{N/m}$

Scale the eigenvectors so that the largest absolute element in each column is set to unity.

(b) Demonstrate that a transformation to modal coordinates using the eigenvectors as modes enables the equations to be written as uncoupled single-DOF systems.

(c) Rescale the eigenvectors so that the mass matrix, in normal mode coordinates, is a unit matrix.

Solution

Part (a)

The equations of motion for this system, in terms of global coordinates z_1 and z_2, and with external forces F_1 and F_2, have already been derived as Eq. (6.10):

$$\begin{bmatrix} m_1 & 0 \\ 0 & m_2 \end{bmatrix} \begin{Bmatrix} \ddot{z}_1 \\ \ddot{z}_2 \end{Bmatrix} + \begin{bmatrix} (k_1 + k_2) & -k_2 \\ -k_2 & k_2 \end{bmatrix} \begin{Bmatrix} z_1 \\ z_2 \end{Bmatrix} = \begin{Bmatrix} F_1 \\ F_2 \end{Bmatrix} \tag{A}$$

Inserting the given numerical values:

$$\begin{bmatrix} 1 & 0 \\ 0 & 2 \end{bmatrix} \begin{Bmatrix} \ddot{z}_1 \\ \ddot{z}_2 \end{Bmatrix} + \begin{bmatrix} 20 & -10 \\ -10 & 10 \end{bmatrix} \begin{Bmatrix} z_1 \\ z_2 \end{Bmatrix} = \begin{Bmatrix} F_1 \\ F_2 \end{Bmatrix} \tag{B}$$

Omitting the applied forces F_1 and F_2 gives the homogeneous equations, from which the eigenvalues and eigenvectors can be found:

$$\begin{bmatrix} 1 & 0 \\ 0 & 2 \end{bmatrix} \begin{Bmatrix} \ddot{z}_1 \\ \ddot{z}_2 \end{Bmatrix} + \begin{bmatrix} 20 & -10 \\ -10 & 10 \end{bmatrix} \begin{Bmatrix} z_1 \\ z_2 \end{Bmatrix} = 0 \tag{C}$$

Arranging into the form of Eq. (6.45): $([K] - \lambda[M])\{\bar{z}\} = 0$, we have

$$\left(\begin{bmatrix} 20 & -10 \\ -10 & 10 \end{bmatrix} - \lambda \begin{bmatrix} 1 & 0 \\ 0 & 2 \end{bmatrix} \right) \begin{Bmatrix} \bar{z}_1 \\ \bar{z}_2 \end{Bmatrix} = 0 \tag{D}$$

or

$$\begin{bmatrix} (20 - \lambda) & -10 \\ -10 & (10 - 2\lambda) \end{bmatrix} \begin{Bmatrix} \bar{z}_1 \\ \bar{z}_2 \end{Bmatrix} = 0 \tag{E}$$

The two eigenvalues, λ_1 and λ_2, are now found by writing the matrix in Eq. (E) as a determinant and equating it to zero:

$$\begin{vmatrix} (20 - \lambda) & -10 \\ -10 & (10 - 2\lambda) \end{vmatrix} = 0,$$

or

$$(20 - \lambda)(10 - 2\lambda) - 100 = 0,$$

giving

$$\lambda^2 - 25\lambda + 50 = 0. \tag{F}$$

Solving this quadratic equation by the formula $\lambda = \frac{25 \pm \sqrt{25^2 - 200}}{2}$ gives the lower eigenvalue, $\lambda_1 = 2.1922$, and the higher eigenvalue, $\lambda_2 = 22.807$.

The two natural frequencies are

$$\omega_1 = \sqrt{\lambda_1} = \sqrt{2.1922} = 1.480\,\text{rad/s, and}$$
$$\omega_2 = \sqrt{\lambda_2} = \sqrt{22.807} = 4.775\,\text{rad/s}$$

The two eigenvectors can now be found by substituting the eigenvalues, one at a time, into Eq. (E). Since the eigenvectors are arbitrary to an overall factor, we can only find the ratio $(\bar{z}_1/\bar{z}_2)$ in each case. Either row of Eq. (E) can be used, giving

$$\left(\frac{\bar{z}_1}{\bar{z}_2} \right) = \left(\frac{10}{20 - \lambda} \right) \text{ or } \left(\frac{10 - 2\lambda}{10} \right)$$

Thus we have

for $\lambda = \lambda_1 = 2.192$, $\left(\dfrac{\bar{z}_1}{\bar{z}_2}\right)_1 = 0.5615$

for $\lambda = \lambda_2 = 22.807$, $\left(\dfrac{\bar{z}_1}{\bar{z}_2}\right)_2 = -3.5615$

The eigenvectors can now be called 'normal modes', and it is usual to write them as vectors:

$$\{\phi\}_1 = \left\{\begin{matrix} \bar{z}_1 \\ \bar{z}_2 \end{matrix}\right\}_1, \quad \{\phi\}_2 = \left\{\begin{matrix} \bar{z}_1 \\ \bar{z}_2 \end{matrix}\right\}_2 \tag{G}$$

Using the first of the standard methods for normalizing the eigenvectors (making the largest absolute element in each column equal to unity), these are

$$\{\phi\}_1 = \left\{\begin{matrix} 0.5615 \\ 1 \end{matrix}\right\} \quad \{\phi\}_2 = \left\{\begin{matrix} 1 \\ -0.2807 \end{matrix}\right\} \tag{H}$$

Part (b)

The procedure for transforming the equations of motion into normal coordinates is now as outlined in Section 6.2.3, which could be applied using any set of modes. What is special about a normal mode transformation, in contrast to an assumed mode transformation, however, is that the matrix $[X]$ in Eq. (6.26):

$$\{z\} = [X]\{q\}$$

is now formed by *using the eigenvectors of the original equations as its columns*. The transformation matrix $[X]$, in this case, is therefore:

$$[X] = [\{\phi\}_1 \{\phi\}_2] = \begin{bmatrix} 0.5615 & 1 \\ 1 & -0.2807 \end{bmatrix} \tag{I}$$

From Eq. (6.31), the new mass matrix $[\underline{M}]$, in modal coordinates, is given by:

$$[\underline{M}] = [X]^T[M][X]$$

or numerically:

$$[\underline{M}] = \begin{bmatrix} 0.5615 & 1 \\ 1 & -0.2807 \end{bmatrix}^T \begin{bmatrix} 1 & 0 \\ 0 & 2 \end{bmatrix} \begin{bmatrix} 0.5615 & 1 \\ 1 & -0.2807 \end{bmatrix} = \begin{bmatrix} 2.315 & 0 \\ 0 & 1.157 \end{bmatrix} \tag{J}$$

From Eq. (6.33), the new stiffness matrix, $[\underline{K}]$, in modal coordinates, is given by:

$$[\underline{K}] = [X]^T[K][X]$$

or numerically:

$$[\underline{K}] = \begin{bmatrix} 0.5615 & 1 \\ 1 & -0.2807 \end{bmatrix}^T \begin{bmatrix} 20 & -10 \\ -10 & 10 \end{bmatrix} \begin{bmatrix} 0.5615 & 1 \\ 1 & -0.2807 \end{bmatrix} = \begin{bmatrix} 5.075 & 0 \\ 0 & 26.40 \end{bmatrix}$$

$$\tag{K}$$

The complete equations in normal mode coordinates are now given by substituting Eqs (J) and (K) into Eq. (6.25):

$$[\underline{M}]\{\ddot{q}\} + [\underline{K}]\{q\} = \{Q\}$$

giving

$$\begin{bmatrix} 2.315 & 0 \\ 0 & 1.157 \end{bmatrix} \begin{Bmatrix} \ddot{q}_1 \\ \ddot{q}_2 \end{Bmatrix} + \begin{bmatrix} 5.075 & 0 \\ 0 & 26.40 \end{bmatrix} \begin{Bmatrix} q_1 \\ q_2 \end{Bmatrix} = \begin{Bmatrix} Q_1 \\ Q_2 \end{Bmatrix} \qquad \text{(L)}$$

It can be seen that the use of the eigenvectors as modes in the transformation has made the mass and stiffness matrices diagonal. Equation (L) now consists of two completely independent single-DOF equations:

$$\underline{m}_{11}\ddot{q}_1 + \underline{k}_{11}q_1 = Q_1 \qquad \text{(M}_1\text{)}$$

and

$$\underline{m}_{22}\ddot{q}_2 + \underline{k}_{22}q_2 = Q_2 \qquad \text{(M}_2\text{)}$$

where the numerical values of $\underline{m}_{11}$, $\underline{m}_{22}$, $\underline{k}_{11}$ and $\underline{k}_{22}$ are clear from Eq. (L).

The quantities $\underline{m}_{11}$ and $\underline{m}_{22}$ are known as the *generalized* (or *modal*) *masses* of modes 1 and 2, respectively, and similarly $\underline{k}_{11}$ and $\underline{k}_{22}$ are the *generalized* (or *modal*) *stiffnesses*. The absolute numerical values of the generalized masses and stiffnesses have no significance, since they depend upon how the eigenvectors are scaled. However, $\underline{k}_{11}/\underline{m}_{11}$ should always be equal to ω_1^2, and $\underline{k}_{22}/\underline{m}_{22}$ should equal ω_2^2. It is easily verified that this is so in this example.

Part (c)

The other standard method for normalizing the eigenvectors is to scale them so that the new mass matrix, $[\underline{M}]$, in modal coordinates, is equal to the unit matrix, $[I]$.

$$[\underline{M}] = \begin{bmatrix} \underline{m}_{11} & 0 \\ 0 & \underline{m}_{22} \end{bmatrix} = \begin{bmatrix} 1 & 0 \\ 0 & 1 \end{bmatrix} = [I] \qquad \text{(N)}$$

The eigenvectors are then described as *weighted normal* or *orthonormal*, and the corresponding stiffness matrix, $[\underline{K}]$, then becomes

$$[\underline{K}] = \begin{bmatrix} \underline{k}_{11} & 0 \\ 0 & \underline{k}_{22} \end{bmatrix} = \begin{bmatrix} \lambda_1 & 0 \\ 0 & \lambda_2 \end{bmatrix} = \begin{bmatrix} \omega_1^2 & 0 \\ 0 & \omega_2^2 \end{bmatrix} \qquad \text{(O)}$$

that is, a diagonal matrix of the squares of the natural frequencies. Scaling the eigenvectors so that the mass and stiffness matrices take these simple forms can be achieved in two ways.

Method(1):

If the diagonal mass and stiffness matrices in modal coordinates have already been evaluated, using some other scaling for the eigenvectors, as in Eqs (J) and (K), they can be re-scaled to orthonormal form by multiplying each eigenvector $\{\phi\}_i$ by a factor α_i, where

$$\alpha_i = \frac{1}{\sqrt{\underline{m}_{ii}}} \qquad \text{(P)}$$

where $\underline{m}_{ii}$ is the diagonal element of the new mass matrix (in modal coordinates) corresponding to eigenvector i. In this case, from Eq. (L):

$$\alpha_1 = \frac{1}{\sqrt{2.315}} \quad \text{and} \quad \alpha_2 = \frac{1}{\sqrt{1.157}}$$

Multiplying the old vectors from Eq. (H) by these factors:

$$\{\phi\}_1^{\text{new}} = \frac{1}{\sqrt{2.315}} \begin{Bmatrix} 0.5615 \\ 1 \end{Bmatrix} = \begin{Bmatrix} 0.3690 \\ 0.6572 \end{Bmatrix} \tag{Q_1}$$

$$\{\phi\}_2^{\text{new}} = \frac{1}{\sqrt{1.15767}} \begin{Bmatrix} 1 \\ -0.2807 \end{Bmatrix} = \begin{Bmatrix} -0.9294 \\ 0.2609 \end{Bmatrix} \tag{Q_2}$$

where Eqs (Q_1) and (Q_2) give the rescaled eigenvectors. The new transformation matrix, now based on orthonormal eigenvectors, is

$$[X]^{\text{new}} = [\{\phi\}_1 \{\phi\}_2] = \begin{bmatrix} 0.3690 & -0.9294 \\ 0.6572 & 0.2609 \end{bmatrix} \tag{R}$$

It is easily verified that using this matrix in the transformation of the mass and stiffness matrices changes them to the forms of Eqs (N) and (O), respectively:

$$[\underline{M}] = [X]^T[M][X] = \begin{bmatrix} 0.3690 & -0.9294 \\ 0.6572 & 0.2609 \end{bmatrix}^T \begin{bmatrix} 1 & 0 \\ 0 & 2 \end{bmatrix} \begin{bmatrix} 0.3690 & -0.9294 \\ 0.6572 & 0.2609 \end{bmatrix} = \begin{bmatrix} 1 & 0 \\ 0 & 1 \end{bmatrix} \tag{S}$$

so $[\underline{M}]$ becomes a unit matrix and

$$[\underline{K}] = [X]^T[K][X] = \begin{bmatrix} 0.3690 & -0.9294 \\ 0.6572 & 0.2609 \end{bmatrix}^T \begin{bmatrix} 20 & -10 \\ -10 & 10 \end{bmatrix} \begin{bmatrix} 0.3690 & -0.9294 \\ 0.6572 & 0.2609 \end{bmatrix}$$

$$= \begin{bmatrix} 2.192 & 0 \\ 0 & 22.807 \end{bmatrix} \tag{T}$$

where the diagonal terms are now seen to be equal to the squares of the natural frequencies, ω_1^2 and ω_2^2.

Method (2):
The eigenvectors in orthonormal form can also be calculated directly from the *original* mass matrix, $[M]$, and the eigenvectors in any form, by noting that in Eq. (P), the generalized mass, $\underline{m}_{ii}$, for mode i, in the modal (transformed) equations, can be calculated from

$$\underline{m}_{ii} = \{\phi\}_i^T[M]\{\phi\}_i \tag{U}$$

where $\{\phi\}_i$ represents eigenvector i in any arbitrary form, for example, in the non-standard form:

$$\{\underline{\phi}\}_1 = \left\{ \begin{array}{c} 0.5615 \\ 1 \end{array} \right\} \quad \text{and} \quad \{\underline{\phi}\}_2 = \left\{ \begin{array}{c} -3.561 \\ 1 \end{array} \right\} \qquad (V_1)(V_2)$$

and $[M]$ is the original mass matrix in global coordinates, in this example given as

$$[M] = \begin{bmatrix} 1 & 0 \\ 0 & 2 \end{bmatrix}$$

Equations (P) and (U) can be combined, giving the more usual expression for α_i:

$$\alpha_i = \left(\frac{1}{\{\phi\}_i^T [M] \{\phi\}_i} \right)^{\frac{1}{2}} \qquad (W)$$

Substituting the numerical values above into Eq. (W) gives $\alpha_1 = 0.6572$ and $\alpha_2 = 0.2609$. The orthonormal eigenvectors are then:

$$\{\phi\}_1 = \alpha_1 \{\underline{\phi}\}_1 = 0.6572 \left\{ \begin{array}{c} 0.5615 \\ 1 \end{array} \right\} = \left\{ \begin{array}{c} 0.3690 \\ 0.6572 \end{array} \right\} \qquad (X_1)$$

$$\{\phi\}_2 = \alpha_2 \{\underline{\phi}\}_2 = 0.260958 \left\{ \begin{array}{c} -3.561 \\ 1 \end{array} \right\} = \left\{ \begin{array}{c} -0.9294 \\ 0.2609 \end{array} \right\} \qquad (X_2)$$

which are seen to be the same as the eigenvectors given in Eqs (Q_1) and (Q_2) by the first method.

The methods used in Example 6.3 above can be generalized to apply to undamped equations of motion of any size, and it will always be true that if we form the transformation from global coordinates to modal coordinates:

$$\{z\} = [X]\{q\} \qquad (6.26)$$

by making the columns of $[X]$ equal to the eigenvalues of the global system:

$$[M]\{\ddot{z}\} + [K]\{z\} = \{F\} \qquad (6.24)$$

then the resulting equations in modal coordinates

$$[\underline{M}]\{\ddot{q}\} + [\underline{K}]\{q\} = \{Q\} \qquad (6.25)$$

will have the desirable property that the matrices $[\underline{M}]$ and $[\underline{K}]$ both become diagonal, and Eq. (6.25) can be written as

$$\underline{m}_{11}\ddot{q}_1 + \underline{k}_{11}q_1 = Q_1$$
$$\underline{m}_{22}\ddot{q}_2 + \underline{k}_{22}q_2 = Q_2$$

$$\vdots$$

$$\underline{m}_{nn}\ddot{q}_n + \underline{k}_{nn}q_n = Q_n \qquad (6.46)$$

that is, as a series of completely independent single-DOF equations.

It can now be seen that this transformation makes calculating the response of even large multi-DOF systems a relatively simple operation. Any of the methods for finding the response of single-DOF systems discussed in Chapters 3 and 4 can be used, and the results are summed. This is known as the *normal mode summation* method, to be discussed in Section 6.5.

Eigenvalues and eigenvectors from the flexibility matrix

In Example 6.3 above, the eigenvalues and eigenvectors were obtained from the mass and stiffness matrices, $[M]$ and $[K]$, respectively. The equations of motion were written in the form of Eq. (6.45):

$$(-\lambda[M] + [K])\{\bar{z}\} = 0 \tag{6.45}$$

where $\lambda = \omega^2$. The eigenvalues, λ_i, were then found as the roots of the characteristic equation given by expanding the determinant $|[K] - \lambda[M]|$, or $|[M]^{-1}[K] - \lambda[I]|$, and the eigenvectors were found by substituting the eigenvalues back into the equations of motion.

Very often, the flexibility matrix, $[K]^{-1}$, is known rather than $[K]$. In this case, there is no need to invert $[K]^{-1}$ to give $[K]$, since the eigenvalues and vectors can be extracted equally well from the mass and flexibility matrices, as follows. Pre-multiplying Eq. (6.45) by $[K]^{-1}$ gives

$$-\lambda[K]^{-1}[M]\{\bar{z}\} + [K]^{-1}[K]\{\bar{z}\} = 0$$

or

$$\left([K]^{-1}[M] - \underline{\lambda}[I]\right)\{\bar{z}\} = 0 \tag{6.47}$$

where $\underline{\lambda} = 1/\lambda = 1/\omega^2$. The eigenvalues $\underline{\lambda}_i$ are then given by the roots of the characteristic equation:

$$\left|[K]^{-1}[M] - \underline{\lambda}[I]\right| = 0 \tag{6.48}$$

and the eigenvectors are obtained by substituting the eigenvalues, one by one, into Eq. (6.47). When using this method, the eigenvalues found, $\underline{\lambda}_i$, are the reciprocals of the squares of the natural frequencies, ω_i, but the eigenvectors are the same as those found using the stiffness matrix.

Orthogonality properties of eigenvectors

The extreme simplicity of Eq. (6.46) comes about due to the orthogonality properties of the eigenvectors with respect to both the mass and stiffness matrices, and this should now be discussed. Returning to Eq. (6.45)

$$(-\lambda[M] + [K])\{\bar{z}\} = 0$$

for one eigenvalue λ_i, and corresponding eigenvector, $\{\bar{z}\}_i$, this is

$$[K]\{\bar{z}\}_i = \lambda_i[M]\{\bar{z}\}_i \tag{6.49}$$

and for a different eigenvalue, λ_j, and eigenvector $\{\bar{z}\}_j$, it is

$$[K]\{\bar{z}\}_j = \lambda_j[M]\{\bar{z}\}_j \tag{6.50}$$

Premultiplying Eq. (6.49) by $\{\bar{z}\}_j^T$:

$$\{\bar{z}\}_j^T[K]\{\bar{z}\}_i = \lambda_i\{\bar{z}\}_j^T[M]\{\bar{z}\}_i \tag{6.51}$$

and transposing Eq. (6.50), using the fact that $[M]$ and $[K]$ are symmetric, then post-multiplying by $\{\bar{z}\}_i$, we have

$$\{\bar{z}\}_j^T[K]\{\bar{z}\}_i = \lambda_j\{\bar{z}\}_j^T[M]\{\bar{z}\}_i \tag{6.52}$$

Subtracting Eq. (6.52) from Eq. (6.51) gives

$$0 = (\lambda_i - \lambda_j)\{\bar{z}\}_j^T[M]\{\bar{z}\}_i \tag{6.53}$$

Assuming that $\lambda_i \neq \lambda_j$, Eq. (6.53) can only be satisfied if

$$\{\bar{z}\}_j^T[M]\{\bar{z}\}_i = 0 \tag{6.54}$$

showing that any two eigenvectors are orthogonal with respect to the mass matrix, $[M]$. From Eq. (6.51) or (6.52) it must also be true that

$$\{\bar{z}\}_j^T[K]\{\bar{z}\}_i = 0 \tag{6.55}$$

showing that they are also orthogonal with respect to the stiffness matrix.
If $i = j$, Eqs (6.51) and (6.52) can both be written as:

$$\{\bar{z}\}_i^T[K]\{\bar{z}\}_i = \lambda_i\{\bar{z}\}_i^T[M]\{\bar{z}\}_i \tag{6.56}$$

or

$$\underline{k}_{ii} = \lambda_i\,\underline{m}_{ii} \tag{6.57}$$

thus defining the generalized mass, $\underline{m}_{ii}$, as:

$$\underline{m}_{ii} = \{\bar{z}\}_i^T[M]\{\bar{z}\}_i \tag{6.58}$$

and similarly defining the generalized stiffness, $\underline{k}_{ii}$, as:

$$\underline{k}_{ii} = \{\bar{z}\}_i^T[K]\{\bar{z}\}_i \tag{6.59}$$

Advantages of normal modes

The advantages of working in normal modes can be listed as follows:
(1) The equations are vastly simplified, and the mass and stiffness properties, at least, are reduced to simple, diagonal matrices.
(2) Provided the damping is not such as to couple the modes significantly, which is true in the majority of cases, the response of a system can be found very easily by summing the responses of the normal modes, each of which can be treated as a single-DOF system.

(3) Since each normal mode has a known natural frequency, it is usually possible to omit those having natural frequencies well above any possible frequency components in the excitation, so reducing the number of equations.
(4) Perhaps most important of all, the easiest method for validating a structural model consists of comparing the calculated normal modes with measured normal modes, and usually it is worth calculating the normal modes of a structure for this reason alone. It will be seen that it is relatively easy to measure the *undamped* normal modes of a *damped* structure, and these are then directly comparable with the undamped calculated modes.

6.4 Damping in multi-DOF systems

6.4.1 The Damping Matrix

The system shown in Fig. 6.4(a) is the simple 2-DOF system considered throughout this chapter, with the addition of the viscous dampers c_1 and c_1.

The equations of motion can be derived using any of the methods used earlier in this chapter for undamped systems: for example, the free body diagram shown in Fig. 6.1(b) now has additional forces due to the dampers, as shown in Fig. 6.4(b). Alternatively, Lagrange's equations, taking z_1 and z_2 as generalized coordinates and $\bar{F}_1$ and $\bar{F}_2$ as generalized forces, can be modified to include dissipation energy, D, as follows:

$$\frac{\mathrm{d}}{\mathrm{d}t}\left(\frac{\partial T}{\partial \dot{z}_i}\right) + \frac{\partial D}{\partial \dot{z}_i} + \frac{\partial U}{\partial z_i} = \bar{F}_i \quad (i = 1, 2) \tag{6.60}$$

where in this case the dissipation energy, D, analogous to the kinetic energy T and the potential energy, U, is given by:

$$D = \tfrac{1}{2}c_1\dot{z}_1^2 + \tfrac{1}{2}c_2(\dot{z}_2 - \dot{z}_1)^2 \tag{6.61}$$

Whichever method is used, the equations of motion are the same as for the undamped system, but with additional terms due to the damping:

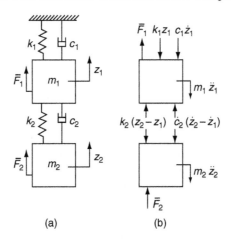

(a)   (b)

Fig. 6.4 Damped 2-DOF system in schematic form.

$$\overline{F}_1 - m_1 \ddot{z}_1 - k_1 z_1 - c_1 \dot{z}_1 + c_2(\dot{z}_2 - \dot{z}_1) + k_2(z_2 - z_1) = 0 \tag{6.62a}$$

$$\overline{F}_2 - m_2 \ddot{z}_2 - c_2(\dot{z}_2 - \dot{z}_1) - k_2(z_2 - z_1) = 0 \tag{6.62b}$$

or in matrix form:

$$\begin{bmatrix} m_1 & 0 \\ 0 & m_2 \end{bmatrix} \begin{Bmatrix} \ddot{z}_1 \\ \ddot{z}_2 \end{Bmatrix} + \begin{bmatrix} (c_1 + c_2) & -c_2 \\ -c_2 & c_2 \end{bmatrix} \begin{Bmatrix} \dot{z}_1 \\ \dot{z}_2 \end{Bmatrix} \begin{bmatrix} (k_1 + k_2) & -k_2 \\ -k_2 & k_2 \end{bmatrix} \begin{Bmatrix} z_1 \\ z_2 \end{Bmatrix} = \begin{bmatrix} \overline{F}_1 \\ \overline{F}_2 \end{bmatrix} \tag{6.63}$$

Equation (6.63) may be generalized to represent any system in global coordinates, with viscous damping and external forces. The equations of motion will always be of the form:

$$\begin{bmatrix} m_{11} & m_{12} & \cdots \\ m_{21} & m_{22} & \cdots \\ \vdots & \vdots & \vdots \end{bmatrix} \begin{Bmatrix} \ddot{z}_1 \\ \ddot{z}_2 \\ \vdots \end{Bmatrix} + \begin{bmatrix} c_{11} & c_{12} & \cdots \\ c_{21} & c_{22} & \cdots \\ \vdots & \vdots & \vdots \end{bmatrix} \begin{Bmatrix} \dot{z}_1 \\ \dot{z}_2 \\ \vdots \end{Bmatrix} \begin{bmatrix} k_{11} & k_{12} & \cdots \\ k_{21} & k_{22} & \cdots \\ \vdots & \vdots & \vdots \end{bmatrix} \begin{Bmatrix} z_1 \\ z_2 \\ \vdots \end{Bmatrix} = \begin{Bmatrix} \overline{F}_1 \\ \overline{F}_2 \\ \vdots \end{Bmatrix} \tag{6.64}$$

or more concisely:

$$[M]\{\ddot{z}\} + [C]\{\dot{z}\} + [K]\{z\} = \{\overline{F}\} \tag{6.65}$$

The damping matrix, $[C]$, is a square, symmetric matrix, comparable to the mass and stiffness matrices.

There are fundamentally two ways of solving an equation such as Eq. (6.65) for particular forms of the applied forces $\{\overline{F}\}$. These are
(1) Solution by the normal mode summation method.
(2) Direct solution of the global equations of motion.

Method (1), the normal mode summation method, is usually preferred, as it is completely analytic. Unfortunately, some forms of the damping matrix, $[C]$, can cause problems when we try to use it, and these are discussed in the next section.

6.4.2 Damped and Undamped Modes

It has been shown that when there is no damping, any system of equations in global coordinates can be transformed into normal mode coordinates, so that the mass and stiffness matrices, $[M]$ and $[K]$, become diagonal. The equations of motion can then be expressed as a set of uncoupled single-DOF equations. A potential problem in extending this idea to damped systems, however, is that the transformation, which consists of real eigenvectors, does not always make the damping matrix, $[C]$, diagonal also, so that the modes can remain coupled by damping terms.

One solution to this problem would be to use the fact that any damping matrix can be made diagonal, together with the mass and stiffness matrices, if complex eigenvalues and eigenvectors are used. This is a considerable extra complication, and is avoided when possible, but there are some systems where this approach has to be used. Some examples are the following:
(1) Rotating systems such as helicopter rotors, where centrifugal and Coriolis forces can lead to damping coupling between the modes;

(2) Aircraft flutter analysis and response calculations using the p method, where large cross-damping terms can arise from aerodynamic forces. The original flutter solution method, now known as the k method, was, in fact, originally devised specifically to avoid the use of complex eigenvalues and eigenvectors.
(3) Systems such as vehicle suspensions where discrete hydraulic dampers may couple the modes.
(4) Some buildings and off-shore structures where the soil or sea-bed damping can be large in relation to that of the structure itself.

Although generally beyond the scope of this text, an indication of how complex eigenvalues and eigenvectors are derived is given in Chapter 7.

Apart from specialized areas such as those listed above, it is found in practice that normal mode summation methods can usually be based on real eigenvalues, especially when the system is a structure. Two arguments can be used to support this common practice:

(1) In a structure, the damping terms tend to be small, and ill-defined, and although the damping may, in theory, couple the modes, the eigenvectors are 'nearly real' and the undamped modes are a good approximation to the actual modes.
(2) The damping can often be considered to be of the *proportional* or *Rayleigh* type. This exists when the damping matrix is of a form that can be diagonalized by the same real eigenvectors used to diagonalize the mass and stiffness matrices. The eigenvalues, however, are complex. This form of damping is discussed in Section 6.4.4.

6.4.3 Damping Inserted from Measurements

For structures, the usual method for introducing damping into a set of equations based on normal modes is, first, to derive the diagonal mass and stiffness matrices, ignoring the damping, and then add an appropriate diagonal damping matrix, on the basis of measurements on the structure, if it exists, or estimates based on similar structures, if it does not.

If there is any doubt about the amount of damping in a given structure, it is usually safer to underestimate it than to overestimate it, since damping almost always tends to reduce vibration response.

The procedure is illustrated by the following example.

Example 6.4

In the modeling of a certain structure, the natural frequencies of the first three normal modes, as calculated, and then verified by tests, were

$$f_1 = 10.1\,\text{Hz}, \quad f_2 = 17.8\,\text{Hz} \quad \text{and} \quad f_3 = 25.3\,\text{Hz}$$

The measured non-dimensional viscous damping coefficients of these modes, expressed as a fraction of critical damping were

$$\gamma_1 = 0.018, \quad \gamma_2 = 0.027, \quad \text{and} \quad \gamma_3 = 0.035$$

respectively. If the modal vectors were scaled to make them orthonormal (so that the mass matrix is a unit matrix), write down the equations of motion.

Solution

It is assumed that the system, in normal mode coordinates, can be treated as three uncoupled, single-DOF, equations as follows:

$$\underline{m}_{ii}\ddot{q}_i + \underline{c}_{ii}\dot{q}_i + \underline{k}_{ii}q_i = Q_i \quad (i = 1, 2, 3) \tag{A}$$

where

$$\underline{m}_{ii} = 1. \tag{B}$$

Using Eq. (2.16) and Eqs (2.18–2.19):

$$\underline{k}_{ii} = \underline{m}_{ii}\omega_i^2 = \underline{m}_{ii}(2\pi f_i)^2 = (2\pi f_i)^2 \tag{C}$$

$$\underline{c}_{ii} = 2\gamma_{ii}\underline{m}_{ii}\omega_i = 4\pi\gamma_{ii}f_i \quad (i = 1, 2, 3) \tag{D}$$

where ω_i is the (undamped) natural frequency of mode i in rad/s, and f_i the corresponding natural frequency in Hz.

Table 6.1 gives the numerical values.

Table 6.1
Calculation of Stiffness and Damping Terms from Measured Data

Mode number i	Mode frequency f_i(Hz)	Generalized mass $\underline{m}_{ii}$	Damping (% Critical)	Damping coefficient (Fraction of critical) γ_{ii}	Stiffness term $\underline{k}_{ii}$	Damping term $\underline{c}_{ii}$
1	10.1	1	1.8	0.018	4 027	2.28
2	17.8	1	2.7	0.027	12 508	6.04
3	25.3	1	3.5	0.035	25 269	11.13

The equations of motion are therefore:

$$\begin{bmatrix} 1 & 0 & 0 \\ 0 & 1 & 0 \\ 0 & 0 & 0 \end{bmatrix} \begin{Bmatrix} \ddot{q}_1 \\ \ddot{q}_2 \\ \ddot{q}_3 \end{Bmatrix} + \begin{bmatrix} 2.28 & 0 & 0 \\ 0 & 6.04 & 0 \\ 0 & 0 & 11.13 \end{bmatrix} \begin{Bmatrix} \dot{q}_1 \\ \dot{q}_2 \\ \dot{q}_3 \end{Bmatrix}$$

$$+ \begin{bmatrix} 4\,027 & 0 & 0 \\ 0 & 12\,508 & 0 \\ 0 & 0 & 25\,269 \end{bmatrix} \begin{Bmatrix} q_1 \\ q_2 \\ q_3 \end{Bmatrix} = \begin{Bmatrix} Q_1 \\ Q_2 \\ Q_3 \end{Bmatrix} \tag{E}$$

6.4.4 Proportional Damping

Viscous proportional damping

In the case of a set of damped equations of motion in global coordinates such as Eq. (6.65),

$$[M]\{\ddot{z}\} + [C]\{\dot{z}\} + [K]\{z\} = \{\bar{F}\}$$

the eigenvalues and eigenvectors of the homogeneous equations

$$[M]\{\ddot{z}\} + [C]\{\dot{z}\} + [K]\{z\} = \{0\} \tag{6.66}$$

are, in general, complex. However, if $[C]$ has the form either $[C] = a[M]$ or $[C] = b[K]$, then the real transformation that diagonalizes both $[M]$ and $[K]$ will also diagonalize $a[M]$ or $b[K]$. Therefore $[C]$ can be made diagonal by the real eigenvectors of the undamped system if it is of the form:

$$[C] = a[M] + b[K] \tag{6.67}$$

where a and b are any constants.

If the matrix $[C]$ in Eq. (6.66) is written in the form of Eq. (6.67), the equations, when transformed into normal mode coordinates, become

$$[\underline{M}]\{\ddot{q}\} + [\underline{C}]\{\dot{q}\} + [\underline{K}]\{q\} = \{0\} \tag{6.68}$$

and all three matrices are diagonal. If a single equation, say the one representing mode j, is taken from the set, Eq. (6.68):

$$\underline{m}_{jj}\ddot{q}_j + \underline{c}_{jj}\dot{q}_j + \underline{k}_{jj}q_j = 0 \tag{6.69}$$

then $\underline{c}_{jj}$ will be given by:

$$\underline{c}_{jj} = a\underline{m}_{jj} + b\underline{k}_{jj} \tag{6.70}$$

and the non-dimensional damping coefficient for that mode, γ_j, will be

$$\gamma_j = \frac{\underline{c}_{jj}}{2\underline{m}_{jj}\omega_j} = \frac{a}{2\omega_j} + \frac{b\omega_j}{2} \tag{6.71}$$

where the relationship $\underline{k}_{jj} = \underline{m}_{jj}\omega_j^2$ has been used, and ω_j is the undamped natural frequency of mode j.

Although a and b can take any arbitrary values, they must apply to the whole system, and cannot be adjusted mode by mode. So the viscous damping coefficient γ_j can be made inversely proportional to the mode frequency, ω_j, or proportional to it, or a combination of these. Since there are only two constants in Eq. (6.71), however, truly proportional damping can only be used to set the damping coefficients for two frequencies, i.e. for two modes. The method described in Section 6.4.3 is therefore more often used in practice.

Hysteretic proportional damping

A more useful form of proportional damping is obtained if it is based on hysteretic, rather than viscous, damping, and it is made proportional to the stiffness matrix only, although hysteretic damping proportional to the mass matrix also is possible. The equations of motion in global coordinates then become

$$[M]\{\ddot{z}\} + (1 + \mathrm{i}g)[K]\{z\} = \{\overline{F}\}$$

or

$$[M]\{\ddot{z}\} + \mathrm{i}g[K]\{z\} + [K]\{z\} = \{\overline{F}\} \tag{6.72}$$

Comparing this with Eq. (6.65), it is seen that the viscous damping forces, $[C]\{\dot{z}\}$, have been replaced by the hysteretic damping forces, $ig[K]\{z\}$, where g is the hysteretic damping coefficient defined in Chapter 5.

Since the transformation that diagonalizes $[M]$ and $[K]$ will also diagonalize $ig[K]$, the equations in normal mode coordinates, will be

$$[\underline{M}]\{\ddot{q}\} + (1+ig)[\underline{K}]\{q\} = \{Q\} \tag{6.73}$$

where $[\underline{M}]$ and $[\underline{K}]$ are diagonal. Now equation j of the set represented by Eq. (6.73) is

$$\underline{m}_{jj}\ddot{q}_j + ig\underline{k}_{jj}\,q_j + \underline{k}_{jj}\,q_j = Q_j \tag{6.74}$$

However, since hysteretic damping is strictly only valid as a concept in the frequency domain, this should be written as:

$$\left(-\omega^2 \underline{m}_{jj} + ig\underline{k}_{jj} + \underline{k}_{ii}\right)\underline{q}_j = \underline{Q}_j \tag{6.75}$$

It can be seen that the same value of hysteretic damping coefficient, g, now applies to all the modes. This gives a particularly simple method for dealing with the elastomeric (rubber-like) materials often used in vibration isolation systems, engine mounts and some vehicle suspensions, that can be formulated to have quite high damping. If the same elastomeric material is used for all the 'springs' in the system, the Specific Damping Capacity (defined in Chapter 5), and hence the hysteretic damping coefficient, g, is likely to be constant throughout, justifying the use of a damping matrix equal to the whole stiffness matrix multiplied by ig.

Of course, hysteretic damping can only be used in the frequency domain, ruling out its use for finding transient responses. However, it can be used in the important cases of periodic and random inputs.

6.5 Response of multi-DOF systems by normal mode summation

We have seen that the equations of motion of a damped system in global coordinates

$$[M]\{\ddot{z}\} + [C]\{\dot{z}\} + [K]\{z\} = \{F\} \tag{6.76}$$

can be transformed into normal mode coordinates using the real eigenvectors of the undamped system, provided that the damping is light, or if not light, of the proportional type, and that this covers the majority of structures, although there are some significant exceptions.

The resulting equations, in modal coordinates, are then assumed to be of the form:

$$[\underline{M}]\{\ddot{q}\} + [\underline{C}]\{\dot{q}\} + [\underline{K}]\{q\} = \{Q\} \tag{6.77}$$

where $[\underline{M}]$, $[\underline{C}]$ and $[\underline{K}]$ are diagonal, and the whole system consists of n completely separate single-DOF equations, each of which is of the form:

$$\underline{m}_{ii}\ddot{q}_i + \underline{c}_{ii}\dot{q}_i + \underline{k}_{ii}q_i = Q_i \quad (i = 1, 2, \dots, n) \tag{6.78}$$

Provided that the system is linear, any of the methods described in Chapters 3 and 4 for finding the response of single-DOF systems can be applied to each of the equations represented by Eq. (6.78), and the results summed to produce the response of the multi-DOF system, a process known as *normal mode summation*. This provides a simple, analytic method for finding the response of even quite large multi-DOF systems, to practically any input, and is therefore almost always the first method considered. It is usually found, in larger systems, that it is not necessary to use all n equations, since those having natural frequencies well above any frequencies present in the excitation will not respond, and can be omitted.

Assuming that the response of the multi-DOF system represented in global coordinates by Eq. (6.76), or in modal coordinates by Eq. (6.77), is required, the procedure is as follows:

(1) The applied external forces will be known in the form of $\{F\}$ in Eq. (6.76), i.e. as individual force time histories applied at the nodes or 'grid points' of the global system. These must be converted to modal forces, $\{Q\}$, using the original transformation, from global to modal forces, Eq. (6.39):

$$\{Q\} = [X]^{\mathrm{T}}\{F\}$$

Since the transformation was to normal mode coordinates, $[X]$ will be the modal matrix.

(2) The modal responses $\{q\}$, $\{\dot{q}\}$ or $\{\ddot{q}\}$ are calculated from Eq. (6.78). This consists of applying the appropriate modal force to each single-DOF equation in the set represented by Eq. (6.78), using any of the methods discussed in Chapters 2 and 3.

(3) The modal responses $\{q\}$, $\{\dot{q}\}$ or $\{\ddot{q}\}$ are converted back to actual responses $\{z\}$, $\{\dot{z}\}$ or $\{\ddot{z}\}$, as required, in the global system, using the same transformation, in the form of Eq. (6.26), say, that was used to form the modal equations:

$$\{z\} = [X]\{q\} \tag{6.26}$$

If global velocity or acceleration responses are required, Eq. (6.26) can be used in differentiated form: $\{\dot{z}\} = [X]\{\dot{q}\}$ and $\{\ddot{z}\} = [X]\{\ddot{q}\}$.

The following example illustrates the normal mode summation method applied to a simple 2-DOF system.

Example 6.5

The simple structure shown in Fig. 6.5 consists of a beam, considered massless, with constant EI, free to bend vertically, with two concentrated masses, m_1 and m_2, located as shown. Numerical values are

$$L = 4\,\mathrm{m}; \quad EI = 2 \times 10^6\,\mathrm{N\,m^2}; \quad m_1 = 10\,\mathrm{kg}; \quad m_2 = 8\,\mathrm{kg}$$

Tests on a similar structure have suggested that the viscous damping coefficient for both normal modes should be 0.02 critical.

(a) Derive the flexibility matrix for the system in terms of the global coordinates z_1 and z_2, and the external forces F_1 and F_2.

(b) Use the flexibility matrix, with mass data, to find the normal modes of the system. Sketch the mode shapes and express them in orthonormal form. Check the results by showing that the orthonormal eigenvectors transform the original mass matrix in global coordinates to a unit matrix in normal mode coordinates.

(c) Write the equations of motion of the system in normal mode coordinates.

(d) Use the normal mode summation method to calculate the time history of z_1 when F_1 is a step force of 1000 N and F_2 is zero, i.e.:

$$F_1 = 1000H(t) \quad F_2 = 0$$

where $H(t)$ is the Heaviside unit step function.
(e) Plot the displacement history of z_1.

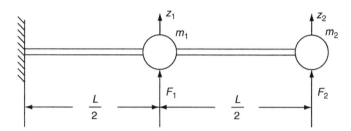

Fig. 6.5   Simple beam structure discussed in Example 6.5.

Solution

Note: The numerical operations in this example were carried out using a larger number of significant figures than indicated.

Part (a):
The flexibility matrix for this problem was derived by the area-moment method in Example B1 of Appendix B:

$$[\alpha] = [K]^{-1} = \begin{bmatrix} \alpha_{11} & \alpha_{12} \\ \alpha_{21} & \alpha_{22} \end{bmatrix} = \frac{L^3}{EI} \begin{bmatrix} \frac{1}{24} & \frac{5}{48} \\ \frac{5}{48} & \frac{1}{3} \end{bmatrix} \tag{A}$$

Part (b):
The equations of motion of the system in global coordinates can be written in the form of Eq. (6.76):

$$[M]\{\ddot{z}\} + [C]\{\dot{z}\} + [K]\{z\} = \{F\} \tag{B}$$

To find the real eigenvalues and eigenvectors, $[C]$ and $\{F\}$ are omitted, leaving the undamped homogeneous equations:

$$[M]\{\ddot{z}\} + [K]\{z\} = \{0\} \tag{C}$$

With the usual substitution, valid for undamped systems: $\{z\} = \{\bar{z}\}e^{i\omega t}$, Eq. (C) can be written as:

$$\big([K] - \omega^2[M]\big)\{\bar{z}\} = 0 \tag{D}$$

In this example, the flexibility matrix, $[K]^{-1}$, is known, rather than the stiffness matrix, $[K]$. Therefore, Eq. (D) is pre-multiplied by $[K]^{-1}$, giving:

$$\left([K]^{-1}[M] - \frac{1}{\omega^2}[I]\right)\{\bar{z}\} = 0 \tag{E}$$

From Eq. (A) we already have

$$[K]^{-1} = \frac{L^3}{EI}\begin{bmatrix} \frac{1}{24} & \frac{5}{48} \\ \frac{5}{48} & \frac{1}{3} \end{bmatrix}$$

The mass matrix, $[M]$, can be derived, as in Section 6.1.2, by noting that the masses, m_1 and m_2, produce inertia forces, $m_1\ddot{z}_1$ and $m_2\ddot{z}_2$, respectively, on the left side of the equation, so the mass matrix simply consists of the masses m_1 and m_2 written as a diagonal matrix:

$$[M] = \begin{bmatrix} m_1 & 0 \\ 0 & m_2 \end{bmatrix} = \begin{bmatrix} 10 & 0 \\ 0 & 8 \end{bmatrix} \tag{F}$$

Equation (E) then becomes

$$\left(\frac{L^3}{48EI}\begin{bmatrix} 2 & 5 \\ 5 & 16 \end{bmatrix}\begin{bmatrix} 10 & 0 \\ 0 & 8 \end{bmatrix} - \frac{1}{\omega^2}\begin{bmatrix} 1 & 0 \\ 0 & 1 \end{bmatrix} \right)\left\{ \begin{matrix} \bar{z}_1 \\ \bar{z}_2 \end{matrix} \right\} = 0. \tag{G}$$

which can be written as:

$$\left(\begin{bmatrix} 20 & 40 \\ 50 & 128 \end{bmatrix} - \Lambda\begin{bmatrix} 1 & 0 \\ 0 & 1 \end{bmatrix} \right)\left\{ \begin{matrix} \bar{z}_1 \\ \bar{z}_2 \end{matrix} \right\} = 0 \tag{H}$$

where Λ is one of the two eigenvalues, defined here as:

$$\Lambda = \frac{48EI}{L^3\omega^2} = \frac{1.5 \times 10^6}{\omega^2} \tag{I}$$

since $EI = 2 \times 10^6$, and $L = 4$, and ω is one of the two natural frequencies.

The two eigenvalues are given by the roots of the characteristic equation, obtained by multiplying out the following determinant, and equating to zero:

$$\begin{vmatrix} (20 - \Lambda) & 40 \\ 50 & (128 - \Lambda) \end{vmatrix} = (20 - \Lambda)(128 - \Lambda) - 2000 = 0$$

$$\Lambda^2 - 148\Lambda + 560 = 0 \tag{J}$$

The two roots, or eigenvalues, are

$$\Lambda = \frac{148 \pm \sqrt{148^2 - 4(560)}}{2} = 74 \pm 70.114$$

that is $\Lambda_1 = 144.11$, and $\Lambda_2 = 3.885$, in ascending order of natural frequency. From Eq. (I), the natural frequencies are $\omega_1 = 102.02$ rad/s and $\omega_2 = 621.30$ rad/s.

The ratios $(\bar{z}_1/\bar{z}_2)_1$ and $(\bar{z}_1/\bar{z}_2)_2$ can be found: by substituting each eigenvalue in turn into Eq. (H):

$$\left(\frac{\bar{z}_1}{\bar{z}_2} \right)_1 = \frac{-40}{20 - \Lambda_1} = \frac{-40}{20 - 144.11} = 0.3222 \tag{K_1}$$

$$\left(\frac{\bar{z}_1}{\bar{z}_2} \right)_2 = \frac{-40}{20 - \Lambda_2} = \frac{-40}{20 - 3.885} = -2.482 \tag{K_2}$$

Temporarily setting $\bar{z}_2$ to 1 in both cases, the two eigenvectors or mode shapes are

$$\{\underline{\phi}\}_1 = \left\{\begin{matrix} \bar{z}_1 \\ \bar{z}_2 \end{matrix}\right\}_1 = \left\{\begin{matrix} 0.3222 \\ 1 \end{matrix}\right\} \tag{L_1}$$

and

$$\{\underline{\phi}\}_2 = \left\{\begin{matrix} \bar{z}_1 \\ \bar{z}_2 \end{matrix}\right\}_2 = \left\{\begin{matrix} -2.482 \\ 1 \end{matrix}\right\} \tag{L_2}$$

These are sketched in Fig. 6.6. To rescale the eigenvectors, or mode shapes, above, so that they are orthonormal, each of them is multiplied by the scalar α_i, given by the following expression, derived in Example 6.3 as Eq. (W):

$$\alpha_i = \left(\frac{1}{\{\underline{\phi}\}_i^{\mathrm{T}}[M]\{\underline{\phi}\}_i}\right)^{\frac{1}{2}} \tag{M}$$

where $\{\underline{\phi}\}_i = \left\{\begin{matrix} \bar{z}_1 \\ \bar{z}_2 \end{matrix}\right\}_i$ is eigenvector i and $[M]$ the mass matrix in global coordinates.

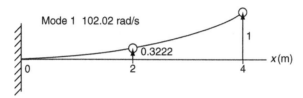

Mode 1 102.02 rad/s

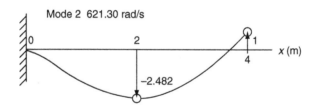

Mode 2 621.30 rad/s

Fig. 6.6

Thus for mode 1:

$$\alpha_1 = \left(\left\{\begin{matrix} \bar{z}_1 \\ \bar{z}_2 \end{matrix}\right\}_1^{\mathrm{T}} \begin{bmatrix} m_1 & 0 \\ 0 & m_2 \end{bmatrix} \left\{\begin{matrix} \bar{z}_1 \\ \bar{z}_2 \end{matrix}\right\}_1\right)^{-\frac{1}{2}} = \left\{\begin{matrix} 0.3222 \\ 1 \end{matrix}\right\}^{\mathrm{T}} \begin{bmatrix} 10 & 0 \\ 0 & 8 \end{bmatrix} \left\{\begin{matrix} 0.3222 \\ 1 \end{matrix}\right\} = 0.3326$$

$$\tag{N_1}$$

and for mode 2:

$$\alpha_2 = \left(\left\{\begin{matrix} \bar{z}_1 \\ \bar{z}_2 \end{matrix}\right\}_2^{\mathrm{T}} \begin{bmatrix} m_1 & 0 \\ 0 & m_2 \end{bmatrix} \left\{\begin{matrix} \bar{z}_1 \\ \bar{z}_2 \end{matrix}\right\}_2\right)^{-\frac{1}{2}} = \left\{\begin{matrix} -2.482 \\ 1 \end{matrix}\right\}^{\mathrm{T}} \begin{bmatrix} 10 & 0 \\ 0 & 8 \end{bmatrix} \left\{\begin{matrix} -2.482 \\ 1 \end{matrix}\right\} = 0.1198$$

$$\tag{N_2}$$

The orthonormal eigenvectors or mode shapes, $\{\phi\}_1$ and $\{\phi\}_2$, are

$$\{\phi\}_1 = \alpha_1\{\underline{\phi}\}_1 = 0.3326\left\{\begin{array}{c} 0.3222 \\ 1 \end{array}\right\} = \left\{\begin{array}{c} 0.1071 \\ 0.3326 \end{array}\right\} \tag{O_1}$$

$$\{\phi\}_2 = \alpha_2\{\underline{\phi}\}_2 = 0.1198\left\{\begin{array}{c} -2.4822 \\ 1 \end{array}\right\} = \left\{\begin{array}{c} -0.2975 \\ 0.1198 \end{array}\right\} \tag{O_2}$$

It can be seen that the scaling applied to the original eigenvectors is immaterial, since the process of converting them to orthonormal form cancels out any overall scaling factor.

The modal matrix, $[X]$, consists of the eigenvectors, in this case the orthonormal set, written as columns:

$$[X] = [\{\phi\}_1\{\phi\}_2] = \begin{bmatrix} 0.1071 & -0.2975 \\ 0.3326 & 0.1198 \end{bmatrix} \tag{P}$$

The matrix $[X]$ will later be used for two purposes:
(1) As the transformation matrix between the global coordinates, $\{z\}$, and the modal coordinates, $\{q\}$:
$$\{z\} = [X]\{q\} \tag{Q}$$

(2) As the transformation matrix between the actual external forces, $\{F\}$, and the modal forces, $\{Q\}$:
$$\{Q\} = [X]^{\mathrm{T}}\{F\} \tag{R}$$

At this point it is as well to check that the transformation:
$$[\underline{M}] = [X]^{\mathrm{T}}[M][X] \tag{S}$$

does, in fact, produce a unit matrix. Using the numerical values from Eqs (P) and (F):

$$[\underline{M}] = \begin{bmatrix} 0.1071 & -0.2975 \\ 0.3326 & 0.1198 \end{bmatrix}^{\mathrm{T}} \begin{bmatrix} 10 & 0 \\ 0 & 8 \end{bmatrix} \begin{bmatrix} 0.1071 & -0.2975 \\ 0.3326 & 0.1198 \end{bmatrix} = \begin{bmatrix} 1 & 0 \\ 0 & 1 \end{bmatrix} = [I] \tag{T}$$

and this is clearly the case.

Part (c):
The essential relationships for forming the undamped equations of motion in modal coordinates, from the equations in global coordinates, were derived in Section 6.2.3. These apply whether the transformation is based on assumed (or arbitrary) modes, or, as in this case, normal modes. However, in the present case, the mass matrix and the stiffness matrix, in modal coordinates, will both be diagonal. Moreover, if the eigenvectors are scaled to orthonormal form, as here, the mass matrix in modal coordinates will become a unit matrix, and the corresponding stiffness matrix will consist of the squares of the undamped natural frequencies in rad/s on the leading diagonal.

The equations in global coordinates, Eq. (B), are now transformed to modal coordinates. The damping terms are initially omitted, and will be based on measured data in the modal equations. Eq. (B) then becomes

$$[M]\{\ddot{z}\} + [K]\{z\} = \{F\} \tag{U}$$

Then the undamped equations of motion in normal mode coordinates will be

$$[\underline{M}]\{\ddot{q}\} + [\underline{K}]\{q\} = \{Q\} \tag{V}$$

Now recalling that the modal matrix $[X]$ was scaled to be orthonormal, i.e. to make $[\underline{M}]$ a unit matrix, we already know that

$$[\underline{M}] = \begin{bmatrix} 1 & 0 \\ 0 & 1 \end{bmatrix}$$

Also, since the two equations represented by Eq. (V) are completely uncoupled, and we know the undamped natural frequencies, ω_1 and ω_2, it follows that

$$[\underline{K}] = \begin{bmatrix} \omega_1^2 & 0 \\ 0 & \omega_2^2 \end{bmatrix}$$

Therefore the undamped equations in normal coordinates are simply:

$$\begin{bmatrix} 1 & 0 \\ 0 & 1 \end{bmatrix} \begin{Bmatrix} \ddot{q}_1 \\ \ddot{q}_2 \end{Bmatrix} + \begin{bmatrix} \omega_1^2 & 0 \\ 0 & \omega_2^2 \end{bmatrix} \begin{Bmatrix} q_1 \\ q_2 \end{Bmatrix} = \begin{Bmatrix} Q_1 \\ Q_2 \end{Bmatrix} \tag{W}$$

The damping matrix can now be inserted. Since it is assumed that the damping does not couple the modes, the equations can continue to be treated as two completely separate single-DOF equations, and the damping term, $\underline{c}_{ii}$, for Mode i, is

$$\underline{c}_{ii} = 2\gamma_i \omega_i \underline{m}_{ii} \tag{X}$$

where γ_i is the non-dimensional damping coefficient for Mode i, ω_i the normal mode undamped natural frequency for Mode i and $\underline{m}_{ii}$ the modal or generalized mass for Mode i, equal to unity, by definition, in this case.

The complete equations of motion are therefore:

$$\begin{bmatrix} 1 & 0 \\ 0 & 1 \end{bmatrix} \begin{Bmatrix} \ddot{q}_1 \\ \ddot{q}_2 \end{Bmatrix} + \begin{bmatrix} 2\gamma_1\omega_1 & 0 \\ 0 & 2\gamma_2\omega_2 \end{bmatrix} \begin{Bmatrix} \dot{q}_1 \\ \dot{q}_2 \end{Bmatrix} + \begin{bmatrix} \omega_1^2 & 0 \\ 0 & \omega_2^2 \end{bmatrix} \begin{Bmatrix} q_1 \\ q_2 \end{Bmatrix} = \begin{Bmatrix} Q_1 \\ Q_2 \end{Bmatrix} \tag{Y}$$

Numerically, since $\omega_1 = 102.02$, $\omega_2 = 621.30$ and $\gamma_1 = \gamma_2 = 0.02$:

$$2\gamma_1\omega_1 = 4.080, \quad 2\gamma_2\omega_2 = 24.85, \quad \omega_1^2 = 10408, \quad \omega_2^2 = 386020$$

In order to put the equations to practical use, we shall also require the relationship between the actual (global) displacements, $\{z\}$, and the modal displacements, $\{q\}$, which is

$$\{z\} = [X]\{q\} \quad \text{or:} \quad \begin{Bmatrix} z_1 \\ z_2 \end{Bmatrix} = \begin{bmatrix} 0.1071 & -0.2975 \\ 0.3326 & 0.1198 \end{bmatrix} \begin{Bmatrix} q_1 \\ q_2 \end{Bmatrix} \tag{Z}$$

and the relationship between the actual external loads, $[F]$, and the modal forces, $[Q]$, which is

$$\{Q\} = [X]^{\mathrm{T}}\{F\}$$

or

$$\begin{Bmatrix} Q_1 \\ Q_2 \end{Bmatrix} = \begin{bmatrix} 0.1071 & -0.2975 \\ 0.3326 & 0.1198 \end{bmatrix}^{\mathrm{T}} \begin{Bmatrix} F_1 \\ F_2 \end{Bmatrix} = \begin{bmatrix} 0.1071 & 0.3326 \\ -0.2975 & 0.1198 \end{bmatrix} \begin{Bmatrix} F_1 \\ F_2 \end{Bmatrix} \tag{a}$$

Part (d):

Equation (Y) can now be treated as two uncoupled, single-DOF equations:

$$\ddot{q}_1 + 2\gamma_1\omega_1\dot{q}_1 + \omega_1^2 q_1 = Q_1 \tag{b_1}$$

and

$$\ddot{q}_2 + 2\gamma_2\omega_2\dot{q}_2 + \omega_2^2 q_2 = Q_2 \tag{b_2}$$

From Eq. (a), we have

$$Q_1 = 0.1071 F_1 + 0.3326 F_2 \tag{c_1}$$

and

$$Q_2 = -0.2975 F_1 + 0.1198 F_2 \tag{c_2}$$

The external applied force F_1 is defined as a step force of 1000 N, and F_2 is zero:

$$F_1 = 1000 H(t), \quad F_2 = 0$$

where $H(t)$ is the unit step or Heaviside function.

Thus,

$$Q_1 = (0.1071 \times 1000) H(t) \tag{d_1}$$

and

$$Q_2 = (-0.2975 \times 1000) H(t) \tag{d_2}$$

From Eq. (B) of Example 3.4 in Section 3.1.3, the response of a single-DOF system to a step force of magnitude a is given as:

$$z = \frac{a}{m\omega_n^2}\left[1 - e^{-\gamma\omega_n t}\left(\cos \omega_d t + \frac{\gamma}{\sqrt{(1-\gamma^2)}}\sin \omega_d t\right)\right] \tag{e}$$

where z is the displacement; m the mass; γ the viscous damping coefficient and ω_n the undamped natural frequency, all of a single-DOF system. Also, ω_d is the damped natural frequency given by $\omega_d = \omega_n\sqrt{1-\gamma^2}$.

In Eq. (e), z is replaced by q_1 and q_2, in turn, and a is replaced by the magnitudes of the step modal forces, Q_1 and Q_2 in turn, to give the responses of the two equations, Eqs (b_1) and (b_2):

$$q_1 = \frac{0.1071 \times 1000}{\omega_1^2}\left[1 - e^{-\gamma_1\omega_1 t}\left(\cos \omega_{1d} t + \frac{\gamma_1}{\sqrt{(1-\gamma_1^2)}}\sin \omega_{1d} t\right)\right] \tag{f_1}$$

$$q_2 = \frac{-0.2975 \times 1000}{\omega_2^2}\left[1 - e^{-\gamma_2\omega_2 t}\left(\cos \omega_{2d} t + \frac{\gamma_2}{\sqrt{(1-\gamma_2^2)}}\sin \omega_{2d} t\right)\right] \tag{f_2}$$

Finally, the required displacement history, z_1, is given by Eq. (Z), which can be written, since z_2 is not required in this case:

$$z_1 = 0.1071 q_1 - 0.2975 q_2 \tag{g}$$

where q_1 and q_2 are given by Eqs (f$_1$) and (f$_2$). The numerical values are ω_1 is the undamped natural frequency of Mode 1 = 102.02 rad/s; ω_2 the undamped natural frequency of Mode 2 = 621.30 rad/s; γ_1 the damping coefficient of Mode 1 = 0.02; γ_2 the damping coefficient of Mode 2 = 0.02; ω_{1d} the damped natural frequency of Mode 1 $= \omega_1\sqrt{1-\gamma_1^2} = 102.02\sqrt{1-0.02^2} = 101.99$ and ω_{2d} the damped natural frequency of Mode 2 $= \omega_2\sqrt{1-\gamma_2^2} = 621.30\sqrt{1-0.02^2} = 621.17$.

The time history of z_1 is plotted in Fig. 6.7.

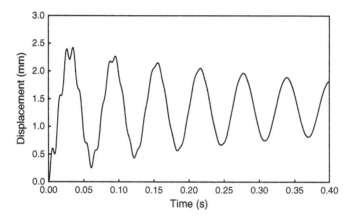

Fig. 6.7

6.6 Response of multi-DOF systems by direct integration

Direct, stepwise solution of the equations of motion in global coordinates can be a useful alternative to the normal mode method described above. When the damping is neither light nor of proportional form, and a time history solution is required anyway, it avoids the complication of complex eigenvectors, and of course it is generally the only viable method for non-linear systems.

The equations of motion to be solved are assumed to be in the form of Eq. (6.76):

$$[M]\{\ddot{z}\} + [C]\{\dot{z}\} + [K]\{z\} = \{F\} \tag{6.76}$$

where the matrices $[M]$, $[C]$ and $[K]$ may all be fully populated. In all methods, Eq. (6.76) is pre-multiplied by $[M]^{-1}$ so that each acceleration depends only on the known input forces and lower derivatives:

$$\{\ddot{z}\} = [M]^{-1}(\{F\} - [C]\{\dot{z}\} - [K]\{z\}) \tag{6.79}$$

The fourth-order Runge–Kutta method, discussed in Chapter 3, can easily be extended to multi-DOF problems, as explained below.

6.6.1 Fourth-order Runge–Kutta Method for Multi-DOF Systems

In Chapter 3, the single-DOF equation

$$m\ddot{z} + c\dot{z} + kz = F$$

was split into two first-order equations

$$u = \dot{z} \tag{3.24a}$$

and, since $\ddot{z} = \dot{u}$,

$$\dot{u} = \frac{F}{m} - \frac{c}{m}u - \frac{k}{m}z \tag{3.24b}$$

To extend Eqs (3.24a) and (3.24b) to cover multi-DOF systems, such as Eq. (6.76), the single variables are written as column vectors:

$$\{u\} = \{\dot{z}\} \tag{6.80a}$$

and, using Eq. (6.79):

$$\{\dot{u}\} = \{\ddot{z}\} = [M]^{-1}(\{F\} - [C]\{u\} - [K]\{z\}) \tag{6.80b}$$

Equations (6.80a) and (6.80b) are usually combined, giving:

$$\left\{ \begin{matrix} \{u\} \\ \{\dot{u}\} \end{matrix} \right\} = \begin{bmatrix} [0] & [I] \\ -[M]^{-1}[K] & -[M]^{-1}[C] \end{bmatrix} \left\{ \begin{matrix} \{z\} \\ \{u\} \end{matrix} \right\} + \begin{bmatrix} 0 \\ [M]^{-1}\{F\} \end{bmatrix} \tag{6.81}$$

The equations defining the average slopes for the single-DOF case, Eqs (3.34) and (3.35),

$$\langle u \rangle = \tfrac{1}{6}\left(u_j^{(1)} + 2u_j^{(2)} + 2u_j^{(3)} + u_j^{(4)} \right)$$

and

$$\langle \dot{u} \rangle = \tfrac{1}{6}\left(\dot{u}_j^{(1)} + 2\dot{u}_j^{(2)} + 2\dot{u}_j^{(3)} + \dot{u}_j^{(4)} \right)$$

now become vectors of average slopes for the multi-DOF version:

$$\{\langle u \rangle\} = \tfrac{1}{6}\left(\{u_j\}^{(1)} + 2\{u_j\}^{(2)} + 2\{u_j\}^{(3)} + \{u_j\}^{(4)} \right) \tag{6.82a}$$

$$\{\langle \dot{u} \rangle\} = \tfrac{1}{6}\left(\{\dot{u}_j\}^{(1)} + 2\{\dot{u}_j\}^{(2)} + 2\{\dot{u}_j\}^{(3)} + \{\dot{u}_j\}^{(4)} \right) \tag{6.82b}$$

Vectors also replace single variables in Table 3.5 as shown in Table 6.2:

Table 6.2
Definition of Average Slopes in Fourth-order Runge–Kutta Method

$\{z_j\}^{(1)} = \{z_j\}$	$\{u_j\}^{(1)} = \{u_j\}$	$\{\dot{u}_j\}^{(1)} = [M]^{-1}\left(\{F_j\} - [C]\{u_j\}^{(1)} - [K]\{z_j\}^{(1)} \right)$
$\{z_j\}^{(2)} = \{z_j\} + \tfrac{h}{2}\{u_j\}^{(1)}$	$\{u_j\}^{(2)} = \{u_j\} + \tfrac{h}{2}\{\dot{u}_j\}^{(1)}$	$\{\dot{u}_j\}^{(2)} = [M]^{-1}\left(\{F_j\} - [C]\{u_j\}^{(2)} - [K]\{z_j\}^{(2)} \right)$
$\{z_j\}^{(3)} = \{z_j\} + \tfrac{h}{2}\{u_j\}^{(2)}$	$\{u_j\}^{(3)} = \{u_j\} + \tfrac{h}{2}\{\dot{u}_j\}^{(2)}$	$\{\dot{u}_j\}^{(3)} = [M]^{-1}\left(\{F_j\} - [C]\{u_j\}^{(3)} - [K]\{z_j\}^{(3)} \right)$
$\{z_j\}^{(4)} = \{z_j\} + h\{u_j\}^{(3)}$	$\{u_j\} = \{u_j\} + h\{\dot{u}_j\}^{(3)}$	$\{\dot{u}_j\}^{(4)} = [M]^{-1}\left(\{F_j\} - [C]\{u_j\}^{(4)} - [K]\{z_j\}^{(4)} \right)$

The recurrence equations for the single-DOF version, Eqs (3.36) and (3.37), are also changed to vectors in the multi-DOF version:

$$\{z_{j+1}\} = \{z_j\} + \frac{h}{6}\left(\{u_j\}^{(1)} + 2\{u_j\}^{(2)} + 2\{u_j\}^{(3)} + \{u_j\}^{(4)}\right) \tag{6.83a}$$

$$\{u_{j+1}\} = \{u_j\} + \frac{h}{6}\left(\{\dot{u}_j\}^{(1)} + 2\{\dot{u}_j\}^{(2)} + 2\{\dot{u}_j\}^{(3)} + \{\dot{u}_j\}^{(4)}\right) \tag{6.83b}$$

Solution then proceeds in the same way as for the single-DOF problem of Example 3.7.

7 Eigenvalues and Eigenvectors

Contents

It was shown in Chapter 6 that provided the eigenvalues and eigenvectors of a system can be found, it is possible to transform the coordinates of the system from local or global coordinates to coordinates consisting of normal or 'principal' modes.

Depending on the damping, the eigenvalues and eigenvectors of a system can be real or complex, as discussed in Chapter 6. However, real eigenvalues and eigenvectors, derived from the undamped equations of motion, can be used in most practical cases, and will be assumed here, unless stated otherwise.

In Example 6.3, we used a very basic 'hand' method to demonstrate the derivation of the eigenvalues and eigenvectors of a simple 2-DOF system; solving the characteristic equation for its roots, and substituting these back into the equations to obtain the eigenvectors. In this chapter, we look at methods that can be used with larger systems.

7.1 The eigenvalue problem in standard form

In Chapter 6, it was shown that the undamped, homogeneous equations of a multi-DOF system can be written as:

$$-\omega^2 \begin{bmatrix} m_{11} & m_{12} & \cdots \\ m_{21} & m_{22} & \cdots \\ \vdots & \vdots & \vdots \end{bmatrix} \begin{Bmatrix} \bar{z}_1 \\ \bar{z}_2 \\ \vdots \end{Bmatrix} + \begin{bmatrix} k_{11} & k_{12} & \cdots \\ k_{21} & k_{22} & \cdots \\ \vdots & \vdots & \vdots \end{bmatrix} \begin{Bmatrix} \bar{z}_1 \\ \bar{z}_2 \\ \vdots \end{Bmatrix} = 0 \tag{7.1}$$

or concisely,

$$\left(-\omega^2 [M] + [K]\right) \{\bar{z}\} = 0 \tag{7.2}$$

where

$[M]$ is the $(n \times n)$, symmetric, mass matrix, $[K]$ the $(n \times n)$, symmetric, stiffness matrix, ω any one of the n natural frequencies and $\{\bar{z}\}$ the corresponding $(n \times 1)$ eigenvector.

Putting $\lambda = \omega^2$, Eq. (7.2) can be written as:

$$(-\lambda[M] + [K])\{\bar{z}\} = 0 \qquad (7.3)$$

where λ is any one of the n eigenvalues.

In order to take advantage of standard mathematical methods for extracting eigenvalues and eigenvectors, Eq. (7.3) must usually be converted to the 'standard form' of the problem. This is the purely mathematical relationship, known as the *algebraic eigenvalue problem*:

$$[A]\{\bar{z}\} = \lambda\{\bar{z}\} \qquad (7.4)$$

where the square *dynamic matrix*, $[A]$, is, as yet, undefined.

Equation (7.4) is often written as:

$$([A] - \lambda[I])\{\bar{z}\} = 0 \qquad (7.5)$$

where the unit matrix $[I]$ is introduced, since a column vector cannot be subtracted from a square matrix.

In Eqs (7.4) and (7.5), although λ is shown as a single scalar number, it can take n different values, where n is the number of degrees of freedom. Similarly, n corresponding eigenvectors, $\{\bar{z}\}$, can be found.

There are now two ways in which Eq. (7.3) can be expressed in the standard form of Eq. (7.5):

(1) The first method is to pre-multiply Eq. (7.2) through by $[M]^{-1}$, giving:

$$-\omega^2[M]^{-1}[M]\{\bar{z}\} + [M]^{-1}[K]\{\bar{z}\} = 0 \qquad (7.6)$$

Since $[M]^{-1}[M] = [I]$, the unit matrix, this is now seen to be in the standard form of Eq. (7.5):

$$([\underline{A}] - \lambda[I])\{\bar{z}\} = 0 \qquad (7.7)$$

where $[\underline{A}] = [M]^{-1}[K]$; and $\lambda = \omega^2$.

(2) The second method is to pre-multiply Eq. (7.2) through by $[K]^{-1}$, giving:

$$-\omega^2[K]^{-1}[M]\{\bar{z}\} + [K]^{-1}[K]\{\bar{z}\} = 0 \qquad (7.8)$$

Since $[K]^{-1}[K] = [I]$, this is also seen to be in the standard form of Eq. (7.5):

$$\left([\underline{\underline{A}}] - \underline{\lambda}[I]\right)\{\bar{z}\} = 0 \qquad (7.9)$$

where $[\underline{\underline{A}}] = [K]^{-1}[M]$; and $\underline{\lambda} = \dfrac{1}{\omega^2}$

The matrices $[\underline{A}]$ and $[\underline{\underline{A}}]$ are not the same, but both can be referred to as the *dynamic matrix*. In general, these matrices are not symmetric, even though both $[M]$ and $[K]$ are. The eigenvectors, $\{\bar{z}\}$, given by the two methods are identical, but it can be seen that the eigenvalues produced by the two methods are *not* the same. In the first case, $\lambda = \omega^2$, and in the second case, $\underline{\lambda} = \frac{1}{\omega^2}$.

Both of the above ways of defining the dynamic matrix, are in common use. It should be noted that the first method uses the stiffness matrix, $[K]$, and the second method uses the flexibility matrix, $[K]^{-1}$.

There are many methods for extracting eigenvalues and eigenvectors from the dynamic matrix, and some of these are discussed in Section 7.2.

7.1.1 The Modal Matrix

When matrix transformation methods are used, it is often convenient to form the eigenvectors into a single matrix known as the *modal matrix*. Then Eq. (7.3),

$$(-\lambda[M] + [K])\{\bar{z}\} = 0$$

can be written as:

$$[K][\Phi] = [M][\Phi][\Lambda] \tag{7.10}$$

where $[\Phi]$ is the *modal matrix* formed by writing the eigenvectors as columns:

$$[\Phi] = [\{\bar{z}\}_1 \{\bar{z}\}_2 \ldots] \tag{7.11}$$

and $[\Lambda]$ is a diagonal matrix of the eigenvalues:

$$[\Lambda] = \begin{bmatrix} \lambda_1 & 0 & \ldots \\ 0 & \lambda_2 & \ldots \\ \vdots & \vdots & \vdots \end{bmatrix} \tag{7.12}$$

Similarly, when the equation is in the form of a dynamic matrix, such as Eq. (7.5),

$$([A] - \lambda[I])\{\bar{z}\} = 0$$

it can be written as:

$$[A][\Phi] = [\Phi][\Lambda] \tag{7.13}$$

where $[\Phi]$ and $[\Lambda]$ are defined by Eqs (7.11) and (7.12), respectively.

In Eq. (7.3), although it is possible to find n eigenvalues and n corresponding eigenvectors, they do not necessarily all have to be extracted. This may be justifiable on physical grounds; for example, we may not be interested in the response of the system above a maximum frequency component in the applied forcing. Therefore, although the matrices $[M]$ and $[K]$ in the foregoing discussion will always be of size $n \times n$, as will the dynamic matrix $[A]$ or $[\underline{A}]$, the modal matrix, $[\Phi]$, may be a rectangular matrix of size $n \times m$ (i.e. n rows and m columns), where $m \leq n$. The diagonal matrix of eigenvalues, $[\Lambda]$, will then be of size $m \times m$. It should be pointed out, however, that some methods for extracting eigenvalues and eigenvectors require all n to be found.

7.2 Some basic methods for calculating real eigenvalues and eigenvectors

Only a brief introduction to this quite complicated subject can be given. The following basic methods are described in this section:

(a) Eigenvalue (only) extraction from the roots of the characteristic equation. The eigenvectors can then be found by substituting the eigenvalues back into the equations, and extracted using one of several methods, of which *Gaussian elimination* is used as an example below.

(b) *Matrix iteration*, giving both the eigenvalues and the eigenvectors.

(c) *Jacobi diagonalization*. The Jacobi method requires a symmetric dynamic matrix as input, and since $[A]$ and $[\underline{A}]$ are not naturally symmetric, this must first be arranged. If both the mass and the stiffness matrices are fully populated, *Choleski factorization* is used to ensure that the dynamic matrix is, in fact, symmetric. If, as is often the case, either the mass matrix or the stiffness matrix is diagonal, a much simpler method can be used.

7.2.1 Eigenvalues from the Roots of the Characteristic Equation and Eigenvectors by Gaussian Elimination

Finding the eigenvalues from the roots of the characteristic equation has already been demonstrated for a 2×2 set of equations in Example 6.3. The following example further illustrates the method, and goes on to find the eigenvectors by Gaussian elimination.

Example 7.1

(a) Find the eigenvalues, and hence the natural frequencies, of the undamped 3-DOF system shown in Fig. 7.1, by solving the characteristic equation for its roots.

(b) Using the eigenvalues, find the eigenvectors by Gaussian elimination.

The numerical values of the masses m_1, m_2 and m_3, and the stiffnesses, k_1, k_2 and k_3, are as follows:

$$m_1 = 2 \text{ kg}; \qquad m_2 = 1 \text{ kg}; \qquad m_3 = 2 \text{ kg};$$
$$k_1 = 1000 \text{ N/m}; \quad k_2 = 2000 \text{ N/m}; \quad k_3 = 1000 \text{ N/m}.$$

Solution

Part (a):

The stiffness matrix for the springs of this system was derived in Example 1.4. Using the method described in Section 6.1.2, the equations for the undamped unforced system are as follows:

$$\begin{bmatrix} m_1 & 0 & 0 \\ 0 & m_2 & 0 \\ 0 & 0 & m_3 \end{bmatrix} \begin{Bmatrix} \ddot{z}_1 \\ \ddot{z}_2 \\ \ddot{z}_3 \end{Bmatrix} + \begin{bmatrix} (k_1 + k_2) & -k_2 & 0 \\ -k_2 & (k_2 + k_3) & -k_3 \\ 0 & -k_3 & k_3 \end{bmatrix} \begin{Bmatrix} z_1 \\ z_2 \\ z_3 \end{Bmatrix} = 0 \qquad \text{(A)}$$

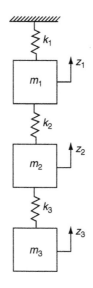

Fig. 7.1

Substituting $\{z\} = \{\bar{z}\}e^{i\omega t}$ in the usual way, and inserting the numerical values gives

$$-\omega^2 \begin{bmatrix} 2 & 0 & 0 \\ 0 & 1 & 0 \\ 0 & 0 & 2 \end{bmatrix} \begin{Bmatrix} \bar{z}_1 \\ \bar{z}_2 \\ \bar{z}_3 \end{Bmatrix} + 10^3 \begin{bmatrix} 3 & -2 & 0 \\ -2 & 3 & -1 \\ 0 & -1 & 1 \end{bmatrix} \begin{Bmatrix} \bar{z}_1 \\ \bar{z}_2 \\ \bar{z}_3 \end{Bmatrix} = 0 \qquad (B)$$

Substituting $\omega^2 = \lambda$ and introducing the scaling $\hat{\lambda} = 10^{-3}\lambda$ to simplify the matrices:

$$\left(\begin{bmatrix} 3 & -2 & 0 \\ -2 & 3 & -1 \\ 0 & -1 & 1 \end{bmatrix} - \hat{\lambda} \begin{bmatrix} 2 & 0 & 0 \\ 0 & 1 & 0 \\ 0 & 0 & 2 \end{bmatrix} \right) \begin{Bmatrix} \bar{z}_1 \\ \bar{z}_2 \\ \bar{z}_3 \end{Bmatrix} = 0. \qquad (C)$$

The determinant D of Eq. (C) is now equated to 0:

$$D = \begin{bmatrix} \left(3 - 2\hat{\lambda}\right) & -2 & 0 \\ -2 & \left(3 - \hat{\lambda}\right) & -1 \\ 0 & -1 & \left(1 - 2\hat{\lambda}\right) \end{bmatrix} = 0 \qquad (D)$$

Using the standard rule for evaluating a determinant:

$$D = (3 - 2\hat{\lambda})[(3 - \hat{\lambda})(1 - 2\hat{\lambda}) - 1] + 2[(-2)(1 - 2\hat{\lambda})] = 0 \qquad (E)$$

which when simplified gives

$$\hat{\lambda}^3 - 5\hat{\lambda}^2 + 4.25\hat{\lambda} - 0.5 = 0 \qquad (F)$$

Using a standard roots solution program, the scaled roots, $\hat{\lambda}_i$, as calculated, are:

$$\hat{\lambda}_1 = 0.1401, \quad \hat{\lambda}_2 = 0.9017, \quad \hat{\lambda}_3 = 3.958$$

However, since the roots, $\hat{\lambda}_i$, were defined as $\hat{\lambda}_i = 10^{-3}\lambda_i$, the eigenvalues of the physical problem, λ_i, are

$$\lambda_1 = 140.1 = \omega_1^2, \quad \lambda_2 = 901.7 = \omega_2^2, \quad \lambda_3 = 3958 = \omega_3^2$$

The natural frequencies, ω_i, in rad/s are

$$\omega_1 = 11.83\,\text{rad/s}, \quad \omega_2 = 30.03\,\text{rad/s}, \quad \omega_3 = 62.91\,\text{rad/s}$$

Part (b):
From Eq. (C):

$$\begin{bmatrix} (3-2\hat{\lambda}) & -2 & 0 \\ -2 & (3-\hat{\lambda}) & -1 \\ 0 & -1 & (1-2\hat{\lambda}) \end{bmatrix} \begin{Bmatrix} \bar{z}_1 \\ \bar{z}_2 \\ \bar{z}_3 \end{Bmatrix} = \begin{Bmatrix} 0 \\ 0 \\ 0 \end{Bmatrix} \qquad (\text{G})$$

The scaled roots $\hat{\lambda}_1$, $\hat{\lambda}_2$ and $\hat{\lambda}_3$ are now substituted, one at a time, into Eq. (G), and Gaussian elimination is used to find the corresponding eigenvectors. Substituting the first root, $\hat{\lambda}_1 = 0.1401$, into Eq. (G) gives

$$\begin{bmatrix} 2.7198 & -2 & 0 \\ -2 & 2.8599 & -1 \\ 0 & -1 & 0.7198 \end{bmatrix} \begin{Bmatrix} \bar{z}_1 \\ \bar{z}_2 \\ \bar{z}_3 \end{Bmatrix}_{(1)} = \begin{Bmatrix} 0 \\ 0 \\ 0 \end{Bmatrix} \qquad (\text{H})$$

The vector has the subscript (1) to indicate that it is the eigenvector corresponding to the first eigenvalue.

Gaussian elimination is a standard method for solving any set of linear equations. The general idea is to eliminate the terms in the matrix below the leading diagonal, then the values of $\bar{z}_1$, $\bar{z}_2$ and $\bar{z}_3$ can be found by back-substitution.

In this case the term -2 in the first column can be removed by multiplying the first row by $(-2/2.7198) = -0.7353$ and subtracting it from the second row, giving

$$\begin{bmatrix} 2.7198 & -2 & 0 \\ 0 & 1.3893 & -1 \\ 0 & -1 & 0.7198 \end{bmatrix} \begin{Bmatrix} \bar{z}_1 \\ \bar{z}_2 \\ \bar{z}_3 \end{Bmatrix}_{(1)} = \begin{Bmatrix} 0 \\ 0 \\ 0 \end{Bmatrix} \qquad (\text{I})$$

Without going further, from Eq. (I) we have the ratios:

$$\frac{\bar{z}_2}{\bar{z}_3} = 0.7198, \quad \frac{\bar{z}_1}{\bar{z}_2} = 0.7353 \text{ and}$$

$$\frac{\bar{z}_1}{\bar{z}_3} = \frac{\bar{z}_1}{\bar{z}_2} \times \frac{\bar{z}_2}{\bar{z}_3} = (0.7353 \times 0.7198) = 0.5293$$

If $\bar{z}_3$ is now arbitrarily assigned the value 1, then the first eigenvector is

$$\begin{Bmatrix} \bar{z}_1 \\ \bar{z}_2 \\ \bar{z}_3 \end{Bmatrix}_{(1)} = \begin{Bmatrix} 0.5293 \\ 0.7198 \\ 1.0000 \end{Bmatrix} \qquad (\text{J})$$

The second eigenvector can be found in precisely the same way, by substituting the second scaled root, $\hat{\lambda}_2 = 0.9017$, into Eq. (G), and repeating the process of eliminating terms below the diagonal and so on. The whole process is then repeated for the third eigenvector.

7.2.2 Matrix Iteration

This method extracts both eigenvalues and eigenvectors from the dynamic matrix. Its use is illustrated by the following example, where the flexibility matrix $[K]^{-1}$ rather than the stiffness matrix $[K]$ is used. Since the matrix iteration method always finds the mode with the largest eigenvalue first, it can be seen from Eq. (7.9) that this will correspond to the lowest natural frequency, and this is usually desirable.

Example 7.2

(a) Basing the calculation on the flexibility matrix, use the matrix iteration method to find the eigenvalue and eigenvector of the lowest frequency mode of the 3-DOF system of Example 7.1.
(b) Indicate how the second and third modes can be found.

Solution

Part (a):
Although we intend to base the calculation on the flexibility matrix, $[K]^{-1}$, suppose for a moment that the homogeneous equations of motion, in terms of the *stiffness* matrix, $[K]$, are

$$[M]\{\ddot{z}\} + [K]\{z\} = \{0\} \tag{A}$$

Substituting $\{z\} = \{\bar{z}\}e^{i\omega t}$ in the usual way:

$$\left(-\omega^2[M] + [K]\right)\{\bar{z}\} = 0 \tag{B}$$

Pre-multiplying by $[K]^{-1}$,

$$-\omega^2[K]^{-1}[M]\{\bar{z}\} + [K]^{-1}[K]\{\bar{z}\} = 0 \tag{C}$$

leads to a standard form

$$[\underline{A}]\{\bar{z}\} = \lambda\{\bar{z}\} \tag{D}$$

where

$$[\underline{A}] = [K]^{-1}[M] \quad \text{and} \quad \underline{\lambda} = \frac{1}{\omega^2}$$

The flexibility matrix $[K]^{-1}$ for the same system was derived in Example 1.4 as:

$$[K]^{-1} = \begin{bmatrix} 1/k_1 & 1/k_1 & 1/k_1 \\ 1/k_1 & (1/k_1 + 1/k_2) & (1/k_1 + 1/k_2) \\ 1/k_1 & (1/k_1 + 1/k_2) & (1/k_1 + 1/k_2 + 1/k_3) \end{bmatrix} \tag{E}$$

Inserting the numerical values, the matrices $[M]$, $[K]^{-1}$ and $[\underline{A}]$ become

$$[M] = \begin{bmatrix} 2 & 0 & 0 \\ 0 & 1 & 0 \\ 0 & 0 & 2 \end{bmatrix}, \quad [K]^{-1} = 10^{-3}\begin{bmatrix} 1 & 1 & 1 \\ 1 & 1.5 & 1.5 \\ 1 & 1.5 & 2.5 \end{bmatrix}, \quad [\underline{A}] = [K]^{-1}[M] = 10^{-3}\begin{bmatrix} 2 & 1 & 2 \\ 2 & 1.5 & 3 \\ 2 & 1.5 & 5 \end{bmatrix}$$

Equation (D), to be solved by matrix iteration, is then,

$$\begin{bmatrix} 2 & 1 & 2 \\ 2 & 1.5 & 3 \\ 2 & 1.5 & 5 \end{bmatrix}\begin{Bmatrix} \bar{z}_1 \\ \bar{z}_2 \\ \bar{z}_3 \end{Bmatrix} = \hat{\lambda}\begin{Bmatrix} \bar{z}_1 \\ \bar{z}_2 \\ \bar{z}_3 \end{Bmatrix} \tag{F}$$

where

$$\hat{\lambda} = \left(10^3 \underline{\lambda}\right) \quad \text{and} \quad \underline{\lambda} = \frac{1}{\omega^2}$$

To find the first eigenvector $\{\bar{z}\}_1$ and first scaled eigenvalue, $\hat{\lambda}_1$, a trial vector, $\{\bar{z}\}$, say,

$$\{\bar{z}\} = \begin{Bmatrix} \bar{z}_1 \\ \bar{z}_2 \\ \bar{z}_3 \end{Bmatrix} = \begin{Bmatrix} 0.4 \\ 0.6 \\ 1.0 \end{Bmatrix} \tag{G}$$

is inserted on the left side of Eq. (F). The equation is then evaluated with this assumption, normalizing the resulting vector by dividing by the largest element:

$$\begin{bmatrix} 2 & 1 & 2 \\ 2 & 1.5 & 3 \\ 2 & 1.5 & 5 \end{bmatrix}\begin{Bmatrix} 0.4 \\ 0.6 \\ 1 \end{Bmatrix} = \begin{Bmatrix} 3.4 \\ 4.7 \\ 6.7 \end{Bmatrix} = 6.7\begin{Bmatrix} 0.5074 \\ 0.7014 \\ 1 \end{Bmatrix} \tag{H}$$

The process is repeated, using the normalized column on the right in place of the trial vector on the left. After four such repeats, we have

$$\begin{bmatrix} 2 & 1 & 2 \\ 2 & 1.5 & 3 \\ 2 & 1.5 & 5 \end{bmatrix}\begin{Bmatrix} 0.5292 \\ 0.7198 \\ 1 \end{Bmatrix} = \begin{Bmatrix} 3.7782 \\ 5.1381 \\ 7.1381 \end{Bmatrix} = 7.1381\begin{Bmatrix} 0.5293 \\ 0.7198 \\ 1 \end{Bmatrix} \tag{I}$$

The final vector on the right is now seen to agree with the first eigenvector from Example 7.1 by the Gaussian elimination method, to four significant figures. The root $\hat{\lambda} = 7.1381 = 10^3 \underline{\lambda} = 10^3/\omega^2$ gives $\omega = 11.83$ rad/s agreeing with the result of Example 7.1.

Part(b):

To find the second eigenvalue and eigenvector, the first eigenvector, already found, must be completely removed from the dynamic matrix $[A]$. This can be achieved by using the orthogonality relationship, Eq. (6.54):

$$\{\bar{z}\}_j^T [M] \{\bar{z}\}_i = 0 \tag{J}$$

At some point in the iteration procedure, the impure vector $\{\bar{z}\}$ must consist of a linear combination of the three pure eigenvectors being sought

$$\{\bar{z}\} = C_1\{\bar{z}\}_1 + C_2\{\bar{z}\}_2 + C_3\{\bar{z}\}_3 \tag{K}$$

where C_1, C_2 and C_3 are constants. Pre-multiplying Eq. (K) by $\{\bar{z}\}_1^T [M]$:

$$\{\bar{z}\}_1^T [M] \{\bar{z}\} = C_1\{\bar{z}\}_1^T [M] \{\bar{z}\}_1 + C_2\{\bar{z}\}_1^T [M] \{\bar{z}\}_2 + C_3\{\bar{z}\}_1^T [M] \{\bar{z}\}_3 \tag{L}$$

Now due to the orthogonality relationship for eigenvectors, Eq. (J), the last two terms in Eq. (L) must be zero, and

$$\{\bar{z}\}_1^T [M] \{\bar{z}\} = C_1\{\bar{z}\}_1^T [M] \{\bar{z}\}_1 \tag{M}$$

To produce a vector $\{\bar{z}\}$ free from $\{\bar{z}\}_1$ we can now set $C_1 = 0$, and since clearly $\{\bar{z}\}_1^T [M] \{\bar{z}\}_1 \neq 0$, it must be true that

$$\{\bar{z}\}_1^T [M] \{\bar{z}\} = 0 \tag{N}$$

or written out, using the notation:

$$\{\bar{z}\}_1 = \begin{Bmatrix} (\bar{z}_1)_1 \\ (\bar{z}_2)_1 \\ (\bar{z}_3)_1 \end{Bmatrix} \quad \text{and} \quad \{\bar{z}\} = \begin{Bmatrix} \bar{z}_1 \\ \bar{z}_2 \\ \bar{z}_3 \end{Bmatrix} :$$

$$(\bar{z}_1)_1 m_1 \bar{z}_1 + (\bar{z}_2)_1 m_2 \bar{z}_2 + (\bar{z}_3)_1 m_3 \bar{z}_3 = 0 \tag{O}$$

from which

$$\bar{z}_1 = s_1 \bar{z}_2 + s_2 \bar{z}_3 \tag{P}$$

where

$$s_1 = \frac{-m_2(\bar{z}_2)_1}{m_1(\bar{z}_1)_1} = \frac{-1(0.7198)}{2(0.5293)} = -0.6799 \tag{Q_1}$$

and

$$s_2 = \frac{-m_3(\bar{z}_3)_1}{m_1(\bar{z}_1)_1} = \frac{-2(1)}{2(0.5293)} = -1.8892 \tag{Q_2}$$

The known values of s_1 and s_2 can now be incorporated into a *sweeping matrix*, $[S]$, where:

$$[S] = \begin{bmatrix} 0 & s_1 & s_2 \\ 0 & 1 & 0 \\ 0 & 0 & 1 \end{bmatrix} = \begin{bmatrix} 0 & -0.6799 & -1.8892 \\ 0 & 1 & 0 \\ 0 & 0 & 1 \end{bmatrix} \tag{R}$$

This is incorporated into the original iteration equation, Eq. (D), as follows:

$$[\underline{A}][S]\{\bar{z}\} = \underline{\lambda}\{\bar{z}\} \tag{S}$$

and Eq. (F) becomes

$$\begin{bmatrix} 2 & 1 & 2 \\ 2 & 1.5 & 3 \\ 2 & 1.5 & 5 \end{bmatrix} \begin{bmatrix} 0 & -0.6799 & -1.8892 \\ 0 & 1 & 0 \\ 0 & 0 & 1 \end{bmatrix} \begin{Bmatrix} \bar{z}_1 \\ \bar{z}_2 \\ \bar{z}_3 \end{Bmatrix} = \hat{\lambda} \begin{Bmatrix} \bar{z}_1 \\ \bar{z}_2 \\ \bar{z}_3 \end{Bmatrix} \tag{T}$$

or

$$\begin{bmatrix} 0 & -0.3598 & 5.7784 \\ 0 & 0.1402 & -0.7784 \\ 0 & 0.1402 & 1.2216 \end{bmatrix} \begin{Bmatrix} \bar{z}_1 \\ \bar{z}_2 \\ \bar{z}_3 \end{Bmatrix} = \hat{\lambda} \begin{Bmatrix} \bar{z}_1 \\ \bar{z}_2 \\ \bar{z}_3 \end{Bmatrix} \tag{U}$$

This is then solved for the second eigenvector and second eigenvalue, in precisely the same way as before, using an estimated vector to start the process.

7.2.3 Jacobi Diagonalization

So far we have considered two forms of the eigenvalue problem:
(1) From Eq. (7.7):

$$([\underline{A}] - \lambda[I])\{\bar{z}\} = 0$$

where

$$[\underline{A}] = [M]^{-1}[K]; \text{ and } \lambda = \omega^2$$

(2) From Eq. (7.9):

$$([\underline{\underline{A}}] - \underline{\lambda}[I])\{\bar{z}\} = 0$$

where

$$[\underline{\underline{A}}] = [K]^{-1}[M]; \text{ and } \underline{\lambda} = \frac{1}{\omega^2}$$

Unfortunately, both $[\underline{A}]$ and $[\underline{\underline{A}}]$ are non-symmetric matrices, and cannot be used directly in the Jacobi diagonalization, which requires the dynamic matrix to be symmetric. It can always be made symmetric, however, by applying a suitable transformation to the eigenvectors:

$$\{\bar{z}\} = [U]^{-1}\{\bar{y}\} \tag{7.14}$$

where $[U]$ is a square matrix, as yet undefined, equal in size to both $[M]$ and $[K]$, $[U]^{-1}$ its inverse and $\{\bar{y}\}$ the transformed eigenvector.

To define the matrix $[U]$, we return to Eq. (7.2):

$$(-\omega^2[M] + [K])\{\bar{z}\} = 0$$

and substitute Eq. (7.14) into it:

$$\left(-\omega^2[M][U]^{-1}+[K][U]^{-1}\right)\{\bar{y}\} = 0 \tag{7.15}$$

Equation (7.15) is now pre-multiplied by the transpose of $[U]^{-1}$, i.e. $\left[[U]^{-1}\right]^{\mathrm{T}}$, which can be written $[U]^{-\mathrm{T}}$:

$$\left(-\omega^2[U]^{-\mathrm{T}}[M][U]^{-1}+[U]^{-\mathrm{T}}[K][U]^{-1}\right)\{\bar{y}\} = 0 \tag{7.16}$$

We now consider two cases, which also define the matrix $[U]$:

Case 1:
We let $[M] = [U]^{\mathrm{T}}[U]$, and substitute it into Eq. (7.16):

$$\left(-\omega^2[U]^{-\mathrm{T}}[U]^{\mathrm{T}}[U][U]^{-1}+[U]^{-\mathrm{T}}[K][U]^{-1}\right)\{\bar{y}\} = 0 \tag{7.17}$$

But since both $[U]^{-\mathrm{T}}[U]^{\mathrm{T}}$ and $[U][U]^{-1}$ are equal to $[I]$, the unit matrix, Eq. (7.17) simplifies to

$$\left(-\omega^2[I] + [U]^{-\mathrm{T}}[K][U]^{-1}\right)\{\bar{y}\} = 0 \tag{7.18}$$

or

$$([\underline{B}] - \lambda[I])\{\bar{y}\} = 0 \tag{7.19}$$

where $\lambda = \omega^2$ and

$$[\underline{B}] = [U]^{-\mathrm{T}}[K][U]^{-1} \tag{7.20}$$

Case 2:
Let $[K] = [U]^{\mathrm{T}}[U]$ and substitute this into Eq. (7.16):

$$\left(-\omega^2[U]^{-\mathrm{T}}[M][U]^{-1}+[U]^{-\mathrm{T}}[U]^{\mathrm{T}}[U][U]^{-1}\right)\{\bar{y}\} = 0 \tag{7.21}$$

which simplifies to

$$\left([\underline{B}] - \underline{\lambda}[I]\right)\{\bar{y}\} = 0 \tag{7.22}$$

where in this case $\underline{\lambda} = \frac{1}{\omega^2}$ and

$$[\underline{B}] = [U]^{-\mathrm{T}}[M][U]^{-1} \tag{7.23}$$

Equations (7.19) and (7.22) are both now in the standard form required, and the new dynamic matrix, whether it is $[\underline{B}]$, given by Eq. (7.20), or $[\underline{B}]$, given by Eq. (7.23), depending upon which of the above methods is used, will be symmetric, as required for the Jacobi method of solution.

The Jacobi method can now be used to find the eigenvalues and eigenvectors of either Eq. (7.19) or Eq. (7.22). To demonstrate the method we can represent either case by:

$$([B] - \lambda[I])\{\bar{y}\} = 0 \tag{7.24}$$

However, if $[B]$ represents $[\underline{B}]$, then $\lambda = \omega^2$, but if $[B]$ represents $[\underline{\underline{B}}]$, then $\lambda = 1/\omega^2$.

For the Jacobi method, Eq. (7.24) is written in terms of the desired modal matrix, $[\Psi]$, i.e. in the form of Eq. (7.13):

$$[B][\Psi] = [\Psi][\Lambda] \tag{7.25}$$

where

$$[\Psi] = [\{\bar{y}\}_1 \{\bar{y}_2\} \ldots] \tag{7.26}$$

and $[\Lambda]$ is the desired diagonal matrix of the eigenvalues, λ_i, these being equal to either ω_i^2 or $1/\omega_i^2$, depending on whether the dynamic matrix was formed by using the method of Case 1 or Case 2 above.

The Jacobi method relies on the fact that, provided $[B]$ is a symmetric, real matrix, then,

$$[\Psi]^{\mathrm{T}}[B][\Psi] = [\Lambda] \tag{7.27}$$

where $[\Psi]$ and $[\Lambda]$ are defined above.

If we apply a series of transformations $[T_1]$, $[T_2]$, $[T_3]$, etc., to $[B]$ in the following way:
first:

$$[B]^{(1)} = [T_1]^{\mathrm{T}}[B][T_1] \tag{7.28a}$$

then:

$$[B]^{(2)} = [T_2]^{\mathrm{T}}[B]^{(1)}[T_2] = [T_2]^{\mathrm{T}}[T_1]^{\mathrm{T}}[B][T_1][T_2] \tag{7.28b}$$

then:

$$[B]^{(3)} = [T_3]^{\mathrm{T}}[B]^{(2)}[T_3] = [T_3]^{\mathrm{T}}[T_2]^{\mathrm{T}}[T_1]^{\mathrm{T}}[B][T_1][T_2][T_3] \tag{7.28c}$$

and so on, then provided the transformations are such that the off-diagonal terms in $[B]$ are removed or reduced with each application, then $[B]$ will eventually be transformed into $[\Lambda]$, and the matrix of eigenvectors, $[\Psi]$, will be given by the product of all the transformation matrices required to achieve this, i.e.,

$$[\Psi] = [T_1][T_2][T_3] \ldots \tag{7.29}$$

Now a 2×2 symmetric matrix, say, $\begin{bmatrix} b_{11} & b_{12} \\ b_{21} & b_{22} \end{bmatrix}$, can be transformed into diagonal form, in one operation, by a transformation matrix:

$$[T] = \begin{bmatrix} \cos\theta & -\sin\theta \\ \sin\theta & \cos\theta \end{bmatrix} \tag{7.30}$$

that is

$$\begin{bmatrix} \cos\theta & -\sin\theta \\ \sin\theta & \cos\theta \end{bmatrix}^{\mathrm{T}} \begin{bmatrix} b_{11} & b_{12} \\ b_{21} & b_{22} \end{bmatrix} \begin{bmatrix} \cos\theta & -\sin\theta \\ \sin\theta & \cos\theta \end{bmatrix} = \begin{bmatrix} \lambda_1 & 0 \\ 0 & \lambda_2 \end{bmatrix} \tag{7.31}$$

It can be shown, by equating the terms in Eq. (7.31), that the value of θ required to achieve this is given by:

$$\tan 2\theta = \frac{2b_{12}}{b_{11} - b_{22}} \tag{7.32}$$

If it is desired to diagonalize a larger matrix, say the following 4×4 symmetric matrix,

$$[B] = \begin{bmatrix} b_{11} & b_{12} & b_{13} & b_{14} \\ b_{21} & b_{22} & b_{23} & b_{24} \\ b_{31} & b_{32} & b_{33} & b_{34} \\ b_{41} & b_{42} & b_{43} & b_{44} \end{bmatrix} \tag{7.33}$$

it must be carried out in stages. The terms in $[B]^{(1)}$ corresponding to b_{12} and b_{21}, i.e. $b_{12}^{(1)}$ and $b_{21}^{(1)}$, where the superscript(1) indicates that these elements belong to $[B]^{(1)}$, can be made zero by the transformation:

$$[B]^{(1)} = [T_1]^{\mathrm{T}}[B][T_1] \tag{7.34}$$

where

$$[T_1] = \begin{bmatrix} \cos\theta_1 & -\sin\theta_1 & 0 & 0 \\ \sin\theta_1 & \cos\theta_1 & 0 & 0 \\ 0 & 0 & 1 & 0 \\ 0 & 0 & 0 & 1 \end{bmatrix} \tag{7.35}$$

and θ_1 is given by

$$\tan 2\theta_1 = \frac{2b_{12}}{b_{11} - b_{22}} \tag{7.36}$$

Similarly, the terms $b_{24}^{(2)}$ and $b_{42}^{(2)}$ in $[B]^{(2)}$ can be made zero by applying:

$$[B]^{(2)} = [T_2]^{\mathrm{T}}[B]^{(1)}[T_2] \tag{7.37}$$

where

$$[T_2] = \begin{bmatrix} 1 & 0 & 0 & 0 \\ 0 & \cos\theta_2 & 0 & -\sin\theta_2 \\ 0 & 0 & 1 & 0 \\ 0 & \sin\theta_2 & 0 & \cos\theta_2 \end{bmatrix} \tag{7.38}$$

and

$$\tan 2\theta_2 = \frac{2b_{24}^{(1)}}{b_{22}^{(1)} - b_{44}^{(1)}} \tag{7.39}$$

It may then be found that elements $b_{12}^{(2)}$ and $b_{21}^{(2)}$ in $[B]^{(2)}$ have become non-zero, but their absolute values will be smaller than those of their original values, b_{12} and b_{21}. Further sweeps may be necessary to reduce them, and all other off-diagonal terms, to some acceptably small tolerance.

This procedure is illustrated numerically in Example 7.3.

It must be remembered that the dynamic matrix was made diagonal by a coordinate change, Eq. (7.14),

$$\{\bar{z}\} = [U]^{-1}\{\bar{y}\}$$

and, of course, if the Jacobi method is used on the matrices $[\underline{B}]$ or $[\underline{B}]$, the eigenvectors found will be the $\{\bar{y}\}$, not the $\{\bar{z}\}$. To obtain the eigenvectors relevant to the physical problem, $\{\bar{z}\}$, they must be transformed back, using the relationship $\{\bar{z}\} = [U]^{-1}\{\bar{y}\}$.

Although we have defined the matrix $[U]$ as being given by either $[M] = [U]^T[U]$ or by $[K] = [U]^T[U]$, we have not yet discussed how it is evaluated. In general, if the matrices $[M]$ and $[K]$ are both non-diagonal, *Choleski decomposition*, described later in Section 7.3, will be required to find the matrix $[U]$, and hence $[U]^T, [U]^{-1}$ and, $[U]^{-T}$. However, if either $[M]$ or $[K]$ is diagonal, $[U]$, $[U]^{-1}$, etc. are of particularly simple form, and can be derived very easily. This is illustrated by the following example.

Example 7.3

Use Jacobi diagonalization to find the eigenvalues and eigenvectors of the 3-DOF system used in Examples 7.1 and 7.2.

Solution

The equations of motion of the system can be written as:

$$\left(-\omega^2[M] + [K]\right)\{\bar{z}\} = 0 \tag{A}$$

or since $\lambda = \omega^2$

$$\left(-\lambda[M] + [K]\right)\{\bar{z}\} = 0 \tag{B}$$

where λ is an eigenvalue and $\{\bar{z}\}$ is an eigenvector. In numerical form, for this example:

$$\left(-\lambda \begin{bmatrix} 2 & 0 & 0 \\ 0 & 1 & 0 \\ 0 & 0 & 2 \end{bmatrix} + 10^3 \begin{bmatrix} 3 & -2 & 0 \\ -2 & 3 & -1 \\ 0 & -1 & 1 \end{bmatrix}\right) \begin{Bmatrix} \bar{z}_1 \\ \bar{z}_2 \\ \bar{z}_3 \end{Bmatrix} = 0 \tag{C}$$

Since the mass matrix is diagonal in this case, decomposing it into $[M] = [U]^T[U]$ is very easy since:

$$[U] = [M]^{\frac{1}{2}} = \begin{bmatrix} \sqrt{m_{11}} & 0 & 0 \\ 0 & \sqrt{m_{22}} & 0 \\ 0 & 0 & \sqrt{m_{33}} \end{bmatrix} = \begin{bmatrix} \sqrt{2} & 0 & 0 \\ 0 & 1 & 0 \\ 0 & 0 & \sqrt{2} \end{bmatrix} \tag{D}$$

The inverse of $[U]$, $[U]^{-1}$ is found by simply inverting the individual terms:

$$[U]^{-1} = \begin{bmatrix} \frac{1}{\sqrt{m_{11}}} & 0 & 0 \\ 0 & \frac{1}{\sqrt{m_{22}}} & 0 \\ 0 & 0 & \frac{1}{\sqrt{m_{33}}} \end{bmatrix} = \begin{bmatrix} \frac{1}{\sqrt{2}} & 0 & 0 \\ 0 & 1 & 0 \\ 0 & 0 & \frac{1}{\sqrt{2}} \end{bmatrix} \tag{E}$$

and

$$[U]^{-T} = [U]^{-1} = \begin{bmatrix} \frac{1}{\sqrt{2}} & 0 & 0 \\ 0 & 1 & 0 \\ 0 & 0 & \frac{1}{\sqrt{2}} \end{bmatrix}$$ (F)

From Eq. (7.19), the eigenvalue problem to be solved is

$$([\underline{B}] - \lambda[I])\{\bar{y}\} = 0$$ (G)

where from Eq. (7.20):

$$[\underline{B}] = [U]^{-T}[K][U]^{-1}$$ (H)

and $\lambda = \omega^2$.

A more convenient form of Eq. (G), using the modal matrix, $[\Psi]$, is

$$[\underline{B}][\Psi] = [\Psi][\Lambda]$$ (I)

where

$$[\Psi] = [\{\bar{y}\}_1 \{\bar{y}\}_2 \{\bar{y}\}_3]$$ (J)

and

$$[\Lambda] = \begin{bmatrix} \lambda_1 & 0 & 0 \\ 0 & \lambda_2 & 0 \\ 0 & 0 & \lambda_3 \end{bmatrix} = \begin{bmatrix} \omega_1^2 & 0 & 0 \\ 0 & \omega_2^2 & 0 \\ 0 & 0 & \omega_3^2 \end{bmatrix}$$ (K)

Evaluating $[\underline{B}]$ from Eq. (H)

$$[\underline{B}] = 10^3 \begin{bmatrix} \frac{1}{\sqrt{2}} & 0 & 0 \\ 0 & 1 & 0 \\ 0 & 0 & \frac{1}{\sqrt{2}} \end{bmatrix} \begin{bmatrix} 3 & -2 & 0 \\ -2 & 3 & -1 \\ 0 & -1 & 1 \end{bmatrix} \begin{bmatrix} \frac{1}{\sqrt{2}} & 0 & 0 \\ 0 & 1 & 0 \\ 0 & 0 & \frac{1}{\sqrt{2}} \end{bmatrix} = \begin{bmatrix} \frac{3}{2} & \frac{-2}{\sqrt{2}} & 0 \\ \frac{-2}{\sqrt{2}} & 3 & \frac{-1}{\sqrt{2}} \\ 0 & \frac{-1}{\sqrt{2}} & \frac{1}{2} \end{bmatrix}$$ (L)

$$[\underline{B}] = 10^3 \begin{bmatrix} 1.5 & -1.4142 & 0 \\ -1.4142 & 3 & -0.7071 \\ 0 & 0 & 0.5 \end{bmatrix}$$ (M)

The end result of the Jacobi diagonalization that we are aiming for is, from Eq. (7.27),

$$[\Psi]^T[\underline{B}][\Psi] = [\Lambda].$$ (N)

So if we can find a series of transformations that eventually change $[\underline{B}]$ into $[\Lambda]$, the product of these transformations must be the modal matrix, $[\Psi]$. The first transformation, aimed at reducing the b_{12} and b_{21} terms to zero, is

$$[B]^{(1)} = [T_1]^T[\underline{B}][T_1]$$ (O)

where

$$[T_1] = \begin{bmatrix} \cos\theta_1 & -\sin\theta_1 & 0 \\ \sin\theta_1 & \cos\theta_1 & 0 \\ 0 & 0 & 1 \end{bmatrix}$$ (P)

and

$$\tan 2\theta_1 = \frac{2b_{12}}{b_{11} - b_{22}} = \frac{2(-1.4142)}{1.5 - 3} = 1.8856$$

giving $\theta_1 = 0.5416$.

Substituting for θ_1,

$$[T_1] = \begin{bmatrix} 0.8569 & -0.5155 & 0 \\ 0.5155 & 0.8569 & 0 \\ 0 & 0 & 1 \end{bmatrix} \tag{Q}$$

Then,

$$[\underline{B}]^{(1)} = [T_1]^{\mathrm{T}}[\underline{B}][T_1] = 10^3 \begin{bmatrix} 0.8569 & -0.5155 & 0 \\ 0.5155 & 0.8569_1 & 0 \\ 0 & 0 & 1 \end{bmatrix}^{\mathrm{T}}$$

$$\begin{bmatrix} 1.5 & -1.4142 & 0 \\ -1.4142 & 3 & -0.7071 \\ 0 & 0 & 0.5 \end{bmatrix} \begin{bmatrix} 0.8569 & -0.5155 & 0 \\ 0.5155 & 0.8569 & 0 \\ 0 & 0 & 1 \end{bmatrix}$$

$$[\underline{B}]^{(1)} = 10^3 \begin{bmatrix} 0.6492 & 0 & -0.3645 \\ 0 & 3.8507 & -0.6059 \\ -0.3645 & -0.6059 & 0.5000 \end{bmatrix} \tag{R}$$

This has reduced the $b_{12}^{(1)}$ and $b_{21}^{(1)}$ terms to zero, as expected. We next reduce the $b_{23}^{(1)}$ and $b_{32}^{(1)}$ terms to zero by the transformation:

$$[\underline{B}]^{(2)} = [T_2]^{\mathrm{T}}[\underline{B}]^{(1)}[T_2] \tag{S}$$

where

$$[T_2] = \begin{bmatrix} 1 & 0 & 0 \\ 0 & \cos\theta_2 & -\sin\theta_2 \\ 0 & \sin\theta_2 & \cos\theta_2 \end{bmatrix} \tag{T}$$

and

$$\tan 2\theta_2 = \frac{2b_{23}^{(1)}}{b_{22}^{(1)} - b_{33}^{(1)}} = \frac{2(-0.6059)}{3.8507 - 0.5000} = 0.36166 \tag{U}$$

giving $\theta_2 = -0.17351$

$$[T_2] = \begin{bmatrix} 1 & 0 & 0 \\ 0 & 0.9849 & 0.1726 \\ 0 & -0.1726 & 0.9849 \end{bmatrix} \tag{V}$$

Applying Eq. (S): $[\underline{B}]^{(2)} = [T_2]^T [\underline{B}]^{(1)} [T_2]$, gives

$$[\underline{B}]^{(2)} = 10^3 \begin{bmatrix} 0.6492 & 0.06291 & -0.3590 \\ 0.06291 & 3.9569 & 0 \\ -0.3590 & 0 & 0.3938 \end{bmatrix} \tag{W}$$

This has reduced the $b_{23}^{(1)}$ and $b_{32}^{(1)}$ terms to zero, but the previously zero $b_{12}^{(1)}$ and $b_{21}^{(1)}$ terms have changed to 0.06291.

Three more transformations were applied, in the same way, with results as follows:
Third transformation, $\theta_3 = -0.6145$:

$$[T_3] = \begin{bmatrix} \cos\theta_3 & 0 & -\sin\theta_3 \\ 0 & 1 & 0 \\ \sin\theta_3 & 0 & \cos\theta_3 \end{bmatrix}; \quad [\underline{B}]^{(3)} = 10^3 \begin{bmatrix} 0.9025 & 0.05140 & 0 \\ 0.05140 & 3.9569 & 0.03626 \\ 0 & 0.03626 & 0.1404 \end{bmatrix}$$

Fourth transformation, $\theta_4 = -0.0168$:

$$[T_4] = \begin{bmatrix} \cos\theta_4 & -\sin\theta_4 & 0 \\ \sin\theta_4 & \cos\theta_4 & 0 \\ \sin\theta_3 & 0 & 1 \end{bmatrix}; \quad [\underline{B}]^{(4)} = 10^3 \begin{bmatrix} 0.9017 & 0 & -0.0006 \\ 0 & 3.9577 & 0.03626 \\ 0 & 0.03626 & 0.1404 \end{bmatrix}$$

Fifth transformation, $\theta_5 = 0.0095$:

$$[T_5] = \begin{bmatrix} 1 & 0 & 0 \\ 0 & \cos\theta_5 & -\sin\theta_5 \\ 0 & \sin\theta_5 & \cos\theta_5 \end{bmatrix}; \quad [\underline{B}]^{(5)} = 10^3 \begin{bmatrix} 0.9017 & 0 & -0.0006 \\ 0 & 3.9581 & 0 \\ -0.0006 & 0 & 0.1401 \end{bmatrix}$$

This could be continued for one or two more transformations, to further reduce the off-diagonal terms, but it can already be seen that $[\underline{B}]^{(5)}$ has practically converged to a diagonal matrix of the eigenvalues. From Example 7.1, representing the same system, these were, in ascending order:

$$\lambda_1 = 140.1, \quad \lambda_2 = 901.7, \quad \lambda_3 = 3958$$

It will be seen that they appear in $[\underline{B}]^{(5)}$ in the different order: $\lambda_2, \lambda_3, \lambda_1$.

The matrix of eigenvectors, $[\Psi]$, is now given by Eq. (7.29). These are the eigenvectors of matrix $[\underline{B}]$, not those of the system, since it will be remembered that the latter were transformed by $[U]^{-1}$ in order to make $[\underline{B}]$ symmetric. Since there were five transformations, $[T_1]$ through $[T_5]$, in this case, the eigenvectors $[\Psi]$ are given by:

$$[\Psi] = [T_1][T_2][T_3][T_4][T_5] \tag{X}$$

or numerically:

$$[\Psi] = \begin{bmatrix} 0.8569 & -0.5155 & 0 \\ 0.5155 & 0.8569 & 0 \\ 0 & 0 & 1 \end{bmatrix} \begin{bmatrix} 1 & 0 & 0 \\ 0 & 0.9849 & 0.1726 \\ 0 & -0.1726 & 0.9849 \end{bmatrix} \begin{bmatrix} 0.8170 & 0 & 0.5766 \\ 0 & 1 & 0 \\ -0.5766 & 0 & 0.8170 \end{bmatrix}$$

$$\times \begin{bmatrix} 0.9998 & 0.01682 & 0 \\ -0.01682 & 0.9998 & 0 \\ 0 & 0 & 1 \end{bmatrix} \begin{bmatrix} 1 & 0 & 0 \\ 0 & 0.9999 & -0.0095 \\ 0 & 0.0095 & 0.9999 \end{bmatrix}$$

$$= \begin{bmatrix} 0.7598 & -0.4910 & 0.4260 \\ 0.3216 & 0.8534 & 0.4099 \\ -0.5649 & -0.1745 & 0.8064 \end{bmatrix}$$

The final matrix product above, in the order in which it emerged from the diagonalization, is the modal matrix:

$$[\Psi] = \left[\{\bar{y}\}_2 \{\bar{y}\}_3 \{\bar{y}\}_1 \right]$$

To find the eigenvectors, $\{\bar{z}\}$, of the physical system, we must use Eq. (7.14):

$$\{\bar{z}\} = [U]^{-1}\{\bar{y}\}$$

All the vectors can be transformed together by writing Eq. (7.14) as:

$$[\Phi] = [U]^{-1}[\Psi] \tag{Y}$$

where $[\Phi]$ is the modal matrix of eigenvectors $\{\bar{z}\}$ and $[\Psi]$ the modal matrix of eigenvectors $\{\bar{y}\}$. From Eq. (Y):

$$[\Phi] = [U]^{-1}[\Psi] = \begin{bmatrix} \frac{1}{\sqrt{2}} & 0 & 0 \\ 0 & 1 & 0 \\ 0 & 0 & \frac{1}{\sqrt{2}} \end{bmatrix} \begin{bmatrix} 0.7598 & -0.4910 & 0.4260 \\ 0.3216 & 0.8534 & 0.4099 \\ -0.5649 & -0.1745 & 0.8064 \end{bmatrix} \tag{Z}$$

$$= \begin{bmatrix} 0.5373 & -0.3472 & 0.3012 \\ 0.3216 & 0.8535 & 0.4099 \\ -0.3994 & -0.1234 & 0.5703 \end{bmatrix}$$

These are, finally, the eigenvectors of the system, $\{\bar{z}\}_i$, in the same order as the eigenvalues, i.e. $\{\bar{z}\}_2$, $\{\bar{z}\}_3$, $\{\bar{z}\}_1$. These would normally be rearranged into the order of ascending frequency.

As a check, we would expect the product $[\Phi]^T[M][\Phi]$ to equal the unit matrix, $[I]$, and this is the case, to within four significant figures, showing that the eigenvectors emerge in orthonormal form, without needing to be scaled.

$$[\Phi]^T[M][\Phi] = \begin{bmatrix} 1.000 & 0 & 0 \\ 0 & 1.000 & 0 \\ 0 & 0 & 1.000 \end{bmatrix}$$

Another check is that the product $[\Phi]^T[K][\Phi]$ should be a diagonal matrix of the eigenvalues, which is also the case to within four significant figures.

$$[\Phi]^T[K][\Phi] = \begin{bmatrix} 901.7 & 0 & 0 \\ 0 & 3958 & 0 \\ 0 & 0 & 140.1 \end{bmatrix}$$

It should be noted that the presentation of the numerical results above, to about four significant figures, does not reflect the accuracy actually used in the computation.

7.3 Choleski factorization

In Example 7.3, advantage was taken of the fact that the mass matrix, $[M]$, was diagonal, which enabled the transformation matrix $[U]$ to be found very easily. It was then easy to form the dynamic matrix $[\underline{B}]$ in symmetric form, by Eq. (7.20), as:

$$[\underline{B}] = [U]^{-T}[K][U]^{-1}$$

This does not work if the matrices $[M]$ and $[K]$ are both non-diagonal, and in this, more general, case, Choleski factorization has to be used to find $[U]$.

We saw in Section 7.2.3 that

(1) If the mass matrix $[M]$ can be decomposed into $[M] = [U]^T[U]$, then the transformation $[U]^{-T}[M][U]^{-1}$ makes the mass matrix into a unit matrix and the transformation $[U]^{-T}[K][U]^{-1}$ makes the stiffness matrix symmetric.
(2) If the stiffness matrix $[K]$ can be decomposed into $[K] = [U]^T[U]$, then the opposite happens, the transformation $[U]^{-T}[K][U]^{-1}$ produces a unit matrix, and $[U]^{-T}[M][U]^{-1}$ is a symmetric mass matrix.

Either way, the dynamic matrix becomes symmetric, as required by many eigenvalue/eigenvector extraction programs.

In the Choleski factorization, $[U]$ is written as an upper triangular matrix of the form:

$$[U] = \begin{bmatrix} u_{11} & u_{12} & u_{13} & \cdots \\ 0 & u_{22} & u_{23} & \cdots \\ 0 & 0 & u_{33} & \cdots \\ \vdots & \vdots & \vdots & \vdots \end{bmatrix} \tag{7.40}$$

and its transpose $[U]^T$ is

$$[U]^T = \begin{bmatrix} u_{11} & 0 & 0 & \cdots \\ u_{12} & u_{22} & 0 & \cdots \\ u_{13} & u_{23} & u_{33} & \cdots \\ \vdots & \vdots & \vdots & \vdots \end{bmatrix} \tag{7.41}$$

This means that, depending on whether $[U]$ is derived from $[M] = [U]^T[U]$ or from $[K] = [U]^T[U]$, the products $[U]^{-T}[K][U]^{-1}$ and $[U]^{-T}[M][U]^{-1}$ will either be equal to the unit matrix, $[I]$, or be symmetric, and in either case a symmetric dynamic matrix will result.

It can now be seen why the matrix $[U]$ is so named; it is because it is an *upper* triangular matrix. It should be pointed out that some authors define the basic matrix as the *lower* triangular matrix, $[L]$. This makes little difference, since for a triangular matrix $[U] = [L]^T$ and $[L] = [U]^T$ and so $[U]^T[U] = [L][L]^T$. Then $[M]$ or $[K]$ can be decomposed into $[L][L]^T$. However, we will stay with the $[U]$ convention.

Decomposing either $[M]$ or $[K]$ into the form $[U]^T[U]$ is fairly straightforward, due to the triangular form assumed for $[U]$. Assuming that $[M]$ is to be decomposed, we can write it as:

$$\begin{bmatrix} m_{11} & m_{12} & m_{13} & \cdots \\ m_{21} & m_{22} & m_{23} & \cdots \\ m_{31} & m_{32} & m_{33} & \cdots \\ \vdots & \vdots & \vdots & \vdots \end{bmatrix} = \begin{bmatrix} u_{11} & 0 & 0 & \cdots \\ u_{12} & u_{22} & 0 & \cdots \\ u_{13} & u_{23} & u_{33} & \cdots \\ \vdots & \vdots & \vdots & \vdots \end{bmatrix} \begin{bmatrix} u_{11} & u_{12} & u_{13} & \cdots \\ 0 & u_{22} & u_{23} & \cdots \\ 0 & 0 & u_{33} & \cdots \\ \vdots & \vdots & \vdots & \vdots \end{bmatrix} \quad (7.42)$$

By equating corresponding elements on the left and right sides of Eq. (7.42), a general expression for the elements of matrix $[U]$ can be derived.

The matrix $[U]^{-1}$, the inverse of $[U]$, will also be required. Since

$$[U][U]^{-1} = [I] \quad (7.43)$$

where $[U]$ is known, $[U]^{-1}$ unknown and $[I]$ the unit matrix, a general expression for the elements of $[U]^{-1}$ can be found relatively easily, again due to the triangular form of $[U]$

7.4 More advanced methods for extracting real eigenvalues and eigenvectors

The foregoing description of the Jacobi method gives some insight into how modern methods for extracting real eigenvalues and eigenvectors work. There are now more efficient methods, and although these are beyond the scope of this book, they should be mentioned.

One of these, the *QR method* is a development of the Jacobi method, with several improvements. The dynamic matrix is decomposed into the product of an orthogonal matrix, $[Q]$, as in the Jacobi method, and an upper triangular matrix, $[R]$. Diagonalization then proceeds as in the Jacobi method, but the zeros are preserved, so that $[R]$ is formed in a finite number of sweeps. The decomposition of the dynamic matrix required for the QR method is achieved using the *Householder transformation*.

The Jacobi and QR methods are used to find complete sets of eigenvalues and eigenvectors. When the set of equations is very large, and only the lower frequency modes are required, alternatives are *sub-space iteration* and the *Lanczos method*. The latter is a method for transforming the mass matrix to a unit matrix, and the stiffness matrix to tri-diagonal form, before using the QR method.

7.5 Complex (damped) eigenvalues and eigenvectors

As discussed in Section 6.4.2, there are some systems where damping couples the modes significantly, and real eigenvectors cannot be used to define the normal modes. Transformation into normal modes is still possible, but the modes are then complex. The use of such modes is generally beyond the scope of this text, and only a brief introduction of the method is included here.

Suppose that the general equations of motion for a damped multi-DOF system at the global coordinate level are

$$[M]\{\ddot{z}\} + [C]\{\dot{z}\} + [K]\{z\} = \{F\} \tag{7.44}$$

where $[M]$, $[C]$ and $[K]$ are of size $n \times n$. If the terms of the damping matrix $[C]$ can not be considered to be either small or proportional to $[M]$ or $[K]$, then the eigenvalues and eigenvectors can still be found, but with greater difficulty.

The initial approach is similar to that already used in Section 6.6.1, to express second-order equations of motion as first-order equations of twice the size. This is referred to as working in the *state space*. First, a column vector, $\{x\}$, of size $2n$ is defined as:

$$\{x\} = \left\{ \begin{array}{c} \{z\} \\ \{\dot{z}\} \end{array} \right\} \tag{7.45}$$

Differentiating with respect to time:

$$\{\dot{x}\} = \left\{ \begin{array}{c} \{\dot{z}\} \\ \{\ddot{z}\} \end{array} \right\} \tag{7.46}$$

and re-arranging Eq. (7.44) to

$$\{\ddot{z}\} = [M]^{-1}(\{F\} - [C]\{\dot{z}\} - [K]\{z\}) \tag{7.47}$$

Equations (7.45)–(7.47) can be combined as:

$$\{\dot{x}\} = [A]\{x\} + [B]\{F\} \tag{7.48}$$

where

$$[A] = \begin{bmatrix} [0] & [I] \\ -[M]^{-1}[K] & -[M]^{-1}[C] \end{bmatrix} \tag{7.49}$$

and

$$[B] = \begin{bmatrix} [0] \\ [M]^{-1} \end{bmatrix} \tag{7.50}$$

Since we only require the eigenvalues and eigenvectors of Eq. (7.48), the external forces, $\{F\}$, can be set to zero, and

$$\{\dot{x}\} = [A]\{x\} \tag{7.51}$$

where the matrix $[A]$ is of size $2n \times 2n$. This first-order set of equations can be solved by the substitution:

$$\{x\} = e^{\lambda t}\{\bar{x}\} \tag{7.52}$$

leading to

$$[A]\{\bar{x}\} = \lambda\{\bar{x}\} \tag{7.53}$$

which is recognizable as an algebraic eigenvalue problem in standard form. Unfortunately, the matrix $[A]$ is not symmetric, the eigenvalues and eigenvectors are, in general, complex, and the eigenvectors are not orthogonal. These difficulties can be overcome, but the reader is referred to more advanced texts [7.1,7.2] for the details.

References

7.1 Meirovitch, L. (2001). *Fundamentals of Vibrations*. McGraw-Hill.
7.2 Petyt, M. (1990). *Introduction to Finite Element Vibration Analysis*. Cambridge University Press.

8 Vibration of Structures

Contents

Traditionally, structures have been analysed either as continuous or as discretized ('lumped') systems. Some structures, such as uniform beams, can still usefully be treated as continuous systems, but most are now regarded as discrete multi-DOF systems. The finite element method, in fact, can be said to combine both approaches: it is continuous within the elements, but discrete at the global coordinate level.

In this chapter, discussion of continuous systems has been limited to uniform beams, and the classical Rayleigh–Ritz method. A description of the latter has been included, even though it is now obsolete, to show the historical link with later methods, such as component mode synthesis, the branch mode method and the finite element method, which are still known as 'Rayleigh–Ritz' methods.

Structures tend to be characterized by low damping, justifying the use of real eigenvalues and eigenvectors in most cases, and this will be assumed in this chapter.

The basic principles outlined in Chapter 6, such as the use of energy methods and normal coordinates, are timeless concepts. However, the application of these fundamental principles to everyday tasks in structural dynamics has changed considerably over the years, due to the development of modern computers. Therefore, to put current methods of analysis into perspective, a brief discussion of their historical development is first presented.

8.1 A historical view of structural dynamics methods

To place the various methods for dealing with the vibration of structures in context, it is helpful to consider how they have developed. It will be seen that the main driving force, over the years, has been the development of computers and computer algorithms rather than of basic mathematical theory. To take the aircraft industry as an example, the classic text by Scanlan and Rosenbaum [8.1] gives a clear snapshot of the methods in use in 1951. These were stated to be, at that time: (1) Rayleigh's method; (2) The Rayleigh–Ritz method; (3) the Stodola method; (4) the Holzer method; (5) the Myklestad method; and (6) Matrix (flexibility) influence coefficient methods. Of these, the Rayleigh, Stodola, Holzer and Myklestad methods are now mainly of historical interest, since they can be replaced by any modern eigenvalue/eigenvector software. Strictly, as noted above, this is also true of the classical Rayleigh–Ritz method, in which a combination of a relatively small number of continuous functions was used, in an optimization process, to find the normal mode shapes of beam-like components. This gave fair accuracy with, above all, small eigenvalue problems, which was very attractive before fast computers were available. However, as we shall see, given enough computing power, each of the functions used in a 'Rayleigh–Ritz' analysis need only represent a part of the structure, and component mode and finite element methods first appeared around 1960. These are still called Rayleigh–Ritz methods, even though they do not now use the original solution process.

The influence coefficient method, the most modern of the methods available in 1951, survived, in everyday use, well into the 1970s. In this method (of which Example 6.5 in Chapter 6 is a simple example), separate flexibility matrices for each major component of a structure, typically derived by using the area-moment method for beams, were calculated. In the case of an aircraft, these would be the wings, fuselage, stabilizers and fin, etc. The flexibility matrices were then used with mass data to find the normal modes of each component. This then allowed normal modes of the complete structure to be found, using component mode methods. Working in several stages, in this way, did not overstretch the computers of the time.

Eventually, more powerful computers, and the wide availability of relatively easy-to-use finite element software, made other methods uncompetitive. Currently, therefore, the analysis of nearly all structures consists of a finite element analysis, giving the equations of motion in global coordinates, usually followed by transformation into normal mode coordinates.

8.2 Continuous systems

8.2.1 Vibration of Uniform Beams in Bending

It is still useful to be able to calculate the natural frequencies and vibration modes of uniform beams, plates, cables, membranes, etc. These are sometimes used in structures, and they are also useful as test cases for unfamiliar programs, since the exact analytic answers are known. As an example, the bending modes of a uniform beam are now derived.

From simple beam theory, for static loading, the following well-known relationships apply:

$$EI\,\frac{d^2y}{dx^2} = M(x) \tag{8.1}$$

$$EI\,\frac{d^3y}{dx^3} = S(x) \tag{8.2}$$

$$EI\,\frac{d^4y}{dx^4} = W(x) \tag{8.3}$$

where

E is the Young's modulus, I is the second moment of area about bending axis, y is the displacement normal to the beam centerline at distance x along the beam, $M(x)$ is the bending moment at x, $S(x)$ is the shear at x, and $W(x)$ is the load per unit length at x.

If there is no applied force or damping, the load distribution $W(x)$ will be the inertia loading due to the mass of the beam:

$$W(x) = -\mu \frac{\partial^2 y}{\partial t^2} \tag{8.4}$$

where μ is the mass of the beam per unit length, constant in this case, and y is now a function of time, t, as well as of x. The negative sign in Eq. (8.4) is, of course, due to D'Alembert's principle.

From Eqs (8.3) and (8.4):

$$EI \frac{\partial^4 y}{\partial x^4} = -\mu \frac{\partial^2 y}{\partial t^2} \tag{8.5}$$

Due to the assumption that there is no applied force, or damping, only simple harmonic motion is possible, i.e.,

$$y = \bar{y} \sin \omega t \tag{8.6}$$

and

$$\frac{\partial^2 y}{\partial t^2} = -\omega^2 \bar{y} \sin \omega t \tag{8.7}$$

where $\bar{y}$ is a mode shape factor, a function of x only. The maximum loading corresponds to $\sin \omega t = 1$, so for this condition, $\partial^2 y / \partial t^2 = -\omega^2 \bar{y}$. Substituting this into Eq. (8.5) gives:

$$\frac{\partial^4 y}{\partial x^4} = \frac{\mu \omega^2}{EI} \bar{y} = \beta^4 \bar{y} \tag{8.8}$$

where

$$\beta^2 = \frac{\omega}{\sqrt{EI/\mu}} \tag{8.9}$$

The general solution of Eq. (8.8) is

$$\bar{y} = A \sin \beta x + B \cos \beta x + C \sinh \beta x + D \cosh \beta x \tag{8.10}$$

The constants A, B, C and D depend upon the boundary conditions, i.e. whether an end is fixed, pinned (simply supported), free, etc.

As an example, a cantilever beam, of length L, has the following boundary conditions, where $x = 0$ at the fixed end and $x = L$ at the free end:

$$\text{At } x = 0, \ \bar{y} = 0; \quad \text{at } x = 0, \ \frac{d\bar{y}}{dx} = 0 \tag{8.11}$$

$$\text{At } x = L, \ \frac{d^2\bar{y}}{dx^2} = 0; \quad \text{at } x = L, \ \frac{d^3\bar{y}}{dx^3} = 0 \tag{8.12}$$

Substituting these conditions into Eq. (8.10), differentiating it as necessary, gives

$$A = -C \quad \text{and} \quad B = -D \tag{8.13}$$

$$\begin{aligned} C(\sin \beta L + \sinh \beta L) + D(\cos \beta L + \cosh \beta L) = 0 \\ C(\cos \beta L + \cosh \beta L) + D(\sinh \beta L - \sin \beta L) = 0 \end{aligned} \tag{8.14}$$

The roots of the simultaneous equations, Eq. (8.14), are given by the determinant:

$$\begin{vmatrix} (\sin \beta L + \sinh \beta L) & (\cos \beta L + \cosh \beta L) \\ (\cos \beta L + \cosh \beta L) & (\sinh \beta L - \sin \beta L) \end{vmatrix} = 0 \tag{8.15}$$

which can be simplified to:

$$1 + \cos \beta L \cosh \beta L = 0 \tag{8.16}$$

Equation (8.16), the 'characteristic equation', can be solved numerically. The first four roots are

$$(\beta L)_1 = 1.87510; \quad (\beta L)_2 = 4.69409; \quad (\beta L)_3 = 7.85476 \quad (\beta L)_4 = 10.99554$$

From Eq. (8.9), the natural frequencies, ω_i, are given by:

$$\omega_i = \beta_i^2 \sqrt{\frac{EI}{\mu}} \tag{8.17}$$

The values of the constants A, B, C and D can be found from Eqs (8.13) and (8.14), and mode shapes are then given by Eq. (8.10).

The procedure for beams with different end conditions is similar to that shown above for the cantilever. Table 8.1 lists the characteristic equation, its roots and the constants, A, B, C and D, for several different configurations. The mode shapes evaluated from the constants given in the table are not normalized to any particular amplitude.

Analytical methods can also be used to find the natural frequencies and mode shapes of beams or rods in torsion or longitudinal vibration, uniform strings or cables, and uniform membranes and plates with rectangular or circular boundaries.

Table 8.1
Natural Frequencies and Mode Shapes for Uniform Beams in Bending

End Conditions	Characteristic equation and roots $\beta_i L$	A	B	C	D
Simply-supported (Pinned-pinned)	$\sin \beta_i L = 0$ $\beta_1 L = \pi$ $\beta_2 L = 2\pi$ $\beta_3 L = 3\pi$ $\beta_4 L = 4\pi$	1	0	0	0
Free-free	$\cos \beta_i L \cdot \cosh \beta_i L = 1$ $\beta_1 L = 4.73004$ $\beta_2 L = 7.85321$ $\beta_3 L = 10.99561$ $\beta_4 L = 14.13717$	1	$\dfrac{\sin \beta_i L - \sinh \beta_i L}{\cosh \beta_i L - \cos \beta_i L}$	1	$\dfrac{\sin \beta_i L - \sinh \beta_i L}{\cosh \beta_i L - \cos \beta_i L}$
Fixed-fixed	$\cos \beta_i L \cdot \cosh \beta_i L = 1$ $\beta_1 L = 4.73004$ $\beta_2 L = 7.85321$ $\beta_3 L = 10.99561$ $\beta_4 L = 14.13717$	−1	$-\dfrac{\sinh \beta_i L - \sin \beta_i L}{\cos \beta_i L - \cosh \beta_i L}$	1	$\dfrac{\sinh \beta_i L - \sin \beta_i L}{\cos \beta_i L - \cosh \beta_i L}$
Cantilever (Fixed-free) (x is measured from the fixed end)	$\cos \beta_i L \cdot \cosh \beta_i L = -1$ $\beta_1 L = 1.87510$ $\beta_2 L = 4.69409$ $\beta_3 L = 7.85475$ $\beta_4 L = 10.99554$	1	$-\dfrac{\sin \beta_i L + \sinh \beta_i L}{\cos \beta_i L + \cosh \beta_i L}$	−1	$\dfrac{\sin \beta_i L + \sinh \beta_i L}{\cos \beta_i L + \cosh \beta_i L}$
Fixed-pinned (x is measured from the fixed end)	$\tan \beta_i L - \tanh \beta_i L = 0$ $\beta_1 L = 3.92660$ $\beta_2 L = 7.06858$ $\beta_3 L = 10.21017$ $\beta_4 L = 13.35177$	1	$-\dfrac{\sin \beta_i L - \sinh \beta_i L}{\cos \beta_i L - \cosh \beta_i L}$	−1	$\dfrac{\sin \beta_i L - \sinh \beta_i L}{\cos \beta_i L - \cosh \beta_i L}$

Notes
In all cases the natural frequencies, in rad/s, are $\omega_i = \beta_i^2 \sqrt{EI/\mu}$, where values of $\beta_i L$, where L is the length of the beam, corresponding to the first four non-zero natural frequencies, are given in the second column of the table. The mode shape is given by:

$$y_i = A \sin \beta_i x + B \cos \beta_i x + C \sinh \beta_i x + D \cosh \beta_i x$$

where A, B, C and D can be found from the table. The mode shapes are not normalized to any particular amplitude. Note that the free–free beam has zero frequency rigid modes.

Example 8.1

A helicopter manufacturer is developing a code of practice for the installation of hydraulic service pipes, and wishes to comply with a 'minimum standard of integrity' vibration standard, which defines, among other requirements, a sinusoidal acceleration level of $\pm 5.0\,g$ from 50 to 500 Hz, at the pipe's attachments to the aircraft structure. One investigation considers straight lengths of steel pipe, installed with spacing, L, between support centers, as shown in Fig. 8.1(a). Only the fundamental vibration mode of the pipe is considered, and this is taken as the exact first bending mode of a simply supported, uniform beam. It may be assumed that a separate investigation has shown the response in higher order modes to be negligible.

Note: The system of units used in this example is the 'British' lbf inch system, in which the acceleration due to gravity, g, is 386 in./s^2, and mass is expressed in lb in^{-1}s^2, a unit sometimes known as the 'mug'.

The properties of the pipe are as follows:

$E =$ Young's modulus for the material $= 30 \times 10^6$ lbf/in.^{2}

$D =$ outer diameter $= 0.3125$ in.

$d =$ inner diameter $= 0.2625$ in.

$I =$ second moment of area of cross-section, $= (\pi/64)\, (D^4 - d^4) = 0.2350 \times 10^{-3}$ in.^{4}

$\mu =$ total mass of pipe per inch $= 0.0214 \times 10^{-3}$ mug/in. (including contained fluid).

The non-dimensional viscous damping coefficient, γ, can be taken as 0.02 of critical.

Find

(a) The vertical single-peak displacement at the center of the pipe span, relative to the supporting structure, if $L = 17$ in.;

(b) The maximum oscillatory stress in the pipe.

Solution

Part (a)

From Table 8.1, for the first normal bending mode of a simply supported uniform beam:

$$\omega_1 = \beta_1^2 \sqrt{\frac{EI}{\mu}} \quad \text{and} \quad \beta_1 L = \pi$$

therefore:

$$\omega_1 = \frac{\pi^2}{L^2} \sqrt{\frac{EI}{\mu}} \tag{A}$$

where ω_1 is the natural frequency in rad/s, and E, I, L, and μ are given above.

Also from Table 8.1, the shape of the first mode, with $i = 1$, is given by $y_1 = \sin \beta_1 x$. Since $\beta_1 = \pi/L$, then $y_1 = \sin(\pi x/L)$. Defining $y_1 = y/y_c$ where y is the actual displacement at distance x along the pipe and y_c is the maximum displacement at the half-span position, then:

$$y = y_c \sin\left(\frac{\pi x}{L}\right) \quad 0 < x < L \tag{B}$$

which is sketched in Fig. 8.1 at (b). The slope, dy/dx, and the curvature, d^2y/dx^2, are given by differentiating Eq. (B) twice with respect to x:

$$\frac{dy}{dx} = y_c \frac{\pi}{L} \cos\left(\frac{\pi x}{L}\right); \quad 0 < x < L \tag{C}$$

$$\frac{d^2y}{dx^2} = y_c \frac{\pi^2}{L^2} \left[-\sin\left(\frac{\pi x}{L}\right)\right] \quad 0 < x < L \tag{D}$$

These are sketched in Fig. 8.1 at (c) and (d).

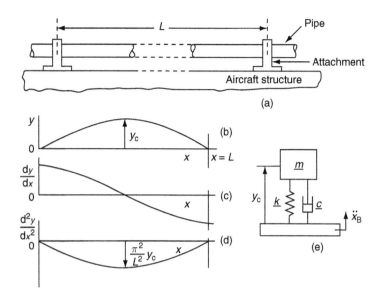

Fig. 8.1   Hydraulic pipe installation discussed in Example 8.1.

We can now derive an equivalent system to describe the vibration of the pipe, lumping the mass, stiffness and damping at the center of the pipe, where the displacement relative to the base is y_c. With the base fixed, the kinetic energy, T, in the pipe is

$$T = \frac{1}{2}\mu \int_0^L \dot{y}^2 \mathrm{d}x \tag{E}$$

Now differentiating Eq. (B) with respect to *time*:

$$\dot{y} = \dot{y}_c \sin \frac{\pi x}{L} \tag{F}$$

which gives the vertical velocity at every point on the pipe.
    Therefore:

$$T = \frac{1}{2}\mu \dot{y}_c^2 \int_0^L \sin^2 \frac{\pi x}{L} \mathrm{d}x = \frac{1}{4}\mu L \dot{y}_c^2 \tag{G}$$

This must be equal to the kinetic energy of the equivalent mass, *m*, which is

$$T = \frac{1}{2}\underline{m}\dot{y}_c^2 \tag{H}$$

Thus,

$$T = \frac{1}{4}\mu L \dot{y}_c^2 = \frac{1}{2}\underline{m}\dot{y}_c^2$$

and so

$$\underline{m} = \frac{1}{2}\mu L \tag{I}$$

The equivalent stiffness, $\underline{k}$, could be found in a similar way by equating the potential energy U in the pipe to that of the equivalent spring stiffness, $\underline{k}$, but a short cut is to use the fact that we already know the first natural frequency, ω_1, from Eq. (A):

$$\omega_1 = \frac{\pi^2}{L^2}\sqrt{\frac{EI}{\mu}},$$

and since $\underline{k} = \underline{m}\omega_1^2$, then:

$$\underline{k} = \underline{m}\omega_1^2 = \frac{1}{2}EI\frac{\pi^4}{L^3} \tag{J}$$

The pipe with damping and base motion can now be represented by the lumped model shown at Fig. 8.1(e). This is seen to be the same as that of the system shown in Fig. 2.3, and the equation of motion is given by Eq. (2.4). In the present notation this is

$$\underline{m}\ddot{y}_c + \underline{c}\dot{y}_c + \underline{k}y_c = -\underline{m}\ddot{x}_B \tag{K}$$

where $\ddot{x}_B$ is the base acceleration input, in this case the acceleration at both pipe attachments.

The modulus of the frequency response function $|y_c|/|x_B|$ of the equivalent system represented by Eq. (K) was given as Eq. (4.26):

$$\frac{|y_c|}{|x_B|} = \frac{\Omega^2}{\sqrt{(1-\Omega^2)^2 + (2\gamma\Omega)^2}} \tag{L}$$

where

$$\Omega = \frac{\omega}{\omega_1} = \frac{f}{f_1};$$

and

ω is the excitation frequency in rad/s, ω_1 the undamped natural frequency in rad/s, f the excitation frequency in Hz, f_1 the undamped natural frequency in Hz and γ the viscous damping coefficient.

Also, from Eq. (4.32), but using the present notation:

$$\frac{|y_c|}{|\ddot{x}_B|} = \frac{1}{\omega^2} \cdot \left(\frac{|y_c|}{|x_B|}\right) \tag{M}$$

From Eqs (L) and (M):

$$\frac{|y_c|}{|\ddot{x}_B|} = \frac{\Omega^2}{\omega^2\sqrt{(1-\Omega^2)^2 + (2\gamma\Omega)^2}} = \frac{1}{\omega_1^2\sqrt{(1-\Omega^2)^2 + (2\gamma\Omega)^2}} \tag{N}$$

Equation (N) now gives the magnitude of the displacement, $|y_c|$ at the center of the pipe in terms of the base acceleration magnitude, $|\ddot{x}_B|$. The maximum value of $|y_c|$ will occur, for practical purposes, when the excitation frequency coincides with the natural frequency, i.e. when $\Omega = 1$. Therefore:

$$|y_c|_{\text{MAX}} = \frac{|\ddot{x}_B|}{2\gamma\omega_1^2} \tag{O}$$

The natural frequency ω_1, in rad/s, is given by Eq. (A). Inserting numerical values:

$$\omega_1 = \frac{\pi^2}{L^2}\sqrt{\frac{EI}{\mu}} = \frac{\pi^2}{17^2}\sqrt{\frac{(30 \times 10^6) \times (0.2350 \times 10^{-3})}{(0.0214 \times 10^{-3})}}$$

$$= 619.8\,\text{rad/s} \quad (\text{or } f_1 = 98.6\,\text{Hz}).$$

Referring to the vibration specification, the required input at this frequency is $\pm 5g$, so $|\ddot{x}_B| = (5 \times 386)\,\text{in./s}^2 = 1930\,\text{in./s}^2$. Substituting these values and $\gamma = 0.02$ into Eq. (O), the required single-peak displacement amplitude at the center of the pipe is

$$|y_c|_{\text{MAX}} = \frac{|\ddot{x}_B|}{2\gamma\omega_1^2} = \frac{1930}{2 \times 0.02 \times (619.8)^2} = 0.126\,\text{in.}$$

Part (b)
From Eq. (D), the curvature of the pipe is given by:

$$\frac{d^2y}{dx^2} = y_c\frac{\pi^2}{L^2}\left[-\sin\left(\frac{\pi x}{L}\right)\right].$$

The curvature at the mid point, $(d^2y/dx^2)_c$, where $x = L/2$, is

$$\left(\frac{d^2y}{dx^2}\right)_c = y_c\frac{\pi^2}{L^2}\left(-\sin\frac{\pi}{2}\right) = -y_c\frac{\pi^2}{L^2} \tag{P}$$

The corresponding bending moment is $M_c = EI(d^2y/dx^2)_c$ and the corresponding stress, s_c, at the center of the span and distance $D/2$ from the bending neutral axis, is

$$s_c = M_c\frac{D}{2I} = \frac{DE}{2}\left(\frac{d^2y}{dx^2}\right)_c = -\frac{\pi^2 DE}{2L^2}y_c \tag{Q}$$

The negative sign in Eq. (Q) is of no importance, since it is arbitrary whether tensile or compressive stress is taken as positive, so we can simply say that

$$|s_c|_{\text{MAX}} = \frac{\pi^2 DE}{2L^2}|y_c|_{\text{MAX}} \tag{R}$$

Inserting numerical values into Eq. (R), $D = 0.3125$ in.; $E = 30 \times 10^6$ lbf/in.^{2}, $L = 17$ in. and $|y_c|_{\text{MAX}} = 0.126$ in., gives

$$|s_c|_{\text{MAX}} = 20\,100\,\text{lbf/in.}^2$$

8.2.2 The Rayleigh–Ritz Method: Classical and Modern

The classical Rayleigh–Ritz method, as devised by Ritz in 1909, and usually applied to continuous beams, is now of historical rather than practical interest. However, the idea behind the method is very much alive, and is the basis for many of today's methods, including all component mode methods and the finite element method. Thus, 'Rayleigh–Ritz method' is now often used as a generic term meaning any method for setting up the equations of motion of a system using assumed modes.

In view of the importance of the Rayleigh–Ritz method, this section includes an explanation of the original procedure, dating from 1909, which is not now used, and the modern equivalent, based on Lagrange's equations. So the 1909 method has been replaced by the 'more modern' method of 1788!

In Example 8.2, the historical Rayleigh–Ritz method is illustrated by an example taken from Scanlan and Rosenbaum's classic text [8.1]. It is then shown that the same result can be produced, more elegantly, using Lagrange's equations.

To understand the classical Rayleigh–Ritz method, we first consider the simpler Rayleigh energy method. This was essentially a method for finding the natural frequency of a single mode, usually of a beam in bending or torsion, the mode shape being assumed. Taking the bending case as an example, the potential energy, U, in the beam is given by:

$$U = \frac{1}{2} \int_0^L EI \left(\frac{\partial^2 y}{\partial x^2} \right)^2 \mathrm{d}x \tag{8.18}$$

where E is Young's modulus, I the second moment of area of the beam cross-section, y the lateral displacement of the beam at a distance x from the end and L the length of the beam. The kinetic energy in the beam, T, is similarly given by:

$$T = \frac{1}{2} \int_0^L \mu \left(\frac{\partial y}{\partial t} \right)^2 \mathrm{d}x \tag{8.19}$$

where μ is the mass of the beam per unit length.

In Eqs (8.18) and (8.19), E, I and μ may vary along the length of the beam.

Rayleigh's method is based on the fact that if the beam is vibrating with simple harmonic motion, the maximum value of the potential energy, U_{max}, must be equal to the maximum value of the kinetic energy, T_{max}. These do not, of course, occur at the same instant of time: U_{max} occurs at the maximum value of y, but T_{max} occurs at the maximum value of $\partial y/\partial t$, which, for simple harmonic motion, we can take as ωy.

Therefore:

$$U_{max} = \frac{1}{2} \int_0^L EI \left(\frac{\mathrm{d}^2 y}{\mathrm{d}x^2} \right)^2 \mathrm{d}x \tag{8.20}$$

and

$$T_{max} = \frac{1}{2} \omega^2 \int_0^L \mu y^2 \mathrm{d}x \tag{8.21}$$

Since $U_{max} = T_{max}$, from Eqs (8.20) and (8.21), an expression giving the natural frequency is

$$\omega^2 = \frac{\int_0^L EI \left(\dfrac{\mathrm{d}^2 y}{\mathrm{d}x^2} \right)^2 \mathrm{d}x}{\int_0^L \mu y^2 \mathrm{d}x} \tag{8.22}$$

where it is understood that y is now the maximum value, at a given value of x, and is a function of x only.

The modification by Ritz, in 1909, assumed that the deflection of the beam was made up of the sum of several functions, rather than just one in the Rayleigh method, i.e.,

$$y = a_1 \phi_1(x) + a_2 \phi_2(x) + \cdots \tag{8.23}$$

where a_1, a_2, etc. are scalar factors, and $\phi_1(x)$, $\phi_2(x)$, etc. are assumed shapes.

It can be shown that the combination of the factors, a_1, a_2, etc., producing the best approximation to a normal mode is that which minimizes the natural frequency, ω. This is found by (partially) differentiating Eq. (8.22) with respect to a_1, a_2, etc. in turn, and equating to zero, producing as many equations as there are factors a_i:

$$\frac{\partial \omega^2}{\partial a_i} = 0 \quad i = 1, 2, \ldots \tag{8.24}$$

Eq. (8.22) may be differentiated by noting that it is a quotient, and it can be shown, after some manipulation, that the required equations are given by:

$$\frac{\partial R}{\partial a_i} = 0 \quad i = 1, 2, \ldots \tag{8.25}$$

where:

$$R = \int_0^L EI\left(\frac{\partial^2 y}{\partial x^2}\right)^2 dx - \omega^2 \int_0^L \mu y^2 dx \tag{8.26}$$

These equations form a dynamic matrix, or eigenvalue problem, which is solved in the usual way, giving approximations to the natural frequencies and normal mode shapes. They are stated here in historical form, and it can be seen that the terms on the right side of Eq. (8.26) are equal to *twice* the maximum potential and kinetic energies, respectively, and when partially differentiated with respect to each factor a_i in turn, they will produce a dynamic matrix in which each term is twice as large as the corresponding term based on Lagrange's equations. This is confusing, but of no great importance, since the partial derivatives are, in any case, equated to zero. However, later writers have sometimes placed the factor $\frac{1}{2}$ before the integral signs in Eqs (8.22) and (8.26), for compatibility with the Lagrange method.

Example 8.2

(a) Use the classical Rayleigh–Ritz method to derive a dynamic matrix enabling the approximate natural frequencies and normal modes of a uniform cantilever beam, in bending, to be found. Use the following two-term series to represent the displacement:

$$y = a_1 x^2 + a_2 x^3 \tag{A}$$

where x is the distance along the beam from the fixed end. The length of the beam is L; Young's modulus is E; the second moment of area of the cross-section is I and the mass per unit length is μ.

(b) Show that the same result can be obtained by the use of Lagrange's equations, taking q_1 and q_2 as generalized coordinates, defined by:

$$y = q_1 x^2 + q_2 x^3 \tag{B}$$

Solution

Part(a)

From Eq. (A):

$$y^2 = a_1^2 x^4 + 2a_1 a_2 x^5 + a_2^2 x^6 \tag{C}$$

$$\frac{\partial^2 y}{\partial x^2} = 2a_1 + 6a_2 x \tag{D}$$

$$\left(\frac{\partial^2 y}{\partial x^2}\right)^2 = 4a_1^2 + 24a_1 a_2 x + 36a_2^2 x^2 \tag{E}$$

From Eq. (8.26), using Eqs (C) and (E):

$$R = EI \int_0^L \left(4a_1^2 + 24a_1 a_2 x + 36a_2^2 x^2\right) dx - m_x \omega^2 \int_0^L \left(a_1^2 x^4 + 2a_1 a_2 x^5 + a_2^2 x^6\right) dx \tag{F}$$

$$= EI\left(4a_1^2 L + 12a_1 a_2 L^2 + 12a_2^2 L^3\right) - m_x \omega^2 \left(\frac{a_1^2 L^5}{5} + \frac{a_1 a_2 L^6}{3} + \frac{a_2^2 L^7}{7}\right) \tag{G}$$

$$\frac{\partial R}{\partial a_1} = EI\left(8a_1 L + 12a_2 L^2\right) - \mu\omega^2 \left(\frac{2}{5}a_1 L^5 + \frac{1}{3}a_2 L^6\right) = 0 \tag{H}$$

$$\frac{\partial R}{\partial a_2} = EI\left(12a_1 L^2 + 24a_2 L^3\right) - \mu\omega^2 \left(\frac{1}{3}a_1 L^6 + \frac{2}{7}a_2 L^7\right) = 0 \tag{I}$$

or combining Eqs (H) and (I) in matrix form:

$$\left(EI\begin{bmatrix} 8L & 12L^2 \\ 12L^2 & 24L^3 \end{bmatrix} - \mu\omega^2 \begin{bmatrix} \frac{2}{5}L^5 & \frac{1}{3}L^6 \\ \frac{1}{3}L^6 & \frac{2}{7}L^7 \end{bmatrix}\right) \begin{Bmatrix} a_1 \\ a_2 \end{Bmatrix} = 0 \tag{J}$$

This can now be seen to be an eigenvalue problem, and it can be solved in any of the usual ways to obtain the two natural frequencies and the two ratios $(a_1/a_2)_1$ and $(a_1/a_2)_2$. The two normal mode shapes are then found from Eq. (A).

Part (b)

The equivalent modern approach is to use Lagrange's equations to derive the equations of motion in terms of the generalized coordinates q_1 and q_2 (which replace a_1 and a_2). With no damping and no external forces, Lagrange's equations are

$$\frac{\mathrm{d}}{\mathrm{d}t}\left(\frac{\partial T}{\partial \dot{q}_1}\right) + \frac{\partial U}{\partial q_1} = 0 \quad \text{and} \quad \frac{\mathrm{d}}{\mathrm{d}t}\left(\frac{\partial T}{\partial \dot{q}_2}\right) + \frac{\partial U}{\partial q_2} = 0 \qquad (\mathrm{K}_1)(\mathrm{K}_2)$$

To find the kinetic energy, T:

From Eq. (8.19),

$$T = \frac{1}{2}\int_0^L \mu\left(\frac{\partial y}{\partial t}\right)^2 \mathrm{d}x \qquad (\mathrm{L})$$

From Eq. (B),

$$y = q_1 x^2 + q_2 x^3$$

and

$$\frac{\partial y}{\partial t} = \dot{q}_1 x^2 + \dot{q}_2 x^3 \qquad (\mathrm{M})$$

so

$$\left(\frac{\partial y}{\partial t}\right)^2 = \dot{q}_1^2 x^4 + 2\dot{q}_1 \dot{q}_2 x^5 + \dot{q}_2^2 x^6 \qquad (\mathrm{N})$$

From Eqs (L) and (N):

$$T = \frac{1}{2}\mu \int_0^L (\dot{q}_1^2 x^4 + 2\dot{q}_1 \dot{q}_2 x^5 + \dot{q}_2^2 x^6)\mathrm{d}x = \frac{1}{2}\mu\left(\frac{L^5}{5}\dot{q}_1^2 + \frac{L^6}{3}\dot{q}_1\dot{q}_2 + \frac{L^7}{7}\dot{q}_2^2\right) \qquad (\mathrm{O})$$

To find the potential energy, U:

From Eq. (8.18):

$$U = \frac{1}{2}\int_0^L EI\left(\frac{\partial^2 y}{\partial x^2}\right)^2 \mathrm{d}x \qquad (\mathrm{P})$$

From Eq. (E), replacing a_1 by q_1 and a_2 by q_2 in each case, we have

$$\left(\frac{\partial^2 y}{\partial x^2}\right)^2 = 4q_1^2 + 24q_1 q_2 x + 36q_2^2 x^2 \qquad (\mathrm{Q})$$

From Eqs (P) and (Q):

$$U = \frac{1}{2}EI \int_0^L (4q_1^2 + 24q_1 q_2 x + 36q_2^2 x^2)\mathrm{d}x = EI(2L^2 q_1^2 + 6L^2 q_1 q_2 + 6L^3 q_2^2) \qquad (\mathrm{R})$$

Using Eqs (O) and (R), the partial derivatives required for Lagrange's equations are

$$\frac{d}{dt}\left(\frac{\partial T}{\partial \dot{q}_1}\right) = \mu\left(\frac{L^5}{5}\ddot{q}_1 + \frac{L^6}{6}\ddot{q}_2\right) \tag{S_1}$$

and

$$\frac{\partial}{\partial t}\left(\frac{\partial T}{\partial \dot{q}_2}\right) = \mu\left(\frac{L^6}{6}\ddot{q}_1 + \frac{L^7}{7}\ddot{q}_2\right) \tag{S_2}$$

$$\frac{\partial U}{\partial q_1} = EI\left(4Lq_1 + 6L^2 q_2\right) \tag{T_1}$$

and

$$\frac{\partial U}{\partial q_2} = EI\left(6L^2 q_1 + 12L^3 q_2\right) \tag{T_2}$$

Substituting $-\omega^2 q_1$ for $\ddot{q}_1$ and $-\omega^2 q_2$ for $\ddot{q}_2$, and writing in matrix form:

$$\left(EI\begin{bmatrix} 4L & 6L^2 \\ 6L^2 & 12L^3 \end{bmatrix} - \mu\omega^2 \begin{bmatrix} \frac{1}{5}L^5 & \frac{1}{6}L^6 \\ \frac{1}{6}L^6 & \frac{1}{7}L^7 \end{bmatrix}\right)\left\{\begin{matrix} q_1 \\ q_2 \end{matrix}\right\} = 0 \tag{U}$$

Comparing Eq. (U) with Eq. (J), it can be seen that each term in Eq. (J) is twice as large as the corresponding term in Eq. (U). As pointed out above, this is due to the use of Eq. (8.26), the historic version of the Rayleigh–Ritz equation, in Part (a) of the solution, and Lagrange's equations in Part (b). It is of no importance, since both sets of equations are equated to zero.

Modern 'Rayleigh–Ritz' methods include all component modes methods, such as all variants of the component mode synthesis method, the branch mode method and the finite element method. All these methods are mathematically similar, differing only in the type and size of the components used as assumed modes. These later developments of the Rayleigh–Ritz method are briefly described in the following sections of this chapter.

8.3 Component mode methods

Two component mode methods are now described. The first of these, known as *component mode synthesis*, is usually associated with Hurty [8.2, 8.3], and further developments have been made by Craig and his associates [8.4, 8.5].

The other method, less well documented, but very useful in practical work, is the *branch mode method*, originally suggested by Hunn [8.6] and further developed by Gladwell [8.7].

As computers have become more powerful in recent years, there has been a tendency to represent even a complicated structure by a single, large, finite element model, rather than to build it up from component modes. These methods therefore now tend to be used less than was the case a few years ago.

8.3.1 Component Mode Synthesis

Component mode synthesis methods enable structures to be analysed as two or more separate substructures, which when joined together become the complete structure. This can have advantages, such as permitting analysis to proceed independently on each substructure, and making each analysis smaller. Each of the individual analyses is now usually carried out by the finite element method.

Component mode synthesis can take many forms, but there are two main variants: (1) the fixed interface method and (2) the free interface method.

In the first of these, the initial modal analysis, on each substructure separately, is made with the interface coordinates fixed. To allow the interface coordinates to move, special constraint modes are introduced. These consist of the *static* displacements of a substructure when unit displacements are imposed at each interface coordinate in turn, a similarity with the finite element method, described later. All the modes, in all the substructures to be joined, are then used in a Rayleigh–Ritz analysis of the complete structure.

In the second method, the free interface method, the initial modal analysis of each substructure is carried out with the interface coordinates free to move. To ensure that coordinates at points subsequently joined move together, constraint equations, making the displacements equal at junctions, are derived. This introduces linear dependencies between the modes of the complete structure, i.e. there are more coordinates than independent degrees of freedom, and Lagrange's equations cannot be used. A special transformation to remove the *superfluous coordinates* must therefore be used before Lagrange's equations are finally used to set up the equations of motion of the complete system.

In both the fixed and the free interface methods, the modes used (other than the constraint modes in the case of the fixed-interface method) are usually normal modes of the substructures. This has several advantages, and is assumed in the following discussion. It is, however, perfectly possible to use assumed polynomial modes, as in Example 8.3.

Both the fixed and the free interface methods are now described. It is assumed here, in both methods, that the equations of motion, expressed in global coordinates, are available for each of two substructures, R and S, which are to be joined. Damping and external forces will be ignored for simplicity, as it is relatively easy to include them later. It will also be assumed that the substructures do not have rigid-body freedoms.

The fixed interface method

The following [8.8] is known as the Craig–Bampton method [8.4]. It is stated by Craig [8.9] to differ only slightly from the Hurty method [8.3].

Let the equations of motion of each of the substructures, R or S, in global coordinates, without damping or external forces, be of the form:

$$[M]\{\ddot{z}\} + [K]\{z\} = 0 \tag{8.27}$$

where the vector, $\{z\}$, contains two kinds of global coordinate:
(1) those *not* at junctions or boundaries, designated $\{z_N\}$. These are always free to move.
(2) those at junctions or boundaries, designated $\{z_B\}$. These are initially fixed, but will later be joined to another substructure.

The vector $\{z\}$ in Eq. (8.27) is now partitioned as follows:

$$\{z\} = \left\{ \begin{matrix} z_N \\ \hline z_B \end{matrix} \right\} \tag{8.28}$$

The matrices $[M]$ and $[K]$ can be similarly partitioned, so Eq. (8.27) can be written as:

$$\begin{bmatrix} M_{NN} & M_{NB} \\ \hline M_{BN} & M_{BB} \end{bmatrix} \left\{ \begin{matrix} \ddot{z}_N \\ \hline \ddot{z}_B \end{matrix} \right\} + \begin{bmatrix} K_{NN} & K_{NB} \\ \hline K_{BN} & K_{BB} \end{bmatrix} \left\{ \begin{matrix} z_N \\ \hline z_B \end{matrix} \right\} = 0 \tag{8.29}$$

Since the boundary coordinates are fixed at this stage, $\{z_B\} = 0$. The natural frequencies and normal modes of the substructure, with junctions fixed, are therefore given by the solution of the 'top left' partitions of Eq. (8.29) only:

$$[M_{NN}]\{\ddot{z}_N\} + [K_{NN}]\{z_N\} = 0 \tag{8.30}$$

With the usual substitution:

$$\{z_N\} = e^{i\omega t}\{\bar{z}_N\} \tag{8.31}$$

$$([K_{NN}] - \omega^2[M_{NN}])\{\bar{z}_N\} = 0 \tag{8.32}$$

which can be solved for natural frequencies, ω_i, and eigenvectors, $\{\bar{z}_N\}_i$, $i = 1, 2, \ldots$ in the usual way.

The resulting modal matrix is then, say:

$$[\phi_N] = [\{\bar{z}_N\}_1 \{\bar{z}_N\}_2 \cdots] \tag{8.33}$$

The modal matrix, $[\phi_N]$, usually has more rows than columns, since often only the first few normal modes are required.

The modal coordinates, $\{q_N\}$, are related to the non-boundary global coordinates, $\{z_N\}$, by:

$$\{z_N\} = [\phi_N]\{q_N\} \tag{8.34}$$

All the above operations are carried out on both substructures, R and S.

Constraint modes are now generated, for each substructure, by applying a unit *static* displacement to each element in $\{z_B\}$, in turn, the other elements in $\{z_B\}$ being fixed. The displacements of each constraint mode are the resulting values of $\{z_N\}$. It can be seen from Eq. (8.29) that since, for static loading, all the accelerations are zero, then $\{z_N\}$ and $\{z_B\}$ are related by:

$$[K_{NN}]\{z_N\} + [K_{NB}]\{z_B\} = 0 \tag{8.35}$$

or

$$\{z_N\} = -[K_{NN}]^{-1}[K_{NB}]\{z_B\} = [\phi_C]\{z_B\} \tag{8.36}$$

where $[\phi_C]$, equal to $-[K_{NN}]^{-1}[K_{NB}]$, can be regarded as a 'modal matrix' giving the displacements of the non-boundary coordinates, $\{z_N\}$, for unit displacements of the boundary displacements $\{z_B\}$. Again, these operations are carried out on both substructures.

The following transformation matrix is then applied to Eq. (8.29), which can be taken to represent either of the substructures, R or S:

$$\left\{\frac{z_N}{z_B}\right\} = \left[\begin{array}{c|c}\phi_N & \phi_C \\ \hline 0 & I\end{array}\right]\left\{\frac{q_N}{z_B}\right\} \tag{8.37}$$

where $[\phi_N]$ and $[\phi_C]$ are defined by Eqs (8.33) and (8.36), respectively and $\{q_N\}$ is defined by Eq. (8.34). This gives, for substructure R:

$$[\bar{M}]^R\{\ddot{q}\}^R + [\bar{K}]^R\{q\}^R = 0 \tag{8.38}$$

where

$$[\bar{M}]^R = \left[\left[\begin{array}{c|c}\phi_N & \phi_C \\ \hline 0 & I\end{array}\right]^R\right]^T\left[\begin{array}{c|c}M_{NN} & M_{NB} \\ \hline M_{BN} & M_{BB}\end{array}\right]^R\left[\begin{array}{c|c}\phi_N & \phi_C \\ \hline 0 & I\end{array}\right]^R = \left[\begin{array}{c|c}\bar{M}_{NN} & \bar{M}_{NB} \\ \hline \bar{M}_{BN} & \bar{M}_{BB}\end{array}\right]^R \tag{8.39}$$

$$[\bar{K}]^R = \left[\left[\begin{array}{c|c}\phi_N & \phi_C \\ \hline 0 & I\end{array}\right]^R\right]^T\left[\begin{array}{c|c}K_{NN} & K_{NB} \\ \hline K_{BN} & K_{BB}\end{array}\right]^R\left[\begin{array}{c|c}\phi_N & \phi_C \\ \hline 0 & I\end{array}\right]^R = \left[\begin{array}{c|c}\bar{K}_{NN} & 0 \\ \hline 0 & \bar{K}_{BB}\end{array}\right]^R \tag{8.40}$$

and

$$\{q\}^R = \left\{\frac{q_N}{z_B}\right\}^R \tag{8.41}$$

In Eqs (8.39) and (8.40), $[\bar{M}_{NN}]$ and $[\bar{K}_{NN}]$ will be diagonal matrices if the substructure concerned is represented by normal modes, as is usually the case. It can be shown that the $[\bar{K}_{NB}]$ and $[\bar{K}_{BN}]$ matrices, missing from Eq. (8.40), are always zero, for any kind of mode.

Equations similar to Eq. (8.38) can be written for subsystem S:

$$[\bar{M}]^S\{\ddot{q}\}^S + [\bar{K}]^S\{q\}^S = 0 \tag{8.38a}$$

where $[\bar{M}]^S, [\bar{K}]^S$ and $\{q\}^S$ are defined in the same way as $[\bar{M}]^R, [\bar{K}]^R$ and $\{q\}^R$.

The equations of motion for the two substructures, R and S, combined into a single system, can now be written as:

$$[\hat{M}]\{\ddot{\hat{q}}\} + [\hat{K}]\{\hat{q}\} = 0 \tag{8.42}$$

where:

$$\{\hat{q}\} = \left\{\begin{array}{c}q_N^R \\ \hline q_N^S \\ \hline z_B\end{array}\right\} \tag{8.43}$$

$$[\hat{M}] = \left[\begin{array}{ccc} \bar{M}^R_{NN} & 0 & \bar{M}^R_{NB} \\ \hline 0 & \bar{M}^S_{NN} & \bar{M}^S_{NB} \\ \hline \bar{M}^R_{BN} & \bar{M}^S_{BN} & \bar{M}^R_{BB} + \bar{M}^S_{BB} \end{array}\right] \tag{8.44}$$

and

$$[\hat{K}] = \left[\begin{array}{ccc} \bar{K}^R_{NN} & 0 & 0 \\ \hline 0 & \bar{K}^S_{NN} & 0 \\ \hline 0 & 0 & \bar{K}^R_{BB} + \bar{K}^S_{BB} \end{array}\right] \tag{8.45}$$

where the superscript R or S indicates the substructure. The boundary coordinates $\{z_B\}$, in Eq. (8.43), are, of course, common to both substructures, when joined, and appear only once. The total number of modes used in the final equations, Eqs (8.43)–(8.45), is therefore the sum of the interface-fixed modes used in the two substructures plus the number of joint displacement coordinates.

Equation (8.42), representing the complete structure, will usually be solved for eigenvalues and eigenvectors, using $\{\hat{q}\}$ in Eq. (8.43) as a set of generalized coordinates. Local displacements, $\{z_N\}$ and $\{z_B\}$, are then found from Eq. (8.37).

The free interface method

In the free interface method, the two subsystems to be joined, R and S, are each analysed separately, as before, but with the junction, or boundary, coordinates free rather than fixed.

The undamped, unforced equations of motion for subsystem R in global coordinates are, say,

$$[M^R]\{\ddot{z}^R\} + [K^R]\{z^R\} = 0 \tag{8.46}$$

and similarly for subsystem S:

$$[M^S]\{\ddot{z}^S\} + [K^S]\{z^S\} = 0 \tag{8.47}$$

Equations (8.46) and (8.47) are now both transformed into modal coordinates, as described in Chapter 6. These are usually *normal* mode coordinates, but assumed modes can also be used (as in Example 8.3). Let the transformations applied to Eqs (8.46) and (8.47) to achieve this be, respectively:

$$\{z^R\} = [\phi^R]\{p^R\} \tag{8.48}$$

and

$$\{z^S\} = [\phi^S]\{p^S\} \tag{8.49}$$

The resulting equations in modal coordinates for the two, still unconnected, substructures are then:

$$[\bar{M}^R]\{\ddot{p}^R\} + [\bar{K}^R]\{p^R\} = 0 \tag{8.50}$$

and

$$[\bar{M}^S]\{\ddot{p}^S\} + [\bar{K}^S]\{p^S\} = 0 \tag{8.51}$$

where:

$$[\bar{M}^R] = [\phi^R]^T[M^R][\phi^R] \tag{8.52a}$$

$$[\bar{K}^R] = [\phi^R]^T[K^R][\phi^R] \tag{8.52b}$$

$$[\bar{M}^S] = [\phi^S]^T[M^S][\phi^S] \tag{8.53a}$$

$$[\bar{K}^S] = [\phi^S]^T[K^S][\phi^S] \tag{8.53b}$$

The matrices $[\bar{M}^R]$, $[\bar{K}^R]$, $[\bar{M}^S]$ and $[\bar{K}^S]$ will be diagonal if, as in the usual case, the modal matrices $[\phi^R]$ and $[\phi^S]$ consist of the eigenvectors of Eqs (8.46) and (8.47), respectively, but not if they are based on assumed modes, as discussed in Chapter 6.

Equations (8.50) and (8.51) are combined, as follows:

$$\begin{bmatrix} \bar{M}^R & 0 \\ \hline 0 & \bar{M}^S \end{bmatrix} \begin{Bmatrix} \ddot{p}^R \\ \ddot{p}^S \end{Bmatrix} + \begin{bmatrix} \bar{K}^R & 0 \\ \hline 0 & \bar{K}^S \end{bmatrix} \begin{Bmatrix} p^R \\ p^S \end{Bmatrix} = 0 \tag{8.54}$$

Equation (8.54) consists of the equations of motion of the two substructures, not yet connected together, and there is still no coupling between them. Joining is, in fact, achieved by applying a transformation to Eq. (8.54), and this is now developed.

The modal matrices $[\phi^R]$ and $[\phi^S]$, in Eqs (8.48) and (8.49) respectively, consist of some rows which correspond to global coordinate displacements at junction (or boundary) nodes, and some which do not. If the global coordinates at junctions are separated out, and designated $\{z_B^R\}$ and $\{z_B^S\}$, for substructures R and S, respectively, then they must be equal when the substructures are joined, i.e.,

$$\{z_B^R\} = \{z_B^S\} \tag{8.55}$$

Using Eqs (8.48) and (8.49):

$$\{z_B^R\} = [\phi_B^R]\{p^R\} \tag{8.56}$$

and

$$\{z_B^S\} = [\phi_B^S]\{p^S\} \tag{8.57}$$

Therefore from Eqs (8.55), (8.56) and (8.57):

$$[\phi_B^R]\{p^R\} = [\phi_B^S]\{p^S\}$$ (8.58)

Each connection in Eq. (8.58), between a mode in subsystem R and a mode in subsystem S, reduces the number of independent degrees of freedom in the system by one. Since the generalized coordinates, to be used in Lagrange's equations, must be independent, the following method is used to eliminate the dependent or superfluous coordinates.

First, Eq. (8.58) is rearranged into the form:

$$\left[\phi_B^R \mid -\phi_B^S\right]\left\{\frac{p^R}{p^S}\right\} = 0$$ (8.59)

or

$$[A]\{p\} = 0$$ (8.60)

The vector $\{p\}$ is of size n_p, where n_p is the number of modal coordinates in both subsystems taken together. The matrix $[A]$ is rectangular and has n_p columns and n_B rows, where n_B is the number of global displacement coordinates at the boundaries of either subsystem, this number obviously being the same for the two subsystems.

Equation (8.60), $[A]\{p\} = 0$, is partitioned, and the order of the terms is changed to:

$$\left[A_1 \mid A_2\right]\left\{\frac{p_d}{p_f}\right\} = 0$$ (8.61)

The matrix $[A_1]$, which must not be singular, is a square matrix, formed from those columns of $[A]$ associated with the dependent coordinates, $\{p_d\}$. The matrix $[A_2]$ consists of the remainder of the columns, associated with the independent coordinates $\{p_f\}$. In theory, it does not matter which n_B coordinates are chosen to be dependent, but some choices may be more convenient than others in practical cases.

Writing Eq. (8.61) as:

$$[A_1]\{p\}_d + [A_2]\{p\}_f = 0$$ (8.62)

then

$$\{p\}_d = -[A_1]^{-1}[A_2]\{p\}_f$$ (8.63)

Introducing the trivial equation $\{p_f\} = [I]\{p_f\}$, Eq. (8.63) can be written as:

$$\left\{\frac{p_d}{p_f}\right\} = \left[\frac{-[A_1]^{-1}[A_2]}{I}\right]\{p\}_f$$ (8.64)

Equation (8.64) is the required transformation, but the individual terms will usually have to be re-arranged to make it compatible with Eq. (8.54), i.e. the elements in the vector $\left\{\frac{p_d}{p_f}\right\}$ must be changed back to the original order, $\left\{\frac{p^R}{p^S}\right\}$, with corresponding changes to the matrix, which then becomes, say, $[\beta]$.

Also, the independent coordinates, $\{p_f\}$, on the right side of Eq. (8.64) are designated as generalized coordinates, $\{q\}$.

Thus, Eq. (8.64) becomes

$$\{p\} = \left\{ \frac{p^R}{p^S} \right\} = [\beta]\{q\} \tag{8.65}$$

The transformation, Eq. (8.65), is now applied to Eq. (8.54), as follows:

$$[\beta]^T \left[\begin{array}{c|c} \bar{M}_R & 0 \\ \hline 0 & \bar{M}_S \end{array} \right] [\beta]\{\ddot{q}\} + [\beta]^T \left[\begin{array}{c|c} \bar{K}_R & 0 \\ \hline 0 & \bar{K}_S \end{array} \right] [\beta]\{q\} = 0 \tag{8.66}$$

or

$$[\hat{M}]\{\ddot{q}\} + [\hat{K}]\{q\} = 0 \tag{8.67}$$

where

$$[\hat{M}] = [\beta]^T \left[\begin{array}{c|c} \bar{M}_R & 0 \\ \hline 0 & \bar{M}_S \end{array} \right] [\beta] \tag{8.68}$$

and

$$[\hat{K}] = [\beta]^T \left[\begin{array}{c|c} \bar{K}_R & 0 \\ \hline 0 & \bar{K}_S \end{array} \right] [\beta] \tag{8.69}$$

Equation (8.67) is the equation of motion of the complete, joined system. It will usually be solved for natural frequencies and normal modes in the usual way, and damping and external forces can then be added.

The following example illustrates the use of the free interface method. It is intended only to demonstrate the mathematical steps involved, and does not represent a typical everyday problem.

Example 8.3

Use the component mode synthesis, free interface, method to find the normal modes of the single system formed when the two uniform cantilever beams, R and S, of lengths L and $2L$, respectively, as shown in Fig. 8.2, are subsequently rigidly joined at the free ends. The two cantilevers have the same mass per unit length, μ, and the same values of Young's modulus, E, and second moment of area of cross-section, I.

Represent the cantilever R by two assumed modes with modal displacements p_1 and p_2, defined as follows:

$$y_1 = \left(\frac{x_1}{L}\right)^2 p_1 + \left(\frac{x_1}{L}\right)^3 p_2 \tag{A}$$

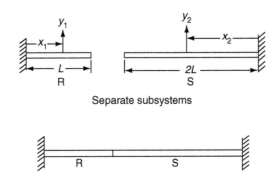

Fig. 8.2 Example showing the joining of two subsystems by the component mode synthesis method.

Represent the cantilever S by three assumed modes with modal displacements p_3, p_4 and p_5 defined by:

$$y_2 = \left(\frac{x_2}{2L}\right)^2 p_3 + \left(\frac{x_2}{2L}\right)^3 p_4 + \left(\frac{x_2}{2L}\right)^4 p_5 \tag{B}$$

where x_1, x_2, y_1, y_2 and L are defined by Fig. 8.2.

Solution

We shall need the following relationships:
From Eq. (A):

$$\frac{dy_1}{dx_1} = \frac{2x_1}{L^2}p_1 + \frac{3x_1^2}{L^3}p_2 \tag{C}$$

$$\frac{d^2y_1}{dx_1^2} = \frac{2}{L^2}p_1 + \frac{6x_1}{L^3}p_2 \tag{D}$$

From Eq. (B):

$$\frac{dy_2}{dx_2} = \frac{x_2}{2L^2}p_3 + \frac{3x_2^2}{8L^3}p_4 + \frac{x_2^3}{4L^4}p_5 \tag{E}$$

$$\frac{d^2y_2}{dx_2^2} = \frac{1}{2L^2}p_3 + \frac{3x_2}{4L^3}p_4 + \frac{3x_2^2}{4L^4}p_5 \tag{F}$$

To derive the equations of motion for the two 'substructures' R and S, separately, we use Lagrange's equations, which require expressions for the kinetic energies T_R and T_S, and the potential energies, U_R and U_S, of R and S, respectively.

$$T_R = \frac{1}{2}\mu \int_0^L \dot{y}_1^2 dx_1 = \frac{1}{2}\mu \int_0^L \left[\left(\frac{x_1}{L}\right)^2 \dot{p}_1 + \left(\frac{x_1}{L}\right)^3 \dot{p}_2\right]^2 dx = \frac{1}{2}\mu L\left(\frac{1}{5}\dot{p}_1^2 + \frac{1}{3}\dot{p}_1\dot{p}_2 + \frac{1}{7}\dot{p}_2^2\right) \tag{G}$$

$$U_R = \frac{1}{2}EI \int_0^L \left(\frac{d^2 y_1}{dx_1}\right)^2 dx_1 = \frac{1}{2}EI \int_0^L \left(\frac{4}{L^4}p_1^2 + \frac{12x_1}{L^5}p_1p_2 + \frac{36x_1^3}{L^6}p_2^2\right)dx_1$$

$$= \frac{EI}{2L^3}\left(4p_1^2 + 12p_1p_2 + 12p_2^2\right) \tag{H}$$

Applying Lagrange's equations

$$\frac{d}{dt}\left(\frac{\partial T_R}{\partial \dot{p}_i}\right) + \frac{\partial U_R}{\partial p_i} = 0 \quad (i = 1, 2)$$

to substructure R gives the two equations:

$$\mu L \left(\frac{1}{5}\ddot{p}_1 + \frac{1}{6}\ddot{p}_2\right) + \frac{EI}{L^3}\left(4p_1 + 6p_2\right) = 0 \tag{I}$$

$$\mu L \left(\frac{1}{6}\ddot{p}_1 + \frac{1}{7}\ddot{p}_2\right) + \frac{EI}{L^3}\left(6p_1 + 12p_2\right) = 0 \tag{J}$$

or in matrix form:

$$\mu L \begin{bmatrix} \frac{1}{5} & \frac{1}{6} \\ \frac{1}{6} & \frac{1}{7} \end{bmatrix} \left\{\begin{matrix} \ddot{p}_1 \\ \ddot{p}_2 \end{matrix}\right\} + \frac{EI}{L^3}\begin{bmatrix} 4 & 6 \\ 6 & 12 \end{bmatrix}\left\{\begin{matrix} p_1 \\ p_2 \end{matrix}\right\} = 0 \tag{K}$$

Similarly, for substructure S:

$$T_S = \frac{1}{2}\mu \int_0^{2L} \dot{y}_2^2 dx_2 = \frac{1}{2}\mu \int_0^{2L} \left[\left(\frac{x_2}{2L}\right)^2 \dot{p}_3 + \left(\frac{x_2}{2L}\right)^3 \dot{p}_4 + \left(\frac{x_2}{2L}\right)^4 \dot{p}_5\right]^2 dx_2 \tag{L}$$

and

$$U_S = \frac{1}{2}EI \int_0^{2L} \left(\frac{d^2 y_2}{dx_2}\right)^2 dx_2 = \frac{1}{2}EI \int_0^{2L} \left(\frac{1}{2L^2}p_3 + \frac{3x_2}{4L^3}p_4 + \frac{3x_2^2}{4L^4}p_5\right)^2 dx_2 \tag{M}$$

Applying Lagrange's equations:

$$\frac{d}{dt}\left(\frac{\partial T_S}{\partial \dot{p}_i}\right) + \frac{\partial U_S}{\partial p_i} = 0 \quad (i = 3, 4, 5)$$

to substructure S leads to:

$$\mu L \begin{bmatrix} \frac{2}{5} & \frac{1}{3} & \frac{2}{7} \\ \frac{1}{3} & \frac{2}{7} & \frac{1}{4} \\ \frac{2}{7} & \frac{1}{4} & \frac{2}{9} \end{bmatrix} \left\{\begin{matrix} \ddot{p}_3 \\ \ddot{p}_4 \\ \ddot{p}_5 \end{matrix}\right\} + \frac{EI}{L^3}\begin{bmatrix} \frac{1}{2} & \frac{3}{4} & 1 \\ \frac{3}{4} & \frac{3}{2} & \frac{9}{4} \\ 1 & \frac{9}{4} & \frac{18}{5} \end{bmatrix}\left\{\begin{matrix} p_3 \\ p_4 \\ p_5 \end{matrix}\right\} = 0 \tag{N}$$

Combining Eq. (K), representing substructure R, with Eq. (N) representing sub-structure S:

$$\mu L \begin{bmatrix} \frac{1}{5} & \frac{1}{6} & 0 & 0 & 0 \\ \frac{1}{6} & \frac{1}{7} & 0 & 0 & 0 \\ 0 & 0 & \frac{2}{5} & \frac{1}{3} & \frac{2}{7} \\ 0 & 0 & \frac{1}{3} & \frac{2}{7} & \frac{1}{4} \\ 0 & 0 & \frac{2}{7} & \frac{1}{4} & \frac{2}{9} \end{bmatrix} \begin{Bmatrix} \ddot{p}_1 \\ \ddot{p}_2 \\ \ddot{p}_3 \\ \ddot{p}_4 \\ \ddot{p}_5 \end{Bmatrix} + \frac{EI}{L^3} \begin{bmatrix} 4 & 6 & 0 & 0 & 0 \\ 6 & 12 & 0 & 0 & 0 \\ 0 & 0 & \frac{1}{2} & \frac{3}{4} & 1 \\ 0 & 0 & \frac{3}{4} & \frac{3}{2} & \frac{9}{4} \\ 0 & 0 & 1 & \frac{9}{4} & \frac{18}{5} \end{bmatrix} \begin{Bmatrix} p_1 \\ p_2 \\ p_3 \\ p_4 \\ p_5 \end{Bmatrix} = 0 \qquad (O)$$

These are still two separate sets of equations, and represent the two substructures not yet joined together.

The constraint equations are now derived by observing that at the junction, where $x_1 = L$ and $x_2 = 2L$, when the substructures are joined, the displacements and slopes must be the same for the two beams, i.e.,

$$y_1(L) = y_2(2L) \qquad (P_1)$$

$$\frac{dy_1}{dx_1}(L) = -\frac{dy_2}{dx_2}(2L) \qquad (P_2)$$

Note: In Eq. (P_2), the negative sign is due to the fact that x_1 and x_2 are defined in opposite directions, affecting slopes but not displacements.

Substituting $x_1 = L$ and $x_2 = 2L$ into Eqs (A), (B), (C) and (E) gives

$$y_1 = p_1 + p_2 \qquad (Q_1)$$

$$\frac{dy_1}{dx_1} = \frac{2}{L}p_1 + \frac{3}{L}p_2 \qquad (Q_2)$$

$$y_2 = p_3 + p_4 + p_5 \qquad (R_1)$$

$$\frac{dy_2}{dx_2} = \frac{1}{L}p_3 + \frac{3}{2L}p_4 + \frac{2}{L}p_5 \qquad (R_2)$$

Combining Eqs (P_1),(P_2),(Q_1),(Q_2),(R_1) and (R_2):

$$p_1 + p_2 - p_3 - p_4 - p_5 = 0 \qquad (S_1)$$

$$2p_1 + 3p_2 + p_3 + \frac{3}{2}p_4 + 2p_5 = 0 \qquad (S_2)$$

or in matrix form, corresponding to Eq. (8.60):

$$\begin{bmatrix} 1 & 1 & -1 & -1 & -1 \\ 2 & 3 & 1 & \frac{3}{2} & 2 \end{bmatrix} \begin{Bmatrix} p_1 \\ p_2 \\ p_3 \\ p_4 \\ p_5 \end{Bmatrix} = 0 \qquad (T)$$

Equation (T) is now partitioned in the same way as Eq. (8.61). Since there are five modal coordinates, and two constraint equations, the number of independent coordinates, q_i, that will finally represent the complete system, is three. These can be any three of the five modes in the system, so choosing:

$$q_1 = p_3, \quad q_2 = p_4, \quad q_3 = p_5,$$

then:

$$[A_1 \mid A_2]\{p\} = \begin{bmatrix} 1 & 1 & \vdots & -1 & -1 & -1 \\ 2 & 3 & \vdots & 1 & \frac{3}{2} & 2 \end{bmatrix} \begin{Bmatrix} p_1 \\ p_2 \\ p_3 \\ p_4 \\ p_5 \end{Bmatrix} = 0 \tag{U}$$

Now applying Eq. (8.64):

$$\begin{Bmatrix} p_d \\ p_f \end{Bmatrix} = \begin{bmatrix} -[A_1]^{-1}[A_2] \\ \hline I \end{bmatrix} \{q\} \tag{V}$$

or numerically, noting that the order of the terms in $\{p\}$ was not changed in this case,

$$\{p\} = \begin{Bmatrix} p_1 \\ p_2 \\ \hline p_3 \\ p_4 \\ p_5 \end{Bmatrix} = \begin{bmatrix} -\begin{bmatrix} 1 & 1 \\ 2 & 3 \end{bmatrix}^{-1}\begin{bmatrix} -1 & -1 & -1 \\ 1 & \frac{3}{2} & 2 \end{bmatrix} \\ \hline \begin{bmatrix} 1 & 0 & 0 \\ 0 & 1 & 0 \\ 0 & 0 & 1 \end{bmatrix} \end{bmatrix} \begin{Bmatrix} q_1 \\ q_2 \\ q_3 \end{Bmatrix} \tag{W}$$

or

$$\{p\} = \begin{Bmatrix} p_1 \\ p_2 \\ p_3 \\ p_4 \\ p_5 \end{Bmatrix} = \begin{bmatrix} 4 & 4.5 & 5 \\ -3 & -3.5 & -4 \\ 1 & 0 & 0 \\ 0 & 1 & 0 \\ 0 & 0 & 1 \end{bmatrix} \begin{Bmatrix} q_1 \\ q_2 \\ q_3 \end{Bmatrix} = [\beta]\{q\} \tag{X}$$

The two separate subsystems in Eq. (O) are now joined together, by applying the transformation $\{p\} = [\beta]\{q\}$. Eq. (O) then becomes

$$[\hat{M}]\{\ddot{q}\} + [\hat{K}]\{q\} = 0 \tag{Y}$$

where, from Eq. (8.68):

$$[\hat{M}] = [\beta]^T \begin{bmatrix} \bar{M}_R & \vdots & 0 \\ \hline 0 & \vdots & M_S \end{bmatrix} [\beta] =$$

$$\mu L \begin{bmatrix} 4 & 4.5 & 5 \\ -3 & -3.5 & -4 \\ 1 & 0 & 0 \\ 0 & 1 & 0 \\ 0 & 0 & 1 \end{bmatrix}^{T} \begin{bmatrix} \frac{1}{5} & \frac{1}{6} & 0 & 0 & 0 \\ \frac{1}{6} & \frac{1}{7} & 0 & 0 & 0 \\ 0 & 0 & \frac{2}{5} & \frac{1}{3} & \frac{2}{7} \\ 0 & 0 & \frac{1}{3} & \frac{2}{7} & \frac{1}{4} \\ 0 & 0 & \frac{2}{7} & \frac{1}{4} & \frac{2}{9} \end{bmatrix} \begin{bmatrix} 4 & 4.5 & 5 \\ -3 & -3.5 & -4 \\ 1 & 0 & 0 \\ 0 & 1 & 0 \\ 0 & 0 & 1 \end{bmatrix}$$

$$= \mu L \begin{bmatrix} 0.8857 & 0.8500 & 0.8333 \\ 0.8500 & 0.8357 & 0.8333 \\ 0.8333 & 0.8333 & 0.8413 \end{bmatrix} \tag{Z_1}$$

and from Eq. (8.69):

$$[\hat{K}] = [\beta]^{T} \begin{bmatrix} \bar{K}_R & 0 \\ 0 & \bar{K}_S \end{bmatrix} [\beta] = \frac{EI}{L^3} \begin{bmatrix} 4 & 4.5 & 5 \\ -3 & -3.5 & -4 \\ 1 & 0 & 0 \\ 0 & 1 & 0 \\ 0 & 0 & 1 \end{bmatrix}^{T} \begin{bmatrix} 4 & 6 & 0 & 0 & 0 \\ 6 & 12 & 0 & 0 & 0 \\ 0 & 0 & \frac{1}{2} & \frac{3}{4} & 1 \\ 0 & 0 & \frac{3}{4} & \frac{3}{2} & \frac{9}{4} \\ 0 & 0 & 1 & \frac{9}{4} & \frac{18}{5} \end{bmatrix}$$

$$\begin{bmatrix} 4 & 4.5 & 5 \\ -3 & -3.5 & -4 \\ 1 & 0 & 0 \\ 0 & 1 & 0 \\ 0 & 0 & 1 \end{bmatrix} = \frac{EI}{L^3} \begin{bmatrix} 28.50 & 33.75 & 39.00 \\ 33.75 & 40.50 & 47.25 \\ 39.00 & 47.25 & 55.60 \end{bmatrix} \tag{Z_2}$$

Thus Eq. (Y), the equation of motion of the joined subsystems, expressed in numerical form is

$$\mu L \begin{bmatrix} 0.8857 & 0.8500 & 0.8333 \\ 0.8500 & 0.8357 & 0.8333 \\ 0.8333 & 0.8333 & 0.8413 \end{bmatrix} \begin{Bmatrix} \ddot{q}_1 \\ \ddot{q}_2 \\ \ddot{q}_3 \end{Bmatrix} + \frac{EI}{L^3} \begin{bmatrix} 28.50 & 33.75 & 39.00 \\ 33.75 & 40.50 & 47.25 \\ 39.00 & 47.25 & 55.60 \end{bmatrix} \begin{Bmatrix} q_1 \\ q_2 \\ q_3 \end{Bmatrix} = 0 \tag{a}$$

To find the natural frequencies and normal modes, since there is no damping, we may substitute $\{q\} = \{\bar{q}\}e^{i\omega t}$ and $\{\ddot{q}\} = -\omega^2\{\bar{q}\}e^{i\omega t}$ in the usual way. Equation (a) can then be written as:

$$\left(\begin{bmatrix} 28.50 & 33.75 & 39.00 \\ 33.75 & 40.50 & 47.25 \\ 39.00 & 47.25 & 55.60 \end{bmatrix} - \lambda \begin{bmatrix} 0.8857 & 0.8500 & 0.8333 \\ 0.8500 & 0.8357 & 0.8333 \\ 0.8333 & 0.8333 & 0.8413 \end{bmatrix} \right) \begin{Bmatrix} \bar{q}_1 \\ \bar{q}_2 \\ \bar{q}_3 \end{Bmatrix} = 0 \tag{b}$$

where:

$$\lambda = \frac{\omega^2 \mu L^4}{EI} \tag{c}$$

Using a standard eigenvalue/eigenvector program, solution of Eq. (b) gives the first two eigenvalues as $\lambda_1 = 6.206$ and $\lambda_2 = 49.44$. The corresponding eigenvectors, arbitrarily scaled to make $\bar{q}_3 = 1$, are

$$\left\{ \begin{array}{c} \bar{q}_1 \\ \bar{q}_2 \\ \bar{q}_3 \end{array} \right\}_1 = \left\{ \begin{array}{c} 2.371 \\ -3.103 \\ 1 \end{array} \right\} \tag{d_1}$$

and

$$\left\{ \begin{array}{c} \bar{q}_1 \\ \bar{q}_2 \\ \bar{q}_3 \end{array} \right\}_2 = \left\{ \begin{array}{c} 0.9267 \\ -1.978 \\ 1 \end{array} \right\} \tag{d_2}$$

From Eq. (c), the two natural frequencies ω_1 and ω_2 are

$$\omega_1 = \frac{1}{L^2} \sqrt{\frac{EI}{\mu}} \sqrt{\lambda_1} = \frac{2.491}{L^2} \sqrt{\frac{EI}{\mu}} \tag{e_1}$$

and

$$\omega_2 = \frac{1}{L^2} \sqrt{\frac{EI}{\mu}} \sqrt{\lambda_2} = \frac{7.031}{L^2} \sqrt{\frac{EI}{\mu}} \tag{e_2}$$

The corresponding mode shapes can be plotted by substituting each eigenvector, from Eqs (d_1) and (d_2), in turn, into Eq. (X), to find the values of p_1 to p_5, for each mode. The local beam displacements, y_1 and y_2, are then given by Eqs (A) and (B). The two mode shapes, scaled to a maximum displacement of unity, are plotted in Fig. 8.3

Since the joined cantilevers form a fixed–fixed uniform beam of length $3L$, it is possible to compare the results above with exact answers from Table 8.1. The first two normal mode frequencies are given by $\beta_1(3L) = 4.7300$ and $\beta_2(3L) = 7.8532$.

Then the exact natural frequencies for a fixed–fixed uniform beam of length $3L$ are

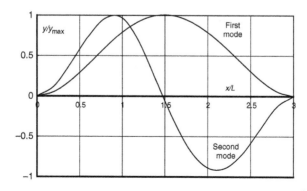

Fig. 8.3   Mode shapes calculated by the component mode synthesis method.

$$\omega_1 = (\beta_1)^2 \sqrt{\frac{EI}{\mu}} = \frac{2.485}{L^2} \sqrt{\frac{EI}{\mu}} \tag{f_1}$$

and

$$\omega_2 = (\beta_2)^2 \sqrt{\frac{EI}{\mu}} = \frac{6.852}{L^2} \sqrt{\frac{EI}{\mu}} \tag{f_2}$$

The natural frequencies, ω_1 and ω_2, calculated by the component mode synthesis method, and given by Eqs (e_1) and (e_2), are 0.24 and 2.6% higher, respectively, than the exact values given by Eqs (f_1) and (f_2).

From Fig. 8.3, the mode shapes are seen to approximate to the first two modes of a uniform fixed–fixed beam.

8.3.2 The Branch Mode Method

The branch mode method appears to have been devised by Hunn [8.6], for finding the normal modes of complete aircraft structures, and was further developed by Gladwell [8.7]. It was mentioned by Hurty, in his 1965 paper [8.3], and described as being similar to the component mode synthesis method. During the period when it was widely used, say from about 1960 until 1980, the influence coefficient (flexibility) method, or, later, the finite element method, was used to calculate the normal modes of the 'branches' of an aircraft structure, for example the fuselage, wings, vertical and horizontal stabilizers, etc., and the branch mode method was then used to combine them into a mathematical model of the complete aircraft. As more powerful computers became available, the finite element method became viable for complete structures, making it unnecessary to use the branch mode method. It remains useful, however, in some circumstances, such as:

(a) when a given parameter, say the stiffness of a particular actuator or joint, is required to appear explicitly in the equations of motion at modal level, either because it is to be varied widely or because it is non-linear. In the latter case it is possible, for example, to separate the linear and non-linear parts in a system of equations.

(b) The method can also be used to make small modifications to models of complete structures at modal level, without having to re-run a large FE model.

The branch mode method uses a special way of defining the shapes of the assumed modes, avoiding the use of constraint modes or constraint equations, as required in component mode synthesis methods, and is therefore much easier to use, essentially being a 'non-mathematical' version of the component mode synthesis method. Although it can be used with completely arbitrary modes, the following describes the more usual procedure, in which normal modes of branches are used.

The method is most easily explained by a simple example. Consider the L-shaped lumped-mass structure shown in Fig. 8.4(a). The normal modes of the two beams, AB

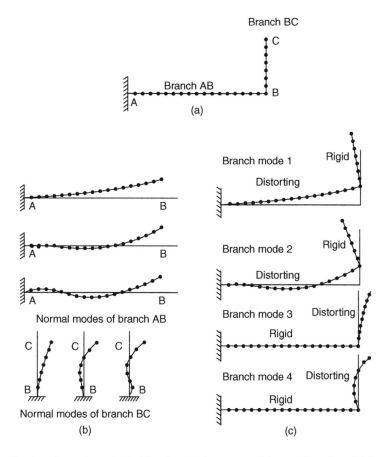

Fig. 8.4. The branch mode method: (a) a branched structure; (b) normal modes of AB and BC as Cantilevers; (c) branch modes

and BC, are first calculated, as separate cantilevers, i.e. fixed at A and B respectively. These are known as the 'branches'. Any suitable method can be used for this first stage, such as the influence coefficient method or the finite element method.

The equations of motion for the two cantilevers, treated as two separate systems, are, at global coordinate level, say,:

for cantilever AB:

$$[M_{AB}]\{\ddot{z}_{AB}\} + [K_{AB}]\{z_{AB}\} = 0 \tag{8.70}$$

and for cantilever BC:

$$[M_{BC}]\{\ddot{z}_{BC}\} + [K_{BC}]\{z_{BC}\} = 0 \tag{8.71}$$

where $[M_{AB}]$ and $[M_{BC}]$ are the mass matrices at global level, and $[K_{AB}]$ and $[K_{BC}]$ are the corresponding stiffness matrices. The vectors $\{z_{AB}\}$ and $\{z_{BC}\}$ are the global coordinates.

These are solved in the usual way, producing two sets of *normal* modes, with modal matrices $[X_{AB}]$ and $[X_{BC}]$. Three of the normal modes of each cantilever are sketched in Fig. 8.4(b). They are now available for use as component modes in a branch mode analysis.

The unique feature of the method is now shown in Fig. 8.4(c), where four examples of the many possible assumed modes, known as branch modes, are sketched. Branch mode 1, for example, consists of the first bending normal mode of beam AB, with beam BC attached to it, but the part BC is *constrained to be rigid*, although free to move with AB. Similarly, branch mode 2 consists of the second bending normal mode of AB, with BC as a rigid extension. In branch mode 3, section AB is assumed rigid, and also fixed, but section BC is allowed to distort in its first cantilever mode. Branch mode 4 is the same, except that the second cantilever mode of BC is used. Of course, when the normal modes of the branches are modified in this way, they become assumed, not normal modes, of the complete structure and cannot be expected to diagonalize the equations of motion as a whole.

It can be seen that forming the branch modes from cantilever modes, as described above, will introduce the need for additional coordinates at some grid points. As an example, the rotation at point B in the cantilever AB must be known, to define the lateral displacements of points in BC, and these points must also have the freedom to move vertically as well as laterally. The need for these extra coordinates must be anticipated when setting up Eqs (8.70) and (8.71).

The branch modes, say the four shown in Fig. 8.4(c), are now used in a straightforward 'Rayleigh–Ritz', assumed modes analysis, which is carried out by applying Lagrange's equations. The four generalized coordinates required are, say, p_1, p_2, p_3 and p_4, corresponding to branch modes 1–4. The third normal mode of each cantilever was not used in this case.

The end result of this will be the 4-DOF equations of motion:

$$[\bar{M}]\{\ddot{p}\} + [\bar{K}]\{p\} = 0 \tag{8.72}$$

where $[\bar{M}]$ and $[\bar{K}]$ are the 4×4 mass and stiffness matrices of the final system. How these matrices are derived is discussed next.

There is, incidentally, no need to account for the constraint forces required to keep one part of the system rigid while another part is allowed to distort. These forces do no work, and therefore do not contribute to the kinetic or potential energies.

The mass matrix

The matrix $[\bar{M}]$ for the complete system is derived using the standard rules for assumed modes methods, as developed in Chapter 6, which are based on Lagrange's equations. Thus,

$$[\bar{M}] = [X]^{T}[M][X] \tag{8.73}$$

where the matrix, $[M]$, is

$$[M] = \begin{bmatrix} M_{AB} & 0 \\ \hline 0 & M_{BC} \end{bmatrix} \tag{8.74}$$

The matrices $[M_{AB}]$ and $[M_{BC}]$ are the mass matrices as used to calculate the original cantilever modes of AB and of BC, respectively, in Eqs (8.70) and (8.71), assuming that provision for the extra coordinates introduced by the formation of the branch modes, as discussed above, has been made.

The modal matrix, $[X]$, for the complete system, is now given by:

$$[X] = \begin{bmatrix} X_{AB} & 0 \\ \hline X_{AC} & X_{BC} \end{bmatrix} \tag{8.75}$$

where:

$[X_{AB}]$ consists of the coordinate displacements in the part AB, due to its distortion, in branch modes 1 and 2, written as one column for each mode. They are a subset of those found for the cantilever AB in isolation, so have already been calculated.

$[X_{BC}]$ consists of the coordinate displacements, due to distortion, of the part BC, in branch modes 3 and 4, also written as one column for each mode. They are a subset of those found for the cantilever BC in isolation.

The matrix $[X_{AC}]$ is new and is a matrix of coordinate displacements in the part BC due to its rigid motion in branch modes 1 and 2. It is formed, using small-angle approximations, using the linear and angular coordinate displacements at point B. The displacements of points in AB, in branch modes 3 and 4, are zero in this example.

The mass matrix $[\bar{M}]$ is now formed using Eqs (8.73), (8.74) and (8.75):

$$
\begin{aligned}
[\bar{M}] = [X]^{\mathrm{T}}[M][X] &= \begin{bmatrix} X_{AB} & 0 \\ \hline X_{AC} & X_{BC} \end{bmatrix}^{\mathrm{T}} \begin{bmatrix} M_{AB} & 0 \\ \hline 0 & M_{BC} \end{bmatrix} \begin{bmatrix} X_{AB} & 0 \\ \hline X_{AC} & X_{BC} \end{bmatrix} \\
&= \begin{bmatrix} \left([X_{AB}]^{\mathrm{T}} M_{AB} X_{AB} + [X_{AC}]^{\mathrm{T}} M_{BC} X_{AC}\right) & [X_{AC}]^{\mathrm{T}} M_{BC} X_{BC} \\ \hline [X_{BC}]^{\mathrm{T}} M_{BC} X_{AC} & [X_{BC}]^{\mathrm{T}} M_{BC} X_{BC} \end{bmatrix}
\end{aligned} \tag{8.76}
$$

Since $[X_{AB}]$ and $[X_{BC}]$ consist of the eigenvectors of the original simple cantilever systems AB and BC, respectively, the products $[X_{AB}]^{\mathrm{T}}[M_{AB}][X_{AB}]$ and $[X_{BC}]^{\mathrm{T}}[M_{BC}][X_{BC}]$ are both 2×2 diagonal matrices. However, the other matrix products are not diagonal, and $[\bar{M}]$, as a whole, is not diagonal.

The stiffness matrix

The overall 4×4 stiffness matrix, $[\bar{K}]$, for the complete system, could be found in a similar way to the mass matrix. However, due to the way the branch modes are formed, a short cut can be taken, as follows.

If the 4×4 stiffness matrix is written out in full:

$$[\bar{K}] = \begin{bmatrix} k_{11} & k_{12} & k_{13} & k_{14} \\ k_{21} & k_{22} & k_{23} & k_{24} \\ k_{31} & k_{32} & k_{33} & k_{34} \\ k_{41} & k_{42} & k_{43} & k_{44} \end{bmatrix} \tag{8.77}$$

Then the following arguments can be used:

(1) The following pairs of branch modes have no distorting sections in common:

Branch modes 1 and 3;
Branch modes 1 and 4;

Branch modes 2 and 3;
Branch modes 2 and 4.

As discussed in more detail by Gladwell [8.7], in order to create cross-terms between these modes, stresses in one mode would have to do work on the corresponding elements in the other mode. However, in this case the elements in the other mode involved are always rigid, and no work can be done. Therefore all the cross stiffness terms between the pairs of branch modes listed above are zero, and

$$k_{13} = k_{14} = k_{23} = k_{24} = k_{31} = k_{41} = k_{32} = k_{42} = 0 \qquad (8.78)$$

(2) In general, stiffness coupling between modes 1 and 2 and modes 3 and 4 is still possible. However, it can be seen from Fig. 8.4(b) and (c) that the potential energy in branch mode 1 must be exactly the same as that in the first cantilever mode of AB, since all the distortions are the same. Similarly, the potential energy in branch mode 2 is the same as that in the second cantilever mode of AB. The stiffness matrix for branch modes 1 and 2 must therefore be the same as for the first two cantilever modes of AB. The latter are normal modes in this case, and their stiffness matrix is diagonal. The same argument applies to branch modes 3 and 4, and so the following cross-terms are also zero:

$$k_{12} = k_{21} = k_{34} = k_{43} = 0$$

Therefore the stiffness matrix $[\bar{K}]$ is given by:

$$[\bar{K}] = \begin{bmatrix} k_{11} & 0 & 0 & 0 \\ 0 & k_{22} & 0 & 0 \\ 0 & 0 & k_{33} & 0 \\ 0 & 0 & 0 & k_{44} \end{bmatrix} \qquad (8.79)$$

where k_{11} and k_{22} are the generalized stiffnesses of the first two normal modes of the original cantilever AB and k_{33} and k_{44} are the generalized stiffnesses of the first two modes of the cantilever BC.

If normal modes of the complete system are required, Eq. (8.72) is solved for eigenvalues and eigenvectors in the usual way, and the following further coordinate transformation is applied:

$$\{p\} = [\Phi]\{q\} \qquad (8.80)$$

where $\{q\}$ are normal mode coordinates and $[\Phi]$ is a modal matrix derived from some or all of the eigenvectors of Eq. (8.72). The equations of motion, without damping or external forces, expressed in terms of the normal mode coordinates, $\{q\}$, are then:

$$[\hat{M}]\{\ddot{q}\} + [\hat{K}]\{q\} = 0 \qquad (8.81)$$

where

$$[\hat{M}] = [\Phi]^{\mathrm{T}}[\bar{M}][\Phi] \qquad (8.82)$$

and

$$[\hat{K}] = [\Phi]^{\mathrm{T}}[\bar{K}][\Phi] \qquad (8.83)$$

The matrices $[\hat{M}]$ and $[\hat{K}]$ are now both diagonal, and with the addition of damping and external forces may be used to represent the complete structure for any purpose, such as its response to external loading by the normal mode summation method.

In the simple example above, the whole structure was restrained at point A. It need not be restrained, and could represent a free–free structure, floating in space or water. This would require the introduction of up to six rigid modes of the whole structure, with appropriate changes to the branch modes used: for example, the two beams used as branches could be considered cantilevered at B.

8.4 The finite element method

The finite element method was originally devised for static stress analysis, but was soon applied to vibration problems also. It is now used to the exclusion of almost all other methods for setting up the equations of motion of structures.

Its development required no new mathematics, since the basic ideas, essentially the Rayleigh–Ritz concept (1909) of using assumed modes and Lagrange's equations (1788), had been around for a very long time. However, because it relies upon carrying out a huge number of calculations, its practical implementation had to wait for the development of high-speed computers, and in practice the use of the method dates from about 1960.

Fundamentally, there are two finite element methods; the *force method*, where forces are assumed and displacements calculated, and the *displacement method* where displacements are assumed, and forces calculated. The latter is mostly used for vibration analysis, and is the only one discussed here.

8.4.1 An Overview

The development of a finite element model for a structure, using the displacement method, consists essentially of the following four stages. These are now briefly described, using Fig. 8.5 to illustrate each stage pictorially.

Stage 1: dividing the structure into finite elements

As is well known, many kinds of finite element can now be used, such as rods, distorting axially or torsionally; beams in bending; membranes; triangular plates; quadrilateral plates; and solids such as tetrahedrons, pentahedrons and hexahedrons.

This stage is shown diagrammatically in Fig. 8.5(a), where a part of a larger structure, consisting of a curved beam, lying in the plane of the paper, is broken down into elements. The elements of the structure are, say, *q, r, s, t, u, ...*, of which *r* and *s* are shown. When discussing elements in isolation, with the ends free, we shall refer to the left end as *a* and the right end as *b*.

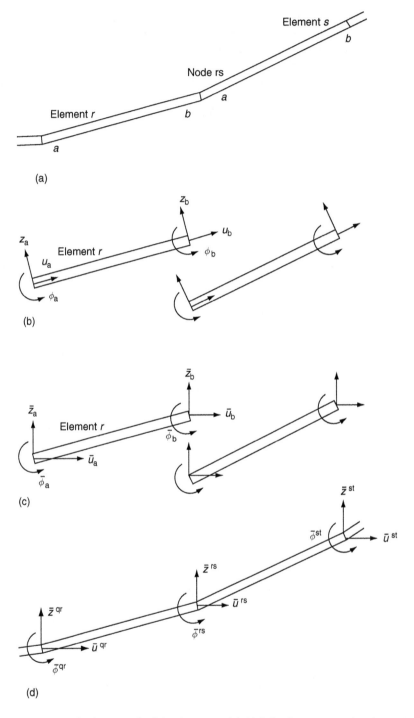

Fig. 8.5 Stages in the development of a finite element model: (a) finite element approximation to structure, (b) separate elements in local coordinates, (c) separate elements transformed to global coordinates, (d) assembled elements in global coordinates

Stage 2: deriving the equations of motion for each element in local axes

The equations of motion for each element separately are usually based on quite simple *static* displacement functions, obtained by imposing a unit static displacement, in turn, to each of the coordinates at the extremities or edges of the element, with the others remaining fixed. The axis systems used for this are whatever is most convenient, and are called *local axes*.

The equations of motion for the element are set up using the static displacement functions as assumed modes, the edge displacements being used as generalized coordinates in a set of Lagrange's equations. This stage is shown pictorially in Fig. 8.5(b), where two of the elements, r and s, are sketched separately. In this example, each element has freedom to bend, and to stretch or compress axially, and has constant properties over its length. Such an element could, in practice, be formed by combining the equations for two of the elements discussed later; an axial rod element and a beam bending element without axial deformation.

Since each element has only a limited number of assumed freedoms, six in this example, it is relatively easy to derive its equations of motion from the kinetic and potential energies in the usual way. Except for constants, the equations of motion of elements of the same type are all of the same standard form, and can be built in to the FE program, once and for all. Element r, for example, shown in Fig. 8.5(b), has a 6-DOF equation of motion of the form:

$$[m]^r\{\ddot{z}\}^r + [k]^r\{z\}^r = \{F\}^r \tag{8.84}$$

where $\{z\}^r$, $\{\ddot{z}\}^r$ and $\{F\}^r$ are, in this particular case:

$$\{z\}^r = \begin{Bmatrix} z_a \\ u_a \\ \phi_a \\ z_b \\ u_b \\ \phi_b \end{Bmatrix}^r \tag{8.85a}$$

$$\{\ddot{z}\}^r = \begin{Bmatrix} \ddot{z}_a \\ \ddot{u}_a \\ \ddot{\phi}_a \\ \ddot{z}_b \\ \ddot{u}_b \\ \ddot{\phi}_b \end{Bmatrix}^r \tag{8.85b}$$

$$\{F\}^r = \begin{Bmatrix} F_{z,a} \\ F_{u,a} \\ M_a \\ F_{z,b} \\ F_{u,b} \\ M_b \end{Bmatrix}^r \tag{8.85c}$$

The end displacements, $\{z\}^r$, are shown in Fig. 8.5(b). The corresponding accelerations, $\{\ddot{z}\}^r$, and forces $\{F\}^r$ (not shown) act in the same sense as $\{z\}^r$. Suffices a and b indicate the left and right ends of the element respectively.

The element mass and stiffness matrices, in this case $[m]^r$ and $[k]^r$, are standard results, and their derivation, for some of the simpler elements, is discussed in Section 8.4.2.

In the derivation shown here, the mass matrix for an element is based upon the same deflection shapes as the stiffness matrix, and this is known as a *consistent mass* approach. It is also possible to use a lumped mass approach, where the stiffness matrix is the same, but all the mass is lumped at the node points. The latter approach has possible computational advantages, in that the mass matrix is diagonal, or more nearly diagonal.

Stage 3: transforming the equations of motion of each element from local axes to global axes

The displacements, accelerations and forces/moments, defined, as above, at the ends or edges of each element, are in local coordinates, and may act at a variety of different angles. Before the elements can be assembled into the overall global system, the displacements, accelerations, and forces, must all be transformed into that system, as shown in Fig. 8.5(c) for elements r and s.

The derivation of the transformation to achieve this is illustrated by Fig. 8.6, where point O is located, say, at the left end of element r. Point O is also taken as the origin of both the global axis system, $\bar{u}$, $\bar{z}$ and of the local axis system, u, z. The local axes are rotated by angle α counter-clockwise from the global axes.

Let the displacements at the end of the element, parallel to global axes, be represented by vectors $\bar{u}_a$ and $\bar{z}_a$, shown in Fig. 8.6 as $\bar{u}_a = OA$ and $\bar{z}_a = OB$. Resolving these in the local axis system, $\bar{u}_a$, produces the components OC and OD, and $\bar{z}_a$ produces components OE and OF, in the local axis system. The total displacement along the u axis, u_a, is

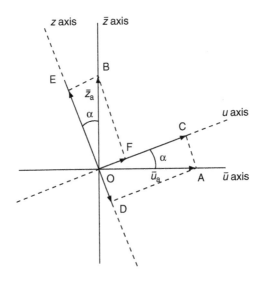

Fig. 8.6   Transforming from local axes to global axes.

$$u_a = \mathrm{OC} + \mathrm{OF} = \bar{u}_a \cos \alpha + \bar{z}_a \sin \alpha \tag{8.86}$$

Similarly, the total displacement along the axis, z_a is

$$z_a = -\mathrm{OD} + \mathrm{OE} = -\bar{u}_a \sin \alpha + \bar{z}_a \cos \alpha \tag{8.87}$$

Writing Eqs (8.86) and (8.87) in matrix form,

$$\begin{Bmatrix} u_a \\ z_a \end{Bmatrix} = \begin{bmatrix} \cos \alpha & \sin \alpha \\ -\sin \alpha & \cos \alpha \end{bmatrix} \begin{Bmatrix} \bar{u}_a \\ \bar{z}_a \end{Bmatrix} \tag{8.88}$$

it can be seen that ϕ_a is unchanged by the transformation, and so $\phi_a = \bar{\phi}_a$. We can therefore write

$$\begin{Bmatrix} u_a \\ z_a \\ \phi_a \end{Bmatrix} = \begin{bmatrix} \cos \alpha & \sin \alpha & 0 \\ -\sin \alpha & \cos \alpha & 0 \\ 0 & 0 & 1 \end{bmatrix} \begin{Bmatrix} \bar{u}_a \\ \bar{z}_a \\ \bar{\phi}_a \end{Bmatrix} \tag{8.89}$$

Although derived for end a of element r, Eq. (8.89) also applies to end b, and, in fact, to either end of any beam-like element where the local axes are rotated by angle α counter-clockwise from the global axes. The transformation matrix for all six coordinates of a complete element is therefore given by:

$$\begin{Bmatrix} z_a \\ u_a \\ \phi_a \\ \hline z_b \\ u_b \\ \phi_b \end{Bmatrix} = \left[\begin{array}{ccc:ccc} \cos \alpha & \sin \alpha & 0 & & & \\ -\sin \alpha & \cos \alpha & 0 & & 0 & \\ 0 & 0 & 1 & & & \\ \hdashline & & & \cos \alpha & \sin \alpha & 0 \\ & 0 & & -\sin \alpha & \cos \alpha & 0 \\ & & & 0 & 0 & 1 \end{array} \right] \begin{Bmatrix} \bar{z}_a \\ \bar{u}_a \\ \bar{\phi}_a \\ \hline \bar{z}_b \\ \bar{u}_b \\ \bar{\phi}_b \end{Bmatrix} \tag{8.90}$$

or in shorter notation:

$$\{z\} = [T]\{\bar{z}\} \tag{8.91}$$

Note: The transformation matrices in Eqs (8.88), (8.89) and (8.90) are *orthogonal matrices*, having the property that $[T]^\mathrm{T}[T] = [T][T]^\mathrm{T} = [I]$, the unit matrix.

Equation (8.90) is the required transformation from local to global coordinates. It has been derived for displacements, but since the accelerations and forces are also vectors, acting in the same directions as the displacements, the same transformation matrix, $[T]$, applies, and

$$\{\ddot{z}\} = [T]\{\ddot{\bar{z}}\} \tag{8.92}$$

$$\{F\} = [T]\{\bar{F}\} \tag{8.93}$$

where quantities in global axes are indicated by bars. Written out in full, the displacement, acceleration and force/moment vectors for element r are as follows.

$$\{\bar{z}\}^{\mathrm{r}} = \begin{Bmatrix} \bar{z}_{\mathrm{a}} \\ \bar{u}_{\mathrm{a}} \\ \bar{\phi}_{\mathrm{a}} \\ \bar{z}_{\mathrm{b}} \\ \bar{u}_{\mathrm{b}} \\ \bar{\phi}_{\mathrm{b}} \end{Bmatrix}^{\mathrm{r}}$$
(8.94a)

$$\{\ddot{\bar{z}}\}^{\mathrm{r}} = \begin{Bmatrix} \ddot{\bar{z}}_{\mathrm{a}} \\ \ddot{\bar{u}}_{\mathrm{a}} \\ \ddot{\bar{\phi}}_{\mathrm{a}} \\ \ddot{\bar{z}}_{\mathrm{b}} \\ \ddot{\bar{u}}_{\mathrm{b}} \\ \ddot{\bar{\phi}}_{\mathrm{b}} \end{Bmatrix}^{\mathrm{r}}$$
(8.94b)

$$\{\bar{F}\}^{\mathrm{r}} = \begin{Bmatrix} \bar{F}_{z,\mathrm{a}} \\ \bar{F}_{u,\mathrm{a}} \\ \bar{M}_{\mathrm{a}} \\ \bar{F}_{z,\mathrm{b}} \\ \bar{F}_{u,\mathrm{b}} \\ \bar{M}_{\mathrm{b}} \end{Bmatrix}^{\mathrm{r}}$$
(8.94c)

The displacements are shown in Fig. 8.5(c), and it should be understood that the accelerations and forces/moments, not shown in the figure, act in the same sense as the displacements.

We can now transform the element equations of motion, Eq. (8.84), which is in local coordinates, into global coordinates. Substituting Eqs (8.91). (8.92) and (8.93) into Eq. (8.84), we have

$$[m][T]\{\ddot{\bar{z}}\} + [k][T]\{\bar{z}\} = [T]\{\bar{F}\}$$
(8.95)

Pre-multiplying through by $[T]^{\mathrm{T}}$:

$$[T]^{\mathrm{T}}[m][T]\{\ddot{\bar{z}}\} + [T]^{\mathrm{T}}[k][T]\{\bar{z}\} = [T]^{\mathrm{T}}[T]\{\bar{F}\}$$
(8.96)

However, since the matrix $[T]$ is orthogonal, $[T]^{\mathrm{T}}[T] = [I]$ and

$$[T]^{\mathrm{T}}[m][T]\{\ddot{\bar{z}}\} + [T]^{\mathrm{T}}[k][T]\{\bar{z}\} = \{\bar{F}\}$$
(8.97)

or

$$[\bar{m}]\{\ddot{\bar{z}}\} + [\bar{k}]\{\bar{z}\} = \{\bar{F}\}$$
(8.98)

where

$$[\bar{m}] = [T]^{\mathrm{T}}[m][T]$$
(8.99a)

and

$$[\bar{k}] = [T]^{\mathrm{T}}[k][T]$$
(8.99b)

Equation (8.98) represents the equation of motion of a single element in global coordinates, and Eqs (8.99a) and (8.99b) give the simple rule for transforming the mass and stiffness matrices from local to global coordinates.

Stage 4: assembling the separate elements into a single set of equations

As a result of Stage 3, the separate elements have been transformed into a common system, i.e. the global system, and it is now required to join the elements together to make a complete system. Using the pictorial example in Fig. 8.5, this corresponds to combining the separate elements shown at (c) into the joined-up system shown at (d).

As an example, let us combine the equations of motion for elements r and s to represent the elements joined together. First, the equations of motion for element r alone are

$$
\begin{bmatrix} \bar{m}^r_{aa} & \bar{m}^r_{ab} \\ \bar{m}^r_{ba} & \bar{m}^r_{bb} \end{bmatrix}
\begin{Bmatrix} \ddot{\bar{z}}^r_a \\ \ddot{\bar{u}}^r_a \\ \ddot{\bar{\phi}}^r_a \\ \ddot{\bar{z}}^r_b \\ \ddot{\bar{u}}^r_b \\ \ddot{\bar{\phi}}^r_b \end{Bmatrix}
+
\begin{bmatrix} \bar{k}^r_{aa} & \bar{k}^r_{ab} \\ \bar{k}^r_{ba} & \bar{k}^r_{bb} \end{bmatrix}
\begin{Bmatrix} \bar{z}^r_a \\ \bar{u}^r_a \\ \bar{\phi}^r_a \\ \bar{z}^r_b \\ \bar{u}^r_b \\ \bar{\phi}^r_b \end{Bmatrix}
=
\begin{Bmatrix} \bar{F}^r_{z,a} \\ \bar{F}^r_{u,a} \\ \bar{M}^r_a \\ \bar{F}^r_{z,b} \\ \bar{F}^r_{u,b} \\ \bar{M}^r_b \end{Bmatrix}
\tag{8.100}
$$

where:

$\bar{z}^r_a$, $\bar{u}^r_a$, $\bar{\phi}^r_a$ are global displacements at the left end of element r, $\ddot{\bar{z}}^r_a$, $\ddot{\bar{u}}^r_a$, $\ddot{\bar{\phi}}^r_a$ the corresponding accelerations, $\bar{F}^r_{z,a}$, $\bar{F}^r_{u,a}$, $\bar{M}^r_a$ the two forces and single moment at the left end of element r, $\bar{z}^r_b$, $\bar{u}^r_b$, and $\bar{\phi}^r_b$ (and similarly for the accelerations) the global displacements at the right end of element r, and $\bar{F}^r_{z,b}$, $\bar{F}^r_{u,b}$, $\bar{M}^r_b$ the two forces and single moment at the right end of element r.

The 6×6 mass and stiffness matrices have each been partitioned into four 3×3 matrices; the matrix $[\bar{k}^r_{ab}]$, for example, relates displacements at the left end to forces at the right end.

Using the same notation, except that the superscripts r are replaced by s, the equations of motion for element s alone are

$$
\begin{bmatrix} \bar{m}^s_{aa} & \bar{m}^s_{ab} \\ \bar{m}^s_{ba} & \bar{m}^s_{bb} \end{bmatrix}
\begin{Bmatrix} \ddot{\bar{z}}^s_a \\ \ddot{\bar{u}}^s_a \\ \ddot{\bar{\phi}}^s_a \\ \ddot{\bar{z}}^s_b \\ \ddot{\bar{u}}^s_b \\ \ddot{\bar{\phi}}^s_b \end{Bmatrix}
+
\begin{bmatrix} \bar{k}^s_{aa} & \bar{k}^s_{ab} \\ \bar{k}^s_{ba} & \bar{k}^s_{bb} \end{bmatrix}
\begin{Bmatrix} \bar{z}^s_a \\ \bar{u}^s_a \\ \bar{\phi}^s_a \\ \bar{z}^s_b \\ \bar{u}^s_b \\ \bar{\phi}^s_b \end{Bmatrix}
=
\begin{Bmatrix} \bar{F}^s_{z,a} \\ \bar{F}^s_{u,a} \\ \bar{M}^s_a \\ \bar{F}^s_{z,b} \\ \bar{F}^s_{u,b} \\ \bar{M}^s_b \end{Bmatrix}
\tag{8.101}
$$

Now, since the right end of element r is joined to left end of element s, the global displacements and accelerations at these locations must be equal, and both must be equal to those at the common node, which will be designated by the superscript rs. Thus:

$$\ddot{z}^r_b = \ddot{z}^s_a = \ddot{z}^{rs} \qquad \ddot{\ddot{z}}^r_b = \ddot{\ddot{z}}^s_a = \ddot{\ddot{z}}^{rs}$$
$$\ddot{u}^r_b = \ddot{u}^s_a = \ddot{u}^{rs} \qquad \ddot{\ddot{u}}^r_b = \ddot{\ddot{u}}^s_a = \ddot{\ddot{u}}^{rs} \qquad (8.102)$$
$$\ddot{\phi}^r_b = \ddot{\phi}^s_a = \ddot{\phi}^{rs} \qquad \ddot{\ddot{\phi}}^r_b = \ddot{\ddot{\phi}}^s_a = \ddot{\ddot{\phi}}^{rs}$$

where $\ddot{z}^{rs}$, $\ddot{u}^{rs}$, $\ddot{\phi}^{rs}$ are the displacements and, $\ddot{\ddot{z}}^{rs}$, $\ddot{\ddot{u}}^{rs}$ and $\ddot{\ddot{\phi}}^{rs}$ are the accelerations, at node rs, with the elements joined. So, for example, $\ddot{z}^{rs}$ can be substituted for both $\ddot{z}^r_b$ and $\ddot{z}^s_a$.

Also, for equilibrium at node *rs*, the sum of the vertical forces acting on the elements must be equal to the external force:

$$\bar{F}^{rs}_z = \bar{F}^r_{z,b} + \bar{F}^s_{z,a} \qquad (8.103)$$

and similarly for horizontal forces and moments:

$$\bar{F}^{rs}_u = \bar{F}^r_{u,b} + \bar{F}^s_{u,a} \qquad (8.104)$$

$$\bar{M}^{rs} = \bar{M}^r_b + \bar{M}^s_a \qquad (8.105)$$

Using Eqs (8.102)–(8.105), Eqs (8.100) and (8.101) can be combined into a single 9×9 set of equations as follows:

$$
\begin{bmatrix}
\bar{m}^r_{aa} & \bar{m}^r_{ab} & 0 \\
\bar{m}^r_{ba} & \bar{m}^r_{bb}+\bar{m}^s_{aa} & \bar{m}^s_{ab} \\
0 & \bar{m}^s_{ba} & \bar{m}^s_{bb}
\end{bmatrix}
\begin{Bmatrix}
\ddot{\ddot{z}}^r_a \\ \ddot{\ddot{u}}^r_a \\ \ddot{\ddot{\phi}}^r_a \\ \ddot{\ddot{z}}^{rs} \\ \ddot{\ddot{u}}^{rs} \\ \ddot{\ddot{\phi}}^{rs} \\ \ddot{\ddot{z}}^s_b \\ \ddot{\ddot{u}}^s_b \\ \ddot{\ddot{\phi}}^s_b
\end{Bmatrix}
+
\begin{bmatrix}
\bar{k}^r_{aa} & \bar{k}^r_{ab} & 0 \\
\bar{k}^r_{ba} & \bar{k}^r_{bb}+\bar{k}^s_{aa} & \bar{k}^s_{ab} \\
0 & \bar{k}^s_{ba} & \bar{k}^s_{bb}
\end{bmatrix}
\begin{Bmatrix}
\bar{z}^r_a \\ \bar{u}^r_a \\ \bar{\phi}^r_a \\ \bar{z}^{rs} \\ \bar{u}^{rs} \\ \bar{\phi}^{rs} \\ \bar{z}^s_b \\ \bar{u}^s_b \\ \bar{\phi}^s_b
\end{Bmatrix}
=
\begin{Bmatrix}
\bar{F}^r_{z,a} \\ \bar{F}^r_{u,a} \\ \bar{M}^r_a \\ \bar{F}^{rs}_z \\ \bar{F}^{rs}_u \\ \bar{M}^{rs} \\ \bar{F}^s_{z,b} \\ \bar{F}^s_{u,b} \\ \bar{M}^s_b
\end{Bmatrix}
\qquad (8.106)
$$

These are the equations of motion for the two elements r and s, joined at node rs, and there are now only nine degrees of freedom rather than twelve. It can be seen that joining the ends of two elements has been achieved very simply by superimposing, or adding, those parts of the mass and stiffness matrices representing the element ends to be joined.

The left end of element r and the right end of element s are still free at this point, and the next steps would be to join these loose ends to their adjacent elements, q and t, respectively, using the same procedure. When all free element ends have been joined in this way, the equations of motion for the whole structure will be in terms of global coordinates at nodes only, as desired.

An easy way to make a computer superimpose selected parts of two matrices is by the use of *locator matrices*. As an example, the 6×6 mass or stiffness matrix for element r can be correctly placed within the 9×9 combined mass or stiffness matrix by using the locator matrix $[L_1]$, where:

$$[L_1] = [I_{6\times6} \mathbin{\vdots} 0_{6\times3}]$$ (8.107)

Then, for the mass matrix, the operation:

$$[L_1]^{\mathrm{T}}[\bar{m}][L_1] = \left[\frac{I_{6\times6}}{0_{3\times6}}\right][\bar{m}_{6\times6}][I_{6\times6} \mathbin{\vdots} 0_{6\times3}] = \left[\frac{\bar{m}_{6\times6} \mathbin{\vdots} 0_{6\times3}}{0_{3\times6} \mathbin{\vdots} 0_{3\times3}}\right]$$ (8.108)

has the desired effect of placing the 6×6 element mass matrix for element r in the top left corner of the combined 9×9 matrix. The same operation is carried out for the stiffness matrix. Another locator matrix

$$L_2 = [0_{6\times3} \mathbin{\vdots} I_{6\times6}]$$ (8.109)

can similarly be used to place the 6×6 mass and stiffness matrices for element s at the bottom right of the 9×9 combined matrix.

The four stages described above are used to create the equations of motion of the structure in global coordinates. These equations may have thousands of degrees of freedom, and although it is sometimes possible to solve them directly, a fifth stage is usually to transform the system into a relatively small number of normal mode coordinates, as discussed in Chapter 6. Damping is usually added at this stage, and the response to any applied input can be found, typically by the normal mode summation method.

The derivation of the equations of motion of the isolated elements is described in the next section.

8.4.2 Equations of Motion for Individual Elements

The equations of motion for three types of element are now derived. These are (1) a rod element with axial deformation, (2) a rod element with torsional deformation and (3) a beam element with bending deformation in one plane.

Rod element with axial deformation

Figure 8.7(a) shows a uniform rod element ab with axial end displacements u_a and u_b, and corresponding axial external end loads F_a and F_b. The rod has the properties: E is Young's modulus; A the cross-sectional area and μ the mass per

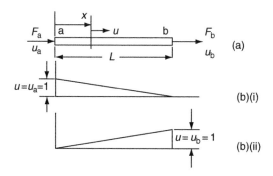

Fig. 8.7   Finite element for a rod with end loading.

unit length, all constant over its length, L. The axial displacement at any point x from the left end is u.

Although it is easy, in this case, to write an expression for the axial displacement, u, in terms of the end displacements u_a and u_b, we shall follow the standard rules, and define the two *displacement functions* by:

(1) applying unit axial displacement at a with b fixed. Then,

$$\begin{pmatrix} u_a = 1 \\ u_b = 0 \end{pmatrix}, \quad \text{and} \quad u = \left(1 - \frac{x}{L}\right), \quad \text{or} \quad u = 1 - \xi, \quad \text{where} \quad \xi = \frac{x}{L} \tag{8.110}$$

(2) applying unit axial displacement at b with a fixed. Then,

$$\begin{pmatrix} u_a = 0 \\ u_b = 1 \end{pmatrix}, \quad \text{and} \quad u = \frac{x}{L} \quad \text{or} \quad u = \xi \tag{8.111}$$

These displacement functions or 'mode shapes' are shown in Fig. 8.7 (b)(i) and (b)(ii), respectively. Since they were produced by applying unit values of u_a and u_b, the axial displacement, u, at any value of ξ $(=x/L)$ can now be written in terms of u_a and u_b as:

$$u = (1 - \xi)u_a + \xi u_b \tag{8.112}$$

The end displacements u_a and u_b are now regarded as generalized coordinates in a set of Lagrange's equations:

$$\frac{d}{dt}\left(\frac{\partial T}{\partial \dot{q}_i}\right) + \frac{\partial U}{\partial q_i} = Q_i \quad i = 1, 2 \tag{8.113}$$

where in this case $q_1 = u_a$, $q_2 = u_b$, $Q_1 = F_a$ and $Q_2 = F_b$.

The following expressions will be needed to find the kinetic energy, T, and the potential energy, U.

Differentiating Eq. (8.112) with respect to time:

$$\dot{u} = (1 - \xi)\dot{u}_a + \xi \dot{u}_b \tag{8.114}$$

Also from Eq. (8.112):

$$\left(\frac{\partial u}{\partial \xi}\right)^2 = (u_a^2 - 2u_a u_b + u_b^2) \tag{8.115}$$

The kinetic energy is

$$T = \frac{1}{2}\mu \int_0^L \dot{u}^2 \cdot dx \tag{8.116}$$

but since $x = L\xi$, then $dx = Ld\xi$, and

$$T = \frac{1}{2}\mu L \int_0^1 \dot{u}^2 . d\xi \tag{8.117}$$

Substituting Eq. (8.114) into Eq. (8.117) gives

$$T = \frac{1}{2}\mu L \int_0^1 [(1 - \xi)\dot{u}_a + \xi \dot{u}_b]^2 d\xi = \frac{1}{6}\mu L \left(\dot{u}_a^2 + \dot{u}_a \dot{u}_b + \dot{u}_b^2\right) \tag{8.118}$$

The potential energy is,

$$U = \frac{1}{2}EA \int_0^L \left(\frac{\partial u}{\partial x}\right)^2 \cdot dx \tag{8.119a}$$

or since $\xi = x/L$

$$U = \frac{EA}{2L} \int_0^1 \left(\frac{\partial u}{\partial \xi}\right)^2 \cdot d\xi \tag{8.119b}$$

Substituting Eq. (8.115) into Eq. (8.119b) gives

$$U = \frac{EA}{2L} \int_0^1 (u_a^2 - 2u_a u_b + u_b^2) d\xi = \frac{EA}{2L} (u_a^2 - 2u_a u_b + u_b^2) \tag{8.120}$$

Now applying Lagrange's equations, Eq. (8.113), term by term:

$$\frac{d}{dt}\left(\frac{\partial T}{\partial \dot{u}_a}\right) = \frac{1}{6}\mu L(2\ddot{u}_a + \ddot{u}_b) \tag{8.121a}$$

$$\frac{d}{dt}\left(\frac{\partial T}{\partial \dot{u}_b}\right) = \frac{1}{6}\mu L(\ddot{u}_a + 2\ddot{u}_b) \tag{8.121b}$$

$$\frac{\partial U}{\partial u_a} = \frac{EA}{2L}(2u_a - 2u_b) \tag{8.122a}$$

$$\frac{\partial U}{\partial u_a} = \frac{EA}{2L}(-2u_a + 2u_b) \tag{8.122b}$$

The equations of motion of the complete element, in matrix form, are thus:

$$\frac{\mu L}{6}\begin{bmatrix} 2 & 1 \\ 1 & 2 \end{bmatrix}\begin{Bmatrix} \ddot{u}_a \\ \ddot{u}_b \end{Bmatrix} + \frac{EA}{L}\begin{bmatrix} 1 & -1 \\ -1 & 1 \end{bmatrix}\begin{Bmatrix} u_a \\ u_b \end{Bmatrix} = \begin{Bmatrix} F_a \\ F_b \end{Bmatrix} \tag{8.123}$$

Rod element with torsional deformation

The equations of motion for an element subjected only to torsion can be found by analogy with the axial element considered above. Figure 8.7 applies to a torsion element also if the axial displacement u is replaced by θ, the angle of twist as a function of x. Similarly, u_a and u_b are replaced by the twist angles at the ends of the element, θ_a and θ_b, and F_a and F_b are replaced by the values of torque, τ_a and τ_b.

The expressions for the element kinetic and potential energies are

$$T = \frac{1}{2}I_x \int_0^L \dot{\theta}^2 dx \tag{8.124}$$

and

$$U = \frac{1}{2}GJ \int_0^L \left(\frac{\partial \theta}{\partial x}\right)^2 dx \tag{8.125}$$

Comparing these with Eqs (8.116) and (8.119), we see that in addition to the changes above, EA must be replaced by GJ and μ by I_x, where G is the shear modulus for the material; J the polar second moment of area of the cross-section, assumed constant and I_x the moment of inertia of the rod per unit length, also assumed constant.

Making these changes, the mathematical steps for deriving the element equations of motion for the torsional element are identical to those for the axial element above, and the element equations of motion are

$$\frac{I_x L}{6}\begin{bmatrix} 2 & 1 \\ 1 & 2 \end{bmatrix}\begin{Bmatrix} \ddot{\theta}_a \\ \ddot{\theta}_b \end{Bmatrix} + \frac{GJ}{L}\begin{bmatrix} 1 & -1 \\ -1 & 1 \end{bmatrix}\begin{Bmatrix} \theta_a \\ \theta_b \end{Bmatrix} = \begin{Bmatrix} \tau_a \\ \tau_b \end{Bmatrix} \qquad (8.126)$$

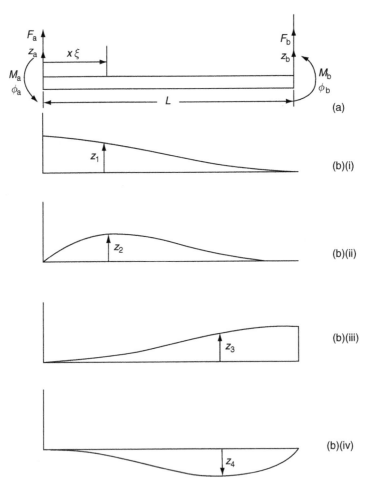

Fig. 8.8   A beam finite element.

Beam element with bending deformation

Figure 8.8, at (a), shows a uniform beam element, free to bend in the plane of the paper. The displacements, z_a, ϕ_a, z_b and ϕ_b, and corresponding forces and moments F_a, M_a, F_b and M_b, at the ends of the element are shown. The length of the beam is L; Young's modulus is E; the second moment of area of the cross-section, constant over the length, is I; the mass per unit length, constant over the length, is μ; and x is the distance along the element from the left end. As before, ξ is the non-dimensional distance along the element, x/L

Figure 8.8 at (b)(i) – b(iv) shows the four static displacement functions, also used as assumed vibration modes, for the element. These are defined by applying unit values of the end displacements, z_a, z_b, ϕ_a and ϕ_b, one at a time, with the other three fixed.

The shapes of the displacement functions can be found by applying standard beam theory, but an easier method is to note that being the static displacements of a uniform beam, with concentrated loads and moments, the displacement shapes are cubic polynomials in x. Therefore, the displacement functions must all be of the form:

$$z(x) = a_1 + a_2 \frac{x}{L} + a_3 \frac{x^2}{L^2} + a_4 \frac{x^3}{L^3} \tag{8.127}$$

or since $\xi = x/L$:

$$z(\xi) = a_1 + a_2\xi + a_3\xi^2 + a_4\xi^3 \tag{8.128}$$

where $a_1 - a_4$ are constants.

The slope at any point, $\phi(x)$, is given by differentiating Eq. (8.127) with respect to x:

$$\phi(x) = \frac{dz}{dx} = \frac{a_2}{L} + \frac{2a_3x}{L^2} + \frac{3a_4x^2}{L^3} \tag{8.129}$$

or, since $\xi = x/L$,

$$\phi(\xi) = \frac{1}{L}\left(a_2 + 2a_3\xi + 3a_4\xi^2\right) \tag{8.130}$$

For each displacement function, the constants $a_1 - a_4$ can be found from the end conditions. For the first function shown in Fig. 8.8(b)(i), these are

$$z(\xi) = z_a = 1 \atop \xi = 0 \quad , \quad \phi(\xi) = \phi_a = 0 \atop \xi = 0 \quad , \quad z(\xi) = z_b = 0 \atop \xi = 1 \quad , \quad \phi(\xi) = \phi_b = 0 \atop \xi = 1 \tag{8.131}$$

Substituting the end conditions, Eq. (8.131), into Eqs (8.128) and (8.130) gives four simultaneous equations, from which the constants, $a_1 - a_4$, for the first displacement function, z_1, can be found.

$$\begin{aligned} 1 &= a_1 \\ 0 &= a_2 \\ 0 &= a_1 + a_2 + a_3 + a_4 \\ 0 &= a_2 + 2a_3 + 3a_4 \end{aligned} \tag{8.132}$$

These are easily solved, giving $a_1 = 1$; $a_2 = 0$; $a_3 = -3$; $a_4 = 2$.

The equation of the first displacement function, z_1, is then, from Eq. (8.128):

$$z_1 = 1 - 3\xi^2 + 2\xi^3 \tag{8.133}$$

The end conditions for the second displacement function shown in Fig. 8.11(b)(ii) are

$$z(\xi) = z_a = 0 \atop \xi = 0 \; ; \quad \phi(\xi) = \phi_a = 1 \atop \xi = 0 \; ; \quad z(\xi) = z_b = 0 \atop \xi = 1 \; ; \quad \phi(\xi) = \phi_b = 0 \atop \xi = 1$$

Inserting these into Eqs (8.128) and (8.130) gives the simultaneous equations:

$$
\begin{aligned}
0 &= a_1 \\
1 &= \frac{1}{L} a_2 \\
0 &= a_1 + a_2 + a_3 + a_4 \\
0 &= \frac{1}{L}(a_2 + 2a_3 + 3a_4)
\end{aligned}
\tag{8.134}
$$

Solving these, $a_1 = 0$; $a_2 = L$; $a_3 = -2L$; $a_4 = L$. The second displacement function, from Eq. (8.128) is therefore:

$$z_2 = L\xi - 2L\xi^2 + L\xi^3 \tag{8.135}$$

In the same way, the remaining two displacement functions, those due to applying unit z_b and ϕ_b, respectively, can be shown to be

$$z_3 = 3\xi^2 - 2\xi^3 \tag{8.136}$$

and

$$z_4 = -L\xi^2 + L\xi^3 \tag{8.137}$$

The displacement, $z(\xi)$, at any point on the element, in terms of end displacements, is

$$z(\xi) = z_1 z_a + z_2 \phi_a + z_3 z_b + z_4 \phi_b \tag{8.138}$$

where z_1, z_2, z_3, and, z_4 are given by Eqs (8.133), (8.135), (8.136) and (8.137).

The corresponding velocity, $\dot{z}(\xi)$, is given by differentiating Eq. (8.138) with respect to time:

$$\dot{z}(\xi) = z_1 \dot{z}_a + z_2 \dot{\phi}_a + z_3 \dot{z}_b + z_4 \dot{\phi}_b \tag{8.139}$$

The equations of motion for the element are given, as always, by Lagrange's equations, which in this case are

$$\frac{d}{dt}\left(\frac{\partial T}{\partial \dot{q}_i}\right) + \frac{\partial U}{\partial q_i} = Q_i \quad i = 1, 2, 3, 4 \tag{8.140}$$

Because the displacement functions $z_1 - z_4$ were chosen to be different from each other, the end displacements z_a, ϕ_a, z_b and, ϕ_b are suitable for use as generalized coordinates, and the corresponding generalized 'forces' (noting that a 'generalized force' can be a moment) are F_a, M_a, F_b and M_b.

Thus,

$$
\begin{aligned}
&q_1 = z_a; && q_2 = \phi_a; && q_3 = z_b; && q_4 = \phi_b; \\
&\dot{q}_1 = \dot{z}_a; && \dot{q}_2 = \dot{\phi}_a; && \dot{q}_3 = \dot{z}_b; && \dot{q}_4 = \dot{\phi}_b; \\
&Q_1 = F_a; && Q_2 = M_a; && Q_3 = F_b; && Q_4 = M_b;
\end{aligned}
\tag{8.141}
$$

The kinetic and potential energies, T and U, are found as follows.

Using a standard result, the kinetic energy, $T = \frac{1}{2}\mu \int_0^L [\dot{z}(x)]^2 dx$, or, since $x = L\xi$, and $dx = L d\xi$:

$$
T = \frac{1}{2}\mu L \int_0^1 [\dot{z}(\xi)]^2 d\xi
\tag{8.142}
$$

The potential energy

$$
U = \frac{1}{2}EI \int_0^L \left(\frac{d^2 z}{dx^2}\right)^2 dx,
$$

or, since $dx = L d\xi$; $dx^2 = L^2 d\xi^2$ and $(dx^2)^2 = L^4 (d\xi^2)^2$:

$$
U = \frac{1}{2}\frac{EI}{L^3} \int_0^1 \left(\frac{d^2 z}{d\xi^2}\right)^2 d\xi
\tag{8.143}
$$

To derive the mass matrix for the element, substituting Eq. (8.139) into Eq. (8.142):

$$
T = \frac{1}{2}\mu L \int_0^1 \left(z_1 \dot{z}_a + z_2 \dot{\phi}_a + z_3 \dot{z}_b + z_4 \dot{\phi}_b\right)^2 d\xi
\tag{8.144}
$$

This will produce sixteen terms within the integral, i.e.,

$$
T = \frac{1}{2}\mu L \int_0^1 \left(z_1^2 \dot{z}_a^2 + z_1 z_2 \dot{z}_a \dot{\phi}_a + z_1 z_3 \dot{z}_a \dot{z}_b + z_1 z_4 \dot{z}_a \dot{\phi}_b + \cdots\right) d\xi
\tag{8.145}
$$

By substituting for z_1, z_2, z_3 and z_4 in Eq. (8.145) and evaluating $d/dt(\partial T/\partial \dot{z}_a)$, $d/dt(\partial T/\partial \dot{\phi}_a)$, $d/dt(\partial T/\partial \dot{z}_b)$ and $d/dt(\partial T/\partial \dot{\phi}_b)$, all the terms of the mass matrix can be found.

To illustrate this process for a single term, for example to find m_{12}, however, only two of the 16 terms within the integral in Eq. (8.145) will be required, i.e.,

$$
T_{12} = \frac{1}{2}\mu L \int_0^1 \left(z_1 z_2 \dot{z}_a \dot{\phi}_a + z_2 z_1 \dot{\phi}_a \dot{z}_a\right) d\xi = \mu L \int_0^1 \left(z_1 z_2 \dot{z}_a \dot{\phi}_a\right) d\xi
\tag{8.146}
$$

where T_{12} is that part of the kinetic energy contributing to the m_{12} term. The following steps will then produce the required m_{12} term.

From Eqs (8.146), (8.133) and (8.135):

$$
T_{12} = \mu L \int_0^1 \left[(1 - 3\xi^2 + 2\xi^3)(L\xi - 2L\xi^2 + L\xi^3) d\xi\right] \dot{z}_a \dot{\phi}_a
\tag{8.147}
$$

$$
T_{12} = \mu L \left(\frac{1}{2}L - \frac{2}{3}L - \frac{1}{2}L + \frac{8}{5}L - \frac{7}{6}L + \frac{2}{7}L\right) \dot{z}_a \dot{\phi}_a = \frac{\mu L}{420}(22L) \dot{z}_a \dot{\phi}_a
\tag{8.148}
$$

$$\frac{\mathrm{d}}{\mathrm{d}t}\left(\frac{\partial T_{12}}{\partial \dot{z}_a}\right) = \frac{\mu L}{420}(22L)\ddot{\phi}_a = m_{12}\ddot{\phi}_a \qquad (8.149)$$

Therefore:

$$m_{12} = \frac{\mu L}{420}(22L) \qquad (8.150)$$

The remaining terms can be found in the same way, and the mass matrix for the complete element is

$$[m] = \frac{\mu L}{420}\begin{bmatrix} 156 & 22L & 54 & -13L \\ 22L & 4L^2 & 13L & -3L^2 \\ 54 & 13L & 156 & -22L \\ -13L & -3L^2 & -22L & 4L^2 \end{bmatrix} \qquad (8.151)$$

which is a well-known standard result.

The stiffness matrix for the element can be found in a similar way, by applying the second term in Eq. (8.140), i.e. $\partial U/\partial q_i$, where q_i is successively: z_a, ϕ_a, z_b and ϕ_b. The expression for the potential energy, U, is Eq. (8.143), and this requires the curvature, $\mathrm{d}^2z/\mathrm{d}\xi^2$, of the element. This is given by differentiating Eq. (8.138) twice, with respect to ξ, thus,

$$\frac{\mathrm{d}^2z}{\mathrm{d}\xi^2}(\xi) = z'' = z_1''z_a + z_2''\phi_a + z_3''z_b + z_4''\phi_b \qquad (8.152)$$

where the notation, z'', for example, is used to indicate the second derivative of z with respect to ξ. So the displacement functions z_1-z_4, from Eqs (8.133), (8.135), (8.136) and (8.137), must now be differentiated twice, giving their second derivatives, $z_1''-z_4''$, as follows:

$$\begin{aligned} z_1 &= 1 - 3\xi^2 + 2\xi^3 & z_1'' &= -6 + 12\xi \\ z_2 &= L\xi - 2L\xi^2 + L\xi^3 & z_2'' &= -4L + 6L\xi \\ z_3 &= 3\xi^2 - 2\xi^3 & z_3'' &= 6 - 12\xi \\ z_4 &= -L\xi^2 + L\xi^3 & z_4'' &= -2L + 6L\xi \end{aligned} \qquad (8.153)$$

Substituting Eq. (8.152) into Eq. (8.143) gives

$$U = \frac{1}{2}\frac{EI}{L^3}\int_0^1 (z_1''z_a + z_2''\phi_a + z_3''z_b + z_4''\phi_b)^2\mathrm{d}\xi$$

$$U = \frac{1}{2}\frac{EI}{L^3}\int_0^1 (z_1''2z_a^2 + z_1''z_2''z_a\phi_a + z_1''z_3''z_az_b + z_1''z_4''z_a\phi_b + \cdots)\mathrm{d}\xi \qquad (8.154)$$

with 16 product terms within the integral, only a few being shown.

Substituting the expressions for $z_1''-z_4''$ from Eq. (8.153) into Eq. (8.154), carrying out the integration, and then taking the partial derivatives $\partial U/\partial z_a$, $\partial U/\partial \phi_a$, etc. gives the stiffness matrix.

To illustrate the process for a single term in the stiffness matrix, for example to find k_{12}, only two of the 16 products in Eq. (8.154) are required. Then if U_{12} is that part of the potential energy contributing to the k_{12} term:

$$
\begin{aligned}
U_{12} &= \frac{1}{2}\frac{EI}{L^3}\int_0^1 \left(z_1''z_2''z_a\phi_a + z_2''z_1''\phi_a z_a\right)\mathrm{d}\xi = \frac{EI}{L^3}\int_0^1 \left(z_1''z_2''\mathrm{d}\xi\right)z_a\phi_a \\
&= \frac{EI}{L^3}\int_0^1 [(-6+12\xi)(-4L+6L\xi)\mathrm{d}\xi]z_a\phi_a = \frac{EI}{L^3}(6L)z_a\phi_a
\end{aligned}
\tag{8.155}
$$

Then the k_{12} term is given by:

$$
\frac{\partial U_{12}}{\partial z_a} = \frac{EI}{L^3}(6L)\phi_a = k_{12}\phi_a
\tag{8.156}
$$

and

$$
k_{12} = \frac{EI}{L^3}(6L)
\tag{8.157}
$$

The remaining terms can be found in the same way, and the complete stiffness matrix for the beam element, a standard result, is

$$
[k] = \frac{EI}{L^3}
\begin{bmatrix}
12 & 6L & -12 & 6L \\
6L & 4L^2 & -6L & 2L^2 \\
-12 & -6L & 12 & -6L \\
6L & 2L^2 & -6L & 4L^2
\end{bmatrix}
\tag{8.158}
$$

The complete equations of motion for the beam element are

$$
\frac{\mu L}{420}
\begin{bmatrix}
156 & 22L & 54 & -13L \\
22L & 4L^2 & 13L & -3L^2 \\
54 & 13L & 156 & -22L \\
-13L & -3L^2 & -22L & 4L^2
\end{bmatrix}
\begin{Bmatrix}
\ddot{z}_a \\
\ddot{\phi}_a \\
\ddot{z}_b \\
\ddot{\phi}_b
\end{Bmatrix}
$$

$$
+ \frac{EI}{L^3}
\begin{bmatrix}
12 & 6L & -12 & 6L \\
6L & 4L^2 & -6L & 2L^2 \\
-12 & -6L & 12 & -6L \\
6L & 2L^2 & -6L & 4L^2
\end{bmatrix}
\begin{Bmatrix}
z_a \\
\phi_a \\
z_b \\
\phi_b
\end{Bmatrix}
=
\begin{Bmatrix}
F_a \\
M_a \\
F_b \\
M_b
\end{Bmatrix}
\tag{8.159}
$$

Example 8.4

Use a finite element model to find the first two natural frequencies and normal modes of the uniform cantilever beam shown in Fig. 8.9(a), and compare the result with the exact answer. Divide the beam into three bending elements each of length $L=1$ meter. Take Young's modulus $= E$; second moment of area of the cross-section $= I$; mass per unit length $= \mu$. The axial displacement of the beam may be assumed to be negligible.

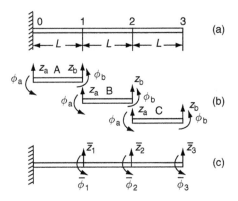

Fig. 8.9   Finite element model used in Example 8.4.

Solution

The equations of motion, in local axes, for each of the unconnected elements, A, B, and C, as shown in Fig. 8.9(b), are given by Eq. (8.159).

Since the global axes are parallel to the local axes, a transformation matrix is not required, and the relationship between global and local displacement coordinates is

$$
\bar{z}_0 = z_a^A; \quad \bar{z}_1 = z_b^A = z_a^B; \quad \bar{z}_2 = z_b^B = z_a^C; \quad \bar{z}_3 = z_b^C;
$$
$$
\bar{\phi}_0 = \phi_a^A; \quad \bar{\phi}_1 = \phi_b^A = \phi_a^B; \quad \bar{\phi}_2 = \phi_b^B = \phi_a^C; \quad \bar{\phi}_3 = \phi_b^C
$$

(A)

where the bars indicate global coordinates. Superscripts A, B, C indicate the element, and subscripts a, b indicate the end (i.e. left or right) in Fig. 8.9(b).

The equations for the three elements, joined at nodes 1 and 2, but still free at the left and right ends, nodes 0 and 3, are formed by superimposing the terms of the mass and stiffness matrices at the joints. With $L = 1$ for all three elements, the mass and stiffness matrices, $[\bar{m}]$ and $[\bar{k}]$, are then given by:

$$
\frac{420}{\mu}[\bar{m}] =
\begin{bmatrix}
156 & 22 & 54 & -13 & 0 & 0 & 0 & 0 \\
22 & 4 & 13 & -3 & 0 & 0 & 0 & 0 \\
54 & 13 & 156 & -22 & 0 & 0 & 0 & 0 \\
-13 & -3 & -22 & 4 & 0 & 0 & 0 & 0 \\
0 & 0 & 0 & 0 & 0 & 0 & 0 & 0 \\
0 & 0 & 0 & 0 & 0 & 0 & 0 & 0 \\
0 & 0 & 0 & 0 & 0 & 0 & 0 & 0 \\
0 & 0 & 0 & 0 & 0 & 0 & 0 & 0
\end{bmatrix}
+
\begin{bmatrix}
0 & 0 & 0 & 0 & 0 & 0 & 0 & 0 \\
0 & 0 & 0 & 0 & 0 & 0 & 0 & 0 \\
0 & 0 & 156 & 22 & 54 & -13 & 0 & 0 \\
0 & 0 & 22 & 4 & 13 & -3 & 0 & 0 \\
0 & 0 & 54 & 13 & 156 & -22 & 0 & 0 \\
0 & 0 & -13 & -3 & -22 & 4 & 0 & 0 \\
0 & 0 & 0 & 0 & 0 & 0 & 0 & 0 \\
0 & 0 & 0 & 0 & 0 & 0 & 0 & 0
\end{bmatrix}
+
$$

$$
\begin{bmatrix}
0 & 0 & 0 & 0 & 0 & 0 & 0 & 0 \\
0 & 0 & 0 & 0 & 0 & 0 & 0 & 0 \\
0 & 0 & 0 & 0 & 0 & 0 & 0 & 0 \\
0 & 0 & 0 & 0 & 0 & 0 & 0 & 0 \\
0 & 0 & 0 & 0 & 156 & 22 & 54 & -13 \\
0 & 0 & 0 & 0 & 22 & 4 & -3 & -3 \\
0 & 0 & 0 & 0 & 54 & 13 & 156 & -22 \\
0 & 0 & 0 & 0 & -13 & -3 & -22 & 4
\end{bmatrix}
=
\begin{bmatrix}
156 & 22 & 54 & -13 & 0 & 0 & 0 & 0 \\
22 & 4 & 13 & -3 & 0 & 0 & 0 & 0 \\
54 & 13 & 312 & 0 & 54 & -13 & 0 & 0 \\
-13 & -3 & 0 & 8 & 13 & -3 & 0 & 0 \\
0 & 0 & 54 & 13 & 312 & 0 & 54 & -13 \\
0 & 0 & -13 & -3 & 0 & 8 & 13 & -3 \\
0 & 0 & 0 & 0 & 54 & 13 & 156 & -22 \\
0 & 0 & 0 & 0 & -13 & -3 & -22 & 4
\end{bmatrix}
$$

(B)

$$\frac{1}{EI}[\bar{k}] = \begin{bmatrix} 12 & 6 & -12 & 6 & 0 & 0 & 0 & 0 \\ 6 & 4 & -6 & 2 & 0 & 0 & 0 & 0 \\ -12 & -6 & 12 & -6 & 0 & 0 & 0 & 0 \\ 6 & 2 & -6 & 4 & 0 & 0 & 0 & 0 \\ 0 & 0 & 0 & 0 & 0 & 0 & 0 & 0 \\ 0 & 0 & 0 & 0 & 0 & 0 & 0 & 0 \\ 0 & 0 & 0 & 0 & 0 & 0 & 0 & 0 \\ 0 & 0 & 0 & 0 & 0 & 0 & 0 & 0 \end{bmatrix} + \begin{bmatrix} 0 & 0 & 0 & 0 & 0 & 0 & 0 & 0 \\ 0 & 0 & 0 & 0 & 0 & 0 & 0 & 0 \\ 0 & 0 & 12 & 6 & -12 & 6 & 0 & 0 \\ 0 & 0 & 6 & 4 & -6 & 2 & 0 & 0 \\ 0 & 0 & -12 & -6 & 12 & -6 & 0 & 0 \\ 0 & 0 & 6 & 2 & -6 & 4 & 0 & 0 \\ 0 & 0 & 0 & 0 & 0 & 0 & 0 & 0 \\ 0 & 0 & 0 & 0 & 0 & 0 & 0 & 0 \end{bmatrix} +$$

$$\begin{bmatrix} 0 & 0 & 0 & 0 & 0 & 0 & 0 & 0 \\ 0 & 0 & 0 & 0 & 0 & 0 & 0 & 0 \\ 0 & 0 & 0 & 0 & 0 & 0 & 0 & 0 \\ 0 & 0 & 0 & 0 & 0 & 0 & 0 & 0 \\ 0 & 0 & 0 & 0 & 12 & 6 & -12 & 6 \\ 0 & 0 & 0 & 0 & 6 & 4 & -6 & 2 \\ 0 & 0 & 0 & 0 & -12 & -6 & 12 & -6 \\ 0 & 0 & 0 & 0 & 6 & 2 & -6 & 4 \end{bmatrix} = \begin{bmatrix} 12 & 6 & -12 & 6 & 0 & 0 & 0 & 0 \\ 6 & 4 & -6 & 2 & 0 & 0 & 0 & 0 \\ -12 & -6 & 24 & 0 & -12 & 6 & 0 & 0 \\ 6 & 2 & 0 & 8 & -6 & 2 & 0 & 0 \\ 0 & 0 & -12 & -6 & 24 & 0 & -12 & 6 \\ 0 & 0 & 6 & 2 & 0 & 8 & -6 & 2 \\ 0 & 0 & 0 & 0 & -12 & -6 & 12 & -6 \\ 0 & 0 & 0 & 0 & 6 & 2 & -6 & 4 \end{bmatrix}$$

(C)

Since we require the beam be a cantilever, fixed at node 0, then the displacements $\bar{z}_0$ and $\bar{\phi}_0$, and the corresponding accelerations, at that node, are zero. The first two columns of the mass and stiffness matrices above can therefore be eliminated, since they would always be multiplied by zero. If, as in this case, the force and moment at the fixed end are not required, the first two rows of the mass and stiffness matrices can also be eliminated. The equations of motion of the uniform, cantilevered beam, in global coordinates, are then:

$$\frac{\mu}{420} \begin{bmatrix} 312 & 0 & 54 & -13 & 0 & 0 \\ 0 & 8 & 13 & -3 & 0 & 0 \\ 54 & 13 & 312 & 0 & 54 & -13 \\ -13 & -3 & 0 & 8 & 13 & -3 \\ 0 & 0 & 54 & 13 & 156 & -22 \\ 0 & 0 & -13 & -3 & -22 & 4 \end{bmatrix} \begin{Bmatrix} \ddot{\bar{z}}_1 \\ \ddot{\bar{\phi}}_1 \\ \ddot{\bar{z}}_2 \\ \ddot{\bar{\phi}}_2 \\ \ddot{\bar{z}}_3 \\ \ddot{\bar{\phi}}_3 \end{Bmatrix}$$

$$+ EI \begin{bmatrix} 24 & 0 & -12 & 6 & 0 & 0 \\ 0 & 8 & -6 & 2 & 0 & 0 \\ -12 & -6 & 24 & 0 & -12 & 6 \\ 6 & 2 & 0 & 8 & -6 & 2 \\ 0 & 0 & -12 & -6 & 12 & -6 \\ 0 & 0 & 6 & 2 & -6 & 4 \end{bmatrix} \begin{Bmatrix} \bar{z}_1 \\ \bar{\phi}_1 \\ \bar{z}_2 \\ \bar{\phi}_2 \\ \bar{z}_3 \\ \bar{\phi}_3 \end{Bmatrix} = \begin{Bmatrix} \bar{F}_1 \\ \bar{M}_2 \\ \bar{F}_2 \\ \bar{M}_2 \\ \bar{F}_3 \\ \bar{M}_3 \end{Bmatrix}$$

(D)

or in shorter form:

$$[\bar{m}]\{\ddot{\bar{z}}\} + [\bar{k}]\{\bar{z}\} = \{\bar{F}\}$$

(E)

In this case, only the natural frequencies and mode shapes of the undamped system are required, so the external forces and moments are set to zero, and the usual substitution $\{z\} = e^{i\omega t}\{\bar{z}\}$ is made, leading to:

$$\left([\bar{k}] - \omega^2[\bar{m}]\right)\{\bar{z}\} = 0 \tag{F}$$

Using Eq. (D), Eq. (F) can be written as:

$$\left(\begin{bmatrix} 24 & 0 & -12 & 6 & 0 & 0 \\ 0 & 8 & -6 & 2 & 0 & 0 \\ -12 & -6 & 24 & 0 & -12 & 6 \\ 6 & 2 & 0 & 8 & -6 & 2 \\ 0 & 0 & -12 & -6 & 12 & -6 \\ 0 & 0 & 6 & 2 & -6 & 4 \end{bmatrix} - \lambda \begin{bmatrix} 312 & 0 & 54 & -13 & 0 & 0 \\ 0 & 8 & 13 & -3 & 0 & 0 \\ 54 & 13 & 312 & 0 & 54 & -13 \\ -13 & -3 & 0 & 8 & 13 & -3 \\ 0 & 0 & 54 & 13 & 156 & -22 \\ 0 & 0 & -13 & -3 & -22 & 4 \end{bmatrix}\right) \begin{Bmatrix} \bar{z}_1 \\ \bar{\phi}_1 \\ \bar{z}_2 \\ \bar{\phi}_2 \\ \bar{z}_3 \\ \bar{\phi}_3 \end{Bmatrix} = 0 \tag{G}$$

where

$$\lambda = \frac{\mu\omega^2}{420EI} \tag{H}$$

Using standard software gives the first two eigenvalues and eigenvectors, as follows:

$$\lambda_1 = 0.3694 \times 10^{-3} \qquad\qquad \lambda_2 = 0.01437$$

$$\begin{Bmatrix} \bar{z}_1 \\ \bar{\phi}_1 \\ \bar{z}_2 \\ \bar{\phi}_2 \\ \bar{z}_3 \\ \bar{\phi}_3 \end{Bmatrix}_1 = \begin{Bmatrix} 0.1664 \\ 0.3118 \\ 0.5472 \\ 0.4439 \\ 1.0000 \\ 0.4470 \end{Bmatrix} \qquad \begin{Bmatrix} \bar{z}_1 \\ \bar{\phi}_1 \\ \bar{z}_2 \\ \bar{\phi}_2 \\ \bar{z}_3 \\ \bar{\phi}_3 \end{Bmatrix}_2 = \begin{Bmatrix} -0.5895 \\ -0.5960 \\ -0.4219 \\ 0.9785 \\ 1.0000 \\ 1.6031 \end{Bmatrix} \tag{I}$$

where the eigenvectors have been scaled to make the tip displacement $\bar{z}_3$ equal to unity in each case.

The natural frequencies are given by Eq. (H):

$$\omega_i = \sqrt{420\lambda_i}\sqrt{\frac{EI}{\mu}} \tag{J}$$

For the first mode, $\lambda_1 = 0.3694 \times 10^{-3}$, and

$$\omega_1 = 0.3939\sqrt{\frac{EI}{\mu}} \tag{K_1}$$

For the second mode,

$$\lambda_2 = 0.01437 \quad \text{and} \quad \omega_2 = 2.456\sqrt{\frac{EI}{\mu}} \tag{K_2}$$

For comparison, the exact natural frequencies of a uniform cantilever are given by Table 8.1, as follows. From the table:

$$\beta_1 L = 1.87510 \quad \beta_2 L = 4.69404 \quad \omega_i = \beta_i^2\sqrt{\frac{EI}{\mu}}$$

Noting that $L = 3\,\text{m}$ in this example,

$$\omega_1 = \beta_1^2\sqrt{\frac{EI}{\mu}} = \left(\frac{1.87510}{3}\right)^2\sqrt{\frac{EI}{\mu}} = 0.39066\sqrt{\frac{EI}{\mu}} \tag{L_1}$$

$$\omega_2 = \beta_2^2\sqrt{\frac{EI}{\mu}} = \left(\frac{4.69409}{3}\right)^2\sqrt{\frac{EI}{\mu}} = 2.4482\sqrt{\frac{EI}{\mu}} \tag{L_2}$$

It can be seen that the first and second natural frequencies from the FE model are about 0.8 and 0.4% higher, respectively, than the exact results, in this case.

The global displacements in each mode are compared with the exact mode shapes of the uniform beam in Fig. 8.10. In both cases the displacements are given as a fraction of the tip displacement, z/z_{TIP}, and the spanwise positions are given as a fraction of the total span. The differences are too small to show at the scale used, being all less than 0.5%.

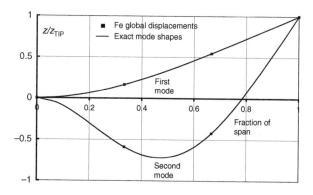

Fig. 8.10   First two modes of a uniform cantilever calculated by a 3-element FE model.

8.5 Symmetrical structures

If a structure has a plane of symmetry, and the normal modes are calculated, or measured, some will be symmetric, and some antisymmetric about that plane. Figure 8.11 represents a structure with left-right symmetry, for example an aircraft. Two symmetrically placed grid points are shown, one on the left and the other on the right, each with three linear displacements and three angular displacements, the directions shown being positive in both cases.

For the symmetric modes, the following is true for all such symmetric pairs of grid points:

$$\begin{Bmatrix} x_L \\ y_L \\ z_L \\ \phi_L \\ \theta_L \\ \psi_L \end{Bmatrix} = \begin{Bmatrix} x_R \\ -y_R \\ z_R \\ -\phi_R \\ \theta_R \\ -\psi_R \end{Bmatrix} \tag{8.160}$$

where suffix L indicates a displacement on the left side and R on the right side.

For the antisymmetric modes the following is true:

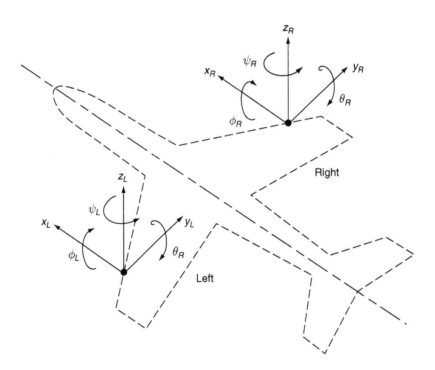

Fig. 8.11

$$
\begin{Bmatrix} x_L \\ y_L \\ z_L \\ \phi_L \\ \theta_L \\ \psi_L \end{Bmatrix} = \begin{Bmatrix} -x_R \\ y_R \\ -z_R \\ \phi_R \\ -\theta_R \\ \psi_R \end{Bmatrix} \tag{8.161}
$$

It is a good idea to sketch these displacements for a few simple structures.

As a result, the symmetric and antisymmetric modes are completely uncoupled, and can be calculated separately, giving two eigenvalue problems of half the size as for the complete system. For each calculation, only half the structure is modeled, with appropriate constraints at the boundary, i.e. the plane of symmetry. This idea is nearly always used when analysing aircraft structures, irrespective of the method used.

The fact that each mode is either symmetric or antisymmetric also simplifies the modal testing of symmetrical aircraft structures (see Chapter 13).

References

8.1 Scanlan, RH and Rosenbaum, R. (1951). *Introduction to the Study of Aircraft Vibration and Flutter.* Macmillan (Reprinted 1968, Dover Publications).

8.2 Hurty, WC (1960). Vibrations of structural systems by component mode synthesis. *Proc Am. Soc. Civil Engrs., J. Eng. Mech. Div.* 86, 51–69.

8.3 Hurty, WC (1965). Dynamic analysis of structural systems using component modes. *AIAA J.* 3, 678–85.

8.4 Craig, RR and Bampton, MCC (1968). Coupling of sub-structures for dynamic analysis *AIAA J.* 6, 1313–19.

8.5 Craig, RR and Chang, CJ (1976). *A review of substructure coupling methods for dynamic analysis.* NASA CP 2001.

8.6 Hunn, BA (1953). A method of calculating space free resonant modes of an aircraft *J. Roy. Aero Soc.* 57, 420.

8.7 Gladwell, GML (1964). Branch mode analysis of vibrating systems. *J. Sound Vib. 1,* 41–59.

8.8 Petyt, M (1990). *Introduction to Finite Element Vibration Analysis.* Cambridge University Press.

8.9 Craig, RR (1981). *Structural Dynamics: An Introduction to Computer Methods.* Wiley.

9 Fourier Transformation and Related Topics

Contents

In this chapter, we introduce some of the basic tools that will be required for the practical analysis of vibration data, preparatory to the discussion of random vibration in Chapter 10. The classical *Fourier series* is first introduced, looking briefly at how the *Fourier integral* and the *Fourier transform* are related to it. We then consider the *discrete Fourier transform*, or *DFT*, the 'computer-friendly' version of the Fourier series, and the basis of most modern digital vibration analysis. The use of digital signal processing brings with it the possibility of serious errors due to the phenomenon of *aliasing*, which is introduced next. Finally, the response of systems to periodic excitation is discussed.

9.1 The Fourier series and its developments

9.1.1 Fourier Series

Figure 9.1 at (a) represents a periodic waveform, $x(t)$, which repeats exactly every T seconds. It was postulated by J. Fourier that this can be represented, subject to some minor exceptions in the case of discontinuities, by a series of the following kind, extending, theoretically, to infinity:

$$
\begin{aligned}
x(t) = a_0 &+ a_1 \cos \omega_0 t + a_2 \cos 2\omega_0 t + a_3 \cos 3\omega_0 t + \cdots \\
&+ b_1 \sin \omega_0 t + b_2 \sin 2\omega_0 t + b_3 \sin 3\omega_0 t + \cdots
\end{aligned}
\tag{9.1}
$$

where

$$
\omega_0 = 2\pi f_0 = \frac{2\pi}{T} \quad \text{and} \quad f_0 = \frac{1}{T}
\tag{9.2}
$$

The frequency ω_0, in rad/s (or f_0 in Hz), is the fundamental frequency of the harmonic series represented by Eq. (9.1), and all the other frequencies are integer multiples of this frequency.

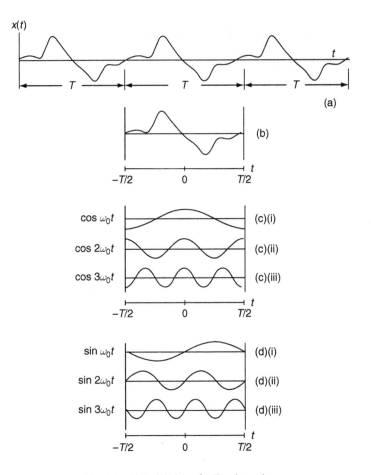

Fig. 9.1   Pictorial view of a Fourier series.

Equation (9.1) can be expressed in the more compact form:

$$x(t) = a_0 + \sum_{n=1}^{\infty} [a_n \cos n\omega_0 t + b_n \sin n\omega_0 t] \quad n = 1, 2, 3, \ldots, \infty \tag{9.3}$$

The constant a_0 is easily found, since it is the average value of $x(t)$ over any complete period, say the period shown in Fig. 9.1 at (b), which extends from $t = -T/2$ to $t = T/2$. Thus, a_0 is given by:

$$a_0 = \frac{1}{T} \int_{-T/2}^{T/2} x(t) \mathrm{d}t \tag{9.4}$$

To find the constants a_n and b_n, for values of n other than zero, both sides of Eq. (9.1), omitting a_0, are multiplied, in turn, by $\cos n\omega_0 t$ and $\sin n\omega_0 t$, and integrated with respect to t, over the same complete period, from $-T/2$ to $T/2$. So in order to find the constants, a_n (not b_n at this stage), we write the equation:

$$\int_{-T/2}^{T/2} [x(t)(\cos n\omega_0 t)]dt$$

$$= \int_{-T/2}^{T/2} (\cos n\omega_0 t)\begin{pmatrix} a_1 \cos \omega_0 t + a_2 \cos 2\omega_0 t + a_3 \cos 3\omega_0 t + \cdots \\ +b_1 \sin \omega_0 t + b_2 \sin 2\omega_0 t + b_3 \sin 3\omega_0 t + \cdots \end{pmatrix}dt \qquad (9.5)$$

$$(n = 1, 2, 3, \ldots, \infty)$$

We now notice that on the right side of Eq. (9.5):
(1) all products of cosine and sine will be zero after integration.
(2) products of two cosines will be zero after integration, except when they are at the same frequency.

It can now be seen why all the frequencies in the series were chosen to be integer multiples of ω_0: all the integrations are then over a whole number of cycles, producing convenient zero values for most of the terms. The one product that does not integrate to zero is $(a_n \cos n\omega_0 t)(\cos n\omega_0 t) = a_n \cos^2 n\omega_0 t$, where n has the same value throughout, and there is one such product for each value of n from 1 to ∞. Equation (9.5) therefore simplifies to:

$$\int_{-T/2}^{T/2} [x(t)(\cos n\omega_0 t)]dt = \int_{-T/2}^{T/2} (a_n \cos^2 n\omega_0 t)dt \quad (n = 1, 2, 3, \ldots, \infty) \qquad (9.6)$$

Since $\int_{-T/2}^{T/2} (a_n \cos^2 n\omega_0 t)dt = a_n T/2$ for any value of the integer n, Eq. (9.6) becomes

$$\int_{-T/2}^{T/2} [x(t)(\cos n\omega_0 t)]dt = a_n \frac{T}{2} \qquad (9.7)$$

Rearranging,

$$a_n = \frac{2}{T}\int_{-T/2}^{T/2} x(t)\cos n\omega_0 t \cdot dt \quad (n = 1, 2, 3, \ldots, \infty) \qquad (9.8)$$

Similarly, to find the coefficients b_n, we multiply both sides of Eq. (9.1), omitting a_0, by $\sin n\omega_0 t$ instead of $\cos n\omega_0 t$, giving

$$\int_{-T/2}^{T/2} [x(t)(\sin n\omega_0 t)]dt$$

$$= \int_{-T/2}^{T/2} (\sin n\omega_0 t)\begin{pmatrix} a_1 \cos \omega_0 t + a_2 \cos 2\omega_0 t + a_3 \cos 3\omega_0 t + \cdots \\ +b_1 \sin \omega_0 t + b_2 \sin 2\omega_0 t + b_3 \sin 3\omega_0 t + \cdots \end{pmatrix}dt \qquad (9.9)$$

$$(n = 1, 2, 3, \ldots, \infty)$$

All the resulting products on the right side of Eq. (9.9) will integrate to zero, except for $(\sin n\omega_0 t)(b_n \sin n\omega_0 t) = b_n \sin^2 n\omega_0 t$. This integrates to $b_n(T/2)$, so the constants b_n are given by:

$$b_n = \frac{2}{T} \int_{-T/2}^{T/2} x(t) \sin n\omega_0 t \cdot \mathrm{d}t \quad (n = 1, 2, 3, \ldots, \infty) \tag{9.10}$$

The process described above is shown pictorially in Fig. 9.1. The single period to be analysed is shown at (b). The first three cosine functions are shown at (c)(i), (c)(ii) and (c)(iii), and the first three sine functions at (d)(i), (d)(ii) and (d)(iii). The coefficient a_1 is twice the average of the product of (b) and (c)(i); the coefficient a_2 is twice the average of the product of (b) and (c)(ii) and so on. The coefficients b_1, b_2 etc. are similarly given by taking twice the average product of (b) and (d)(i); (b) and (d)(ii) and so on.

Example 9.1

(a) Derive a Fourier series to represent the voltage waveform shown in Fig. 9.2(a), a square wave with amplitude $\pm 1\mathrm{V}$, and period T seconds, by representing one period as an even function.
(b) Repeat (a) using an odd function to represent one period.
(c) Compare the results.
(d) If the period of the square wave is 1 second, plot the sums of each of the two series derived in (a) and (b) above, against time, t, showing that the original square wave is reproduced approximately.

Solution

Part (a)
Figure 9.2(b) shows one complete period, of duration T, chosen to be an *even function*, defined as a function where $x(t) = x(-t)$. For analysis purposes this period will be assumed to extend from $t = -T/2$ to $t = T/2$. Since the waveform has zero mean value, $a_0 = 0$, in this case.

The coefficients a_n $(n = 1, 2, 3, \ldots, \infty)$ are given by Eq. (9.8), which, using Eq. (9.2), $\omega_0 = (2\pi/T)$, can be written as:

$$a_n = \frac{2}{T} \int_{-T/2}^{T/2} x(t) \cos n\omega_0 t \cdot \mathrm{d}t = \frac{2}{T} \int_{-T/2}^{T/2} x(t) \cos n\left(\frac{2\pi}{T}\right)t \cdot \mathrm{d}t \tag{A}$$

The coefficients b_n $(n = 1, 2, 3, \ldots, \infty)$ are given by Eq. (9.10), which can similarly be written as:

$$b_n = \frac{2}{T} \int_{-T/2}^{T/2} x(t) \sin n\omega_0 t \cdot \mathrm{d}t = \frac{2}{T} \int_{-T/2}^{T/2} x(t) \sin n\left(\frac{2\pi}{T}\right)t \cdot \mathrm{d}t \tag{B}$$

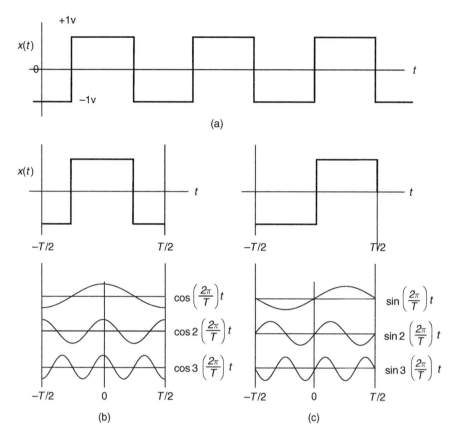

Fig. 9.2   Derivation of a Fourier series for a square wave.

It can be seen from Fig. 9.2 that the values of a_n, given by Eq. (A), are zero for even values of n, and the values of b_n, given by Eq. (B), are all zero. Therefore, the complete Fourier series consists only of cosine terms with odd values of n:

$$x(t) = a_1 \cos\left(\frac{2\pi}{T}\right)t + a_3 \cos3\left(\frac{2\pi}{T}\right)t + a_5 \cos5\left(\frac{2\pi}{T}\right)t + a_7 \cos7\left(\frac{2\pi}{T}\right)t + \cdots \qquad (C)$$

From Eq. (A), the numerical value of a_1 is given by:

$$a_1 = \frac{2}{T}\int_{-T/2}^{T/2} x(t)\cos\left(\frac{2\pi}{T}\right)t \cdot dt = 4\cdot\frac{2}{T}\int_{0}^{T/4} \cos\left(\frac{2\pi}{T}\right)t \cdot dt = 4\cdot\frac{2}{T}\frac{T}{2\pi}\left[\sin\left(\frac{2\pi}{T}\right)t\right]_{0}^{T/4} = \frac{4}{\pi}$$
$$(D)$$

In the same way, it can be shown that

$$a_3 = -\frac{1}{3}\cdot\left(\frac{4}{\pi}\right); \quad a_5 = \frac{1}{5}\cdot\left(\frac{4}{\pi}\right); \quad a_7 = -\frac{1}{7}\cdot\left(\frac{4}{\pi}\right) + \cdots$$

and Eq. (C) becomes

$$x(t) = \frac{4}{\pi}\left[\cos\left(\frac{2\pi}{T}\right)t - \frac{1}{3}\cos 3\left(\frac{2\pi}{T}\right)t + \frac{1}{5}\cos 5\left(\frac{2\pi}{T}\right)t - \frac{1}{7}\cos 7\left(\frac{2\pi}{T}\right)t + \cdots\right] \quad \text{(E)}$$

noting that the terms are alternately positive and negative.

Part (b)

Alternatively, the single period to be analysed can be taken as the *odd function* shown in Fig. 9.2 at (c), an odd function being defined mathematically as one where $x(t) = -x(-t)$. It can be seen that, in this case, all values of a_n are zero, as are values of b_n, when n is an even number. The complete series therefore consists only of sine terms with odd values of n, and from Eq. (B):

$$b_n = \frac{2}{T}\int_{-T/2}^{T/2} x(t)\sin n\left(\frac{2\pi}{T}\right)t \cdot dt \quad (n = 1, 3, 5, 7, \ldots, \infty) \quad \text{(F)}$$

Numerically:

$$b_1 = \frac{2}{T}\int_{-T/2}^{T/2} x(t)\sin\left(\frac{2\pi}{T}\right)t \cdot dt = 2 \cdot \frac{2}{T}\int_{0}^{T/2} \sin\left(\frac{2\pi}{T}\right)dt$$

$$= 2 \cdot \frac{2}{T} \cdot \frac{T}{2\pi}\left[-\cos\left(\frac{2\pi}{T}\right)t\right]_{0}^{T/2} = \frac{4}{\pi}. \quad \text{(G)}$$

Similarly,

$$b_3 = \frac{1}{3}\cdot\left(\frac{4}{\pi}\right); \quad b_5 = \frac{1}{5}\cdot\left(\frac{4}{\pi}\right); \quad b_7 = \frac{1}{7}\cdot\left(\frac{4}{\pi}\right); \quad \cdots$$

The series is therefore:

$$x(t) = \frac{4}{\pi}\left[\sin\left(\frac{2\pi}{T}\right)t + \frac{1}{3}\sin 3\left(\frac{2\pi}{T}\right)t + \frac{1}{5}\sin 5\left(\frac{2\pi}{T}\right)t + \frac{1}{7}\sin 7\left(\frac{2\pi}{T}\right)t + \cdots\right] \quad \text{(H)}$$

noting that, in this case, the terms are all positive.

Part (c)

Comparing the two series in Eqs (E) and (H), it can be seen that the amplitude components are the same, but the phases are apparently different. This is due to the fact that the period chosen as an even function, shown in Fig. 9.2 at (b), leads that chosen as an odd function, in Fig. 9.2 at (c), by $T/4$ seconds, which is equivalent to a phase shift of $\pi/2$ radians at a frequency of $(2\pi/T)$ rad./s; or $(3\pi/2)$ radians at frequency $3(2\pi/T)$ rad/s and so on. Thus the magnitudes of the terms in the two series are identical, but their apparent phases naturally depend upon the arbitrary point in the waveform chosen as $t = 0$.

Part (d)

Plotting $x(t)$, as given by either Eq. (E) or Eq. (H), versus t, should reproduce the original waveform. Taking T equal to 1 second, Fig. 9.3(a) is a plot of Eq. (E), using only the first four terms of the series. Figure 9.3(b) is a similar plot, from Eq. (H), also

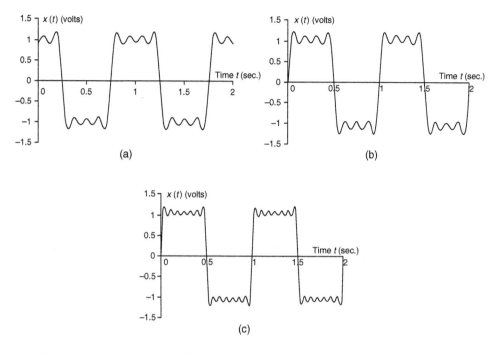

Fig. 9.3   (a) Square wave reconstituted from first four cosine terms, (b) square wave reconstituted from first four sine terms, (c) square wave reconstituted from first seven sine terms.

using only the first four terms, and it can be seen that the the waveforms reproduced are identical, apart from the shift of $T/4$ seconds horizontally, equal to 0.25 s. in this case, as would be expected. The square wave is only approximately reproduced, due to the small number of terms included. However, this improves as the number of terms is increased, as shown in Fig. 9.3(c), which is a repeat of Fig. 9.3(b), using seven terms.

9.1.2 Fourier Coefficients in Magnitude and Phase Form

From Eq. (9.3),

$$x(t) = a_0 + \sum_{n=1}^{\infty} [a_n \cos n\omega_0 t + b_n \sin n\omega_0 t]$$

it can be seen that for values of n other than zero, the content of $x(t)$ at any one frequency, say $n\omega_0$ rad/s, consists of the sum of a cosine component and a sine component. It is sometimes useful to express this in the form of a series of single components, each with magnitude and phase, instead. The single component at each harmonic frequency can be taken as either a sine or a cosine.

Figure 9.4 shows one typical pair of coefficients, on the left as rotating vectors, of length a_n and b_n, and on the right as their projections $a_n \cos n\omega_0 t$ and $b_n \sin n\omega_0 t$, respectively. The vector d_n is the resultant of a_n and b_n, and its length, representing the

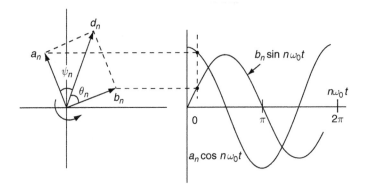

Fig. 9.4   A pair of Fourier components expressed as rotating vectors.

modulus of the component of $x(t)$ at frequency $n\omega_0$, is $\sqrt{a_n^2 + b_n^2}$. The vector d_n leads the vector b_n by angle $\theta_n = \tan^{-1}(a_n/b_n)$. Therefore, for a sine series representation in terms of modulus and phase, we can write

$$a_n \cos n\omega_0 t + b_n \sin n\omega_0 t = d_n \sin(n\omega_0 t + \theta_n) \tag{9.11}$$

where

$$d_n = \sqrt{a_n^2 + b_n^2} \tag{9.12}$$

and

$$\theta_n = \tan^{-1}\left(\frac{a_n}{b_n}\right) \tag{9.13}$$

Since these equations apply for all values of n other than zero, an alternative form of Eq. (9.3) is

$$x(t) = a_0 + \sum_{n=1}^{\infty} [d_n \sin(n\omega_0 t + \theta_n)] \tag{9.14}$$

Similarly, a cosine-only series can be formed by noting that vector d_n *lags* vector a_n by angle $\psi_n = \tan^{-1}(b_n/a_n)$, leading to:

$$x(t) = a_0 + \sum_{n=1}^{\infty} [d_n \cos(n\omega_0 t - \psi_n)] \tag{9.15}$$

where

$$\psi_n = \tan^{-1}\left(\frac{b_n}{a_n}\right) \tag{9.16}$$

It can easily be verified that Eqs (9.14) and (9.15) are equivalent. Using the identity $\sin A = \cos(A - \pi/2)$, then:

$$d_n \sin(n\omega_0 t + \theta_n) = d_n \cos\left(n\omega_0 + \theta_n - \frac{\pi}{2}\right) \tag{9.17}$$

But from Fig. 9.4:

$$\theta_n + \psi_n = \frac{\pi}{2}\text{rad} \tag{9.18}$$

Combining Eqs (9.17) and (9.18):

$$d_n \sin(n\omega_0 t + \theta_n) = d_n \cos(n\omega_0 t - \psi_n) \tag{9.19}$$

showing that Eqs (9.14) and (9.15) are, in fact, equivalent.

9.1.3 The Fourier Series in Complex Notation

Using the standard results:

$$\cos n\omega_0 t = \frac{1}{2}\left(e^{in\omega_0 t} + e^{-in\omega_0 t}\right) \quad \text{and} \quad \sin n\omega_0 t = -i\frac{1}{2}\left(e^{in\omega_0 t} - e^{-in\omega_0 t}\right)$$

Equation (9.3) can be written as:

$$x(t) = a_0 + \sum_{n=1}^{\infty}\left[\frac{1}{2}a_n\left(e^{in\omega_0 t} + e^{-in\omega_0 t}\right) - i\frac{1}{2}b_n\left(e^{in\omega_0 t} - e^{-in\omega_0 t}\right)\right] \tag{9.20}$$

or

$$x(t) = a_0 + \sum_{n=1}^{\infty}\left[\frac{1}{2}(a_n - ib_n)e^{in\omega_0 t} + \frac{1}{2}(a_n + ib_n)e^{-in\omega_0 t}\right] \tag{9.21}$$

or

$$x(t) = c_0 + \sum_{n=1}^{\infty}\left[c_n e^{in\omega_0 t} + c_n^* e^{-in\omega_0 t}\right] \tag{9.22}$$

where

$$c_0 = a_0 \tag{9.23a}$$

$$c_n = \frac{1}{2}(a_n - ib_n) \tag{9.23b}$$

$$c_n^* = \frac{1}{2}(a_n + ib_n) \tag{9.23c}$$

where c_n^* is the *complex conjugate* of c_n, formed, as for any complex number, by reversing the sign of the imaginary part.

Now since terms with negative n are complex conjugates of terms with positive n, then $\sum_{n=1}^{\infty} c_n^* e^{-in\omega_0 t} = \sum_{n=-1}^{-\infty} c_n e^{in\omega_0 t}$ and Eq. (9.22) can now be written as:

$$x(t) = c_0 + \sum_{n=1}^{\infty} c_n e^{in\omega_0 t} + \sum_{n=-1}^{-\infty} c_n e^{in\omega_0 t} \tag{9.24}$$

Noting the trivial relationship, $c_0 = c_0 e^{in\omega_0 t}$, for $n = 0$, we see that the whole of Eq. (9.24) can be replaced by the simple equation,

$$x(t) = \sum_{n=-\infty}^{\infty} c_n e^{in w_0 t} \tag{9.25}$$

which will correctly represent all the possible integer values of n, whether negative, zero or positive. Equation (9.25) is the double-sided Fourier series expressed in complex notation. The inverse relationship can be shown to be

$$c_n = \frac{1}{T} \int_{-T/2}^{T/2} x(t) e^{-in w_0 t} dt \quad (n = 0, \pm 1, \pm 2, \ldots) \tag{9.26}$$

It will be seen that this has introduced negative frequencies, which should not be given physical significance.

9.1.4 The Fourier Integral and Fourier Transforms

In principle, the Fourier integral can be developed from Eqs (9.25) and (9.26) by supposing that the period of the waveform, T, approaches infinity, and the frequency interval, w_0, between adjacent components approaches zero, so that the function becomes non-periodic. The coefficients, c_n, then also approach zero, but this problem can be overcome [9.1] by working with the product $T c_n$, which remains finite, rather than with c_n. We now introduce the notation $w_n = n w_0$, so that w_n represents the actual frequency, in rad/s, of component n. Then, if we regard the frequency interval, w_0, as a small change of w_n equal to δw_n, so that $w_0 = \delta w_n$, we can write

$$\frac{1}{T} = \frac{w_0}{2\pi} = \frac{1}{2\pi} \cdot \delta w_n \tag{9.27}$$

where the relationship $w_0 = 2\pi/T$, from Eq. (9.2), has also been used.

Using Eq. (9.27), Eq. (9.25) can be written in the form:

$$x(t) = \sum_{n=-\infty}^{\infty} \frac{1}{T}(T c_n) e^{i w_n t} = \frac{1}{2\pi} \sum_{n=-\infty}^{\infty} (T c_n) e^{i w_n t} \delta w_n \tag{9.28}$$

Multiplying both sides of Eq. (9.26) by T:

$$T c_n = \int_{-T/2}^{T/2} x(t) e^{-i w_n t} dt \tag{9.29}$$

Now taking the limit as $T \to \infty$, the discrete frequencies, w_n, become the continuous frequency, w, and δw_n becomes dw. The summation, Eq. (9.28), then becomes the integral:

$$x(t) = \frac{1}{2\pi} \int_{-\infty}^{\infty} X(w) e^{i w t} dw \tag{9.30}$$

where

$$X(w) = T c_n$$

Also, Eq. (9.29) becomes

$$X(\omega) = \int_{-\infty}^{\infty} x(t)\,e^{-i\omega t}dt \tag{9.31}$$

Equation (9.30) is known as the *Fourier integral*, and Eqs (9.30) and (9.31), taken together, are a *Fourier transform pair*. The quantity $X(\omega)$ is the *Fourier transform* of $x(t)$, and $x(t)$ is the *inverse Fourier transform* of $X(\omega)$.

It should be mentioned that Eqs (9.30) and (9.31) can be expressed in several alternate ways, and the factor $1/2\pi$ may be found in the transform instead of in the integral. If frequency is expressed as f, in Hz, rather than as ω, in rad/s, since $\omega = 2\pi f$, the factor $1/2\pi$ disappears altogether, and the Fourier transform pair becomes

$$x(t) = \int_{-\infty}^{\infty} X(f)\,e^{i2\pi ft}df \tag{9.32}$$

and

$$X(f) = \int_{-\infty}^{\infty} x(t)\,e^{-i2\pi ft}dt \tag{9.33}$$

These equations are used in the derivation of many of the expressions we shall encounter in random vibration, which is discussed in Chapter 10.

The Fourier transform can also be used in much the same way as the Laplace transform, discussed in Chapter 3, to find the response of systems represented by their equations of motion. The Fourier transform is, in fact, a special case of the Laplace transform, where the real part, σ, of the Laplace operator, $s = \sigma + i\omega$, is zero. The Fourier transform can only be used in this way for a limited range of forcing functions, however, and in practice the Laplace transform is more widely used.

9.2 The discrete Fourier transform

Of course, Eqs (9.32) and (9.33) can only be used as they stand when $x(t)$ or $X(f)$ is known analytically. This is not the case in practical vibration analysis, where a discrete series of data points is obtained by digitizing the analog outputs of transducers, such as accelerometers or strain gages, at constant time intervals. The *discrete Fourier transform*, or *DFT*, is essentially a way of implementing Eq. (9.33) in a stripwise, numerical way. The *inverse discrete Fourier transform* (*IDFT*) is similarly a way of solving Eq. (9.32) numerically. The following derivation of the DFT retains the complex form, as is normal practice. A trigonometric form also exists.

After the DFT has been used to transform the time series data into complex Fourier coefficients, further analysis will depend upon the nature of the vibration problem. In the case of periodic vibration, we are likely to require only the magnitude, and perhaps the phase, of the harmonic components of the original time data. In the case of random vibration, a wider range of functions may be required, including power spectra, correlation functions, etc. These will later be seen to be relatively easy to compute, using the DFT.

9.2.1 Derivation of the Discrete Fourier Transform

The DFT can be derived from Eqs (9.3), (9.8) and (9.10), but in the standard form, the factor 2, which appears in the last two of these equations, is omitted. These three equations are therefore re-defined as follows.

First, Eq. (9.3), which was originally

$$x(t) = a_0 + \sum_{n=1}^{\infty} [a_n \cos n\omega_0 t + b_n \sin n\omega_0 t] \quad (n = 1, 2, \ldots)$$

is now written in the form:

$$x(t) = a_0 + 2\sum_{k=1}^{\infty} \left[a_k \cos \frac{2\pi kt}{T} + b_k \sin \frac{2\pi kt}{T} \right] \quad (k = 1, 2, \ldots) \tag{9.34}$$

where n has been changed to k and the relationship $\omega_0 = 2\pi/T$, from Eq. (9.2), has been used.

Secondly, Eqs (9.8) and (9.10), which were originally

$$a_n = \frac{2}{T} \int_{-T/2}^{T/2} x(t) \cos n\omega_0 t \cdot dt \quad \text{and} \quad b_n = \frac{2}{T} \int_{-T/2}^{T/2} x(t) \sin n\omega_0 t \cdot dt, \quad (n = 1, 2 \ldots)$$

are now written as

$$a_k = \frac{1}{T} \int_0^T x(t) \cos \frac{2\pi kt}{T} \cdot dt \quad (k = 0, 1, 2, \ldots) \tag{9.35}$$

and

$$b_k = \frac{1}{T} \int_0^T x(t) \sin \frac{2\pi kt}{T} \cdot dt \quad (k = 1, 2 \ldots) \tag{9.36}$$

where the period covered by the integrals is now from 0 to T instead of from $-T/2$ to $T/2$. It can be seen that in order to make Eqs (9.34), (9.35) and (9.36) compatible with Eqs (9.3), (9.8) and (9.10), the values of the coefficients have been changed so that

$$a_k = \frac{a_n}{2} \quad \text{and} \quad b_k = \frac{b_n}{2} \tag{9.37}$$

The value of a_0 is unchanged, and it has been absorbed into Eq (9.35), which now includes a term for $k = 0$, whereas Eq. (9.36) does not, since there is, of course, no term b_0.

To combine Eqs (9.35) and (9.36) into a single equation, in complex notation, since

$$e^{-i\left(\frac{2\pi kt}{T}\right)} = \cos\left(\frac{2\pi kt}{T}\right) - i \sin\left(\frac{2\pi kt}{T}\right) \tag{9.38}$$

we can write, using Eqs (9.35), (9.36) and (9.38):

$$X_k = \frac{1}{T} \int_0^T x(t) e^{-i\left(\frac{2\pi kt}{T}\right)} dt \tag{9.39}$$

where

$$X_k = a_k - ib_k \tag{9.40}$$

X_k is seen to be the complex discrete Fourier coefficient for a given value of k, and its real and imaginary parts are a_k and $(-b_k)$, respectively.

In practical measurements, $x(t)$ becomes a series of discrete sampled values at equal intervals of time, Δ, where

$$\Delta = \frac{T}{N} \tag{9.41}$$

where N is the total number of values of $x(t)$ and, of course, T is the length of the period being transformed, in seconds. The time, t, corresponding to any sampled value, x_j, is

$$t = j\Delta \tag{9.42a}$$

or, using Eq. (9.41):

$$t = \frac{jT}{N} \quad (j = 0, 1, 2, \ldots, N-1) \tag{9.42b}$$

Equation (9.42b) defines the relationship between time, t, in seconds, and the integer j.

The integer k, which is equal to the integer n in the original Fourier series, is related to frequency in a similar way. If frequency is measured in hertz, then from Eq. (9.2), $f_0 = 1/T$, where f_0 is the fundamental frequency of the transform. Thus, values of X_k will be computed at frequencies, in Hz, given by:

$$f = \frac{k}{T} \quad (k = 0, 1, 2, \ldots, N-1) \tag{9.43a}$$

If frequency is measured in rad/s, however, we also have from Eq. (9.2);

$$\omega_0 = 2\pi f_0 = \frac{2\pi}{T}$$

and the frequencies in rad/s at which values of X_k will be computed will be given by:

$$\omega = \frac{2\pi k}{T} \quad (k = 0, 1, 2, \ldots, N-1) \tag{9.43b}$$

A stripwise solution to Eq. (9.39) is

$$X_k = \frac{1}{T} \sum_{j=0}^{N-1} x_j e^{-i\left(\frac{2\pi k}{T}\right)(j\Delta)} \Delta \tag{9.44}$$

or using Eq. (9.41):

$$X_k = \frac{1}{N} \sum_{j=0}^{N-1} x_j e^{-i\left(\frac{2\pi jk}{N}\right)} \tag{9.45}$$

Equation (9.45) is the *DFT* of the discrete time series:

$$x_j = (x_0, x_1, x_2, \ldots, x_{N-1})$$

The corresponding *IDFT* can be shown to be

$$x_j = \sum_{k=0}^{N-1} X_k e^{i\left(\frac{2\pi j k}{N}\right)} \qquad (9.46)$$

Some authorities place the factor $1/N$ in the IDFT, Eq. (9.46), instead of in the DFT, Eq. (9.45), and another possibility is to place the factor $1/\sqrt{N}$ in both. Whatever arrangement is adopted, however, the factors will always be arranged so that the IDFT will reverse the effect of the DFT exactly.

In the specific case of the derivation above, it can be seen from Eqs (9.37) and (9.40) that the real and imaginary parts of X_k are related to the original Fourier coefficients, a_n and b_n as follows:

$$X_k = \frac{a_n}{2} - i\left(\frac{b_n}{2}\right) \qquad (9.47a)$$

or

$$\mathrm{Re}(X_k) = \frac{a_n}{2} \quad \text{and} \quad \mathrm{Im}(X_k) = -\frac{b_n}{2} \qquad (9.47b)$$

These relationships apply when X_k is defined by Eq. (9.45), but since the definition of the DFT is to some extent arbitrary, other relationships are possible. When using particular software for the first time, therefore, the associated 'small print' should be read carefully. It is always a good idea, also, when using unfamiliar software, to run a simple case to which the exact answer is known. It should always be remembered that the basic Fourier coefficients, a_n and b_n, being the amplitudes of the cosine and sine components that make up the original time history, have a physical significance. Careful examination of the definition of the DFT will show, in a particular case, how the actual output of a DFT program is related to these basic coefficients.

The following example, by using a small value for N, and a known time history as input, can be worked through easily 'by hand', demonstrating the practical use, and some of the peculiarities, of the DFT. The input time series would not, of course, generally be known analytically in this way, but in this example it permits the output from the DFT calculation to be compared with known correct answers.

Example 9.2

(a) Using a 10-point DFT, i.e. $N = 10$, show how Eq. (9.45) can be used to find the real and imaginary parts of the discrete Fourier transform, X_k, for $k = 3$, for the sampled periodic time history shown in Fig. 9.5(a), where the period, $T = 0.5\,\mathrm{s}$, is represented by ten values of x_j, ($j = 0, 1, 2, \ldots, 9$), corresponding to $t = 0, 0.05, 0.10, \ldots, 0.45\,\mathrm{s}$.

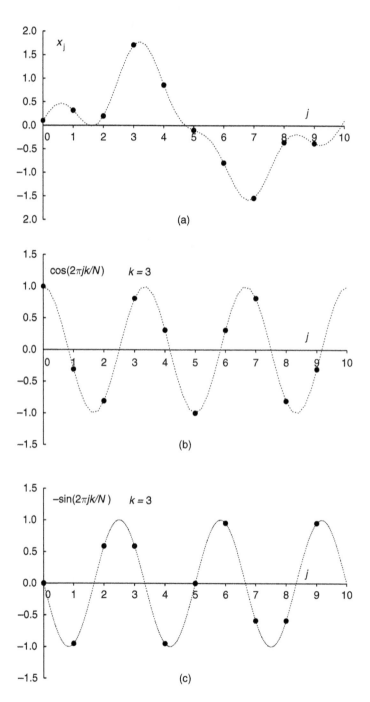

Fig. 9.5   Illustrating a simple 10-point DFT (Example 9.2).

Note: The test input shown in Fig. 9.5(a) was generated from Eq. (9.3), using the following Fourier coefficients: $a_3 = 0.1$; $b_1 = 1.0$; $b_2 = -0.5$; $b_4 = 0.4$; all others being zero. Using Eq. (9.2): $\omega_0 = 2\pi f_0 = 2\pi/T$, and Eq. (9.42): $t = jT/N$, the discretized values of $x(t)$, that is, x_j ($j = 0, 1, 2, \ldots, 9$) were then given by:

$$x_j = 1.0 \sin \frac{2\pi j}{10} - 0.5 \sin \frac{4\pi j}{10} + 0.1 \cos \frac{6\pi j}{10} + 0.4 \sin \frac{8\pi j}{10} \quad (j = 0, 1, 2, \cdots, 9) \quad \text{(A)}$$

(b) Using a simple spreadsheet calculation, evaluate X_k for $k = 0, 1, 2, \ldots, 9$, and compare the results with the Fourier coefficients used to derive the discretized values of x_j, determined as in the note above. Also show the values of frequency, f, in Hz, corresponding to the values of k.

(c) Discuss the apparently spurious terms generated for values of k greater than $N/2$, i.e. $k > 5$ in this example.

Solution

Part (a)

Figure 9.5, (a), (b) and (c) illustrate a possible, but inefficient, implementation of Eq. (9.45). The sampled values x_0, x_1, x_2, ... x_9 are shown at (a) as dots. Since the whole period, from $j = 0$ to 10, is 0.5 s long, these values are at 0.05 s intervals. Writing Eq. (9.45) in the form:

$$X_k = \frac{1}{N} \sum_{j=0}^{N-1} x_j e^{-i\left(\frac{2\pi jk}{N}\right)} = \frac{1}{N} \sum_{j=0}^{N-1} \left[x_j \cos\left(\frac{2\pi jk}{N}\right) - i x_j \sin\left(\frac{2\pi jk}{N}\right) \right] \quad \text{(B)}$$

discretized values of the functions $\cos(2\pi jk/N)$ and $-\sin(2\pi jk/N)$ are shown in Fig. 9.5 at (b) and (c), respectively, for $k = 3$ in this case.

From Eq. (B), the real part of $x_j e^{-i\left(\frac{2\pi jk}{N}\right)}$, for $k = 3$, is given by multiplying together corresponding discrete values in Fig. 9.5, at (a) and (b), for all values of j, from 0 to 9, summing the products, and finally multiplying by $1/N$. A similar process, multiplying the discrete values in (a) and (c), summing and multiplying by $1/N$ gives the imaginary part of X_k.

These operations are shown in Table 9.1, from which the real and imaginary parts of X_3 (i.e. X_k for $k = 3$) are seen to be 0.05 and 0, respectively.

Part (b)

Repeating Table 9.1 for all the other values of k, we can produce Table 9.2, showing the real and imaginary parts of X_k for $k = 0, 1, 2, \ldots, 9$. For comparison, the Fourier coefficients used to calculate the original values of x_j are also shown.

We can convert the discrete values of k to discrete frequencies, f, in hertz, using Eq. (9.43a): $f = k/T$ ($k = 0, 1, 2, \ldots, N - 1$). Since $T = 0.5$ s, then $f = 2k$.

It can be seen from Table 9.2 that the DFT results are, in fact, related to the Fourier coefficients, a_n and b_n, where the latter exist, as predicted by Eq. (9.47b):

$$\text{Re}(X_k) = \frac{a_n}{2} \quad \text{and} \quad \text{Im}(X_k) = -\frac{b_n}{2}$$

It can also be seen from Table 9.2, however, that the DFT has introduced 'spurious' values of X_k at values of k greater than 5, or $N/2$, corresponding to $f > 10$ Hz. These

Table 9.1
Calculation of one Complex Fourier Component, Example 9.2.

t	j	x_j	$\cos(2\pi jk/N)$ $k=3\ N=10$	$-\sin(2\pi jk/N)$ $k=3\ N=10$	$x_j[\cos(2\pi jk/N)]$ $k=3\ N=10$	$x_j[-\sin(2\pi jk/N)]$ $k=3\ N=10$
0	0	0.1000	1	0	0.1000	0.0000
0.05	1	0.3165	−0.3090	−0.9511	−0.0978	−0.3010
0.10	2	0.1958	−0.8090	0.5878	−0.1584	0.1151
0.15	3	1.7063	0.8090	0.5878	1.3804	1.0029
0.20	4	0.8591	0.3090	−0.9511	0.2655	−0.8171
0.25	5	−0.1000	−1	0	0.1000	0.0000
0.30	6	−0.7973	0.3090	0.9511	−0.2464	−0.7583
0.35	7	−1.5445	0.8090	−0.5878	−1.2495	0.9078
0.40	8	−0.3576	−0.8090	−0.5878	0.2893	0.2102
0.45	9	−0.3783	−0.3090	0.9511	0.1169	−0.3598

$$\sum = 0.5$$
$$\mathrm{Re}(X_3) =$$
$$(1/N)0.5 = 0.05$$

$$\sum = 0$$
$$\mathrm{Im}(X_3) = 0$$

Table 9.2
Complex Fourier Coefficients, Example 9.2.

k	f(Hz)	n	DFT results		Fourier coeffs	
			$\mathrm{Re}(X_k)$	$\mathrm{Im}(X_k)$	a_n	b_n
0	0	0	0	0	0	—
1	2	1	0	−0.5	0	1.0
2	4	2	0	0.25	0	−0.5
3	6	3	0.05	0	0.1	0
4	8	4	0	−0.2	0	0.4
5	10	5	0	0	—	—
6	12	6	0	0.2	—	—
7	14	7	0.05	0	—	—
8	16	8	0	−0.25	—	—
9	18	9	0	0.5	—	—

have been produced by the DFT itself, and have nothing to do with the original input data, which clearly contained no components above $n = k = 4$. This phenomenon is discussed next.

Part (c)

The spurious values of X_k can be explained by Fig. 9.6(a) and (b). These show (as solid lines) the multiplying functions $\cos(2\pi jk/N)$ and $-\sin(2\pi jk/N)$, respectively, for $k = 7$. When sampled at $j = 0, 1, 2, \ldots, 9$, however, these become *aliased*, and the sampled values, joined by a dotted line, are, for the cosine function, Fig. 9.6(a), precisely the same as those shown in Fig. 9.5(b), which were computed for $k = 3$. The DFT cannot distinguish between the genuine cosine function for $k = 3$ and the aliased cosine function for $k = 7$, and so produces the same output.

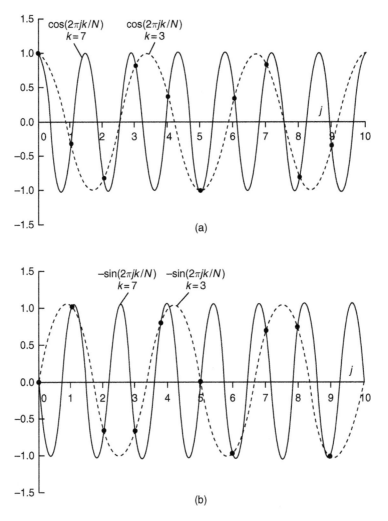

Fig. 9.6   Explanation of 'spurious' components in a DFT.

The (− sine) function, for $k = 7$, as shown in Fig. 9.6(b), is similarly aliased, but in this case it will be seen that the sampled version, also joined by a dotted line, is reversed in sign with respect to Figure 9.5(c).

The same effect occurs for all values of k greater than five, in this example, or, in general, for values of k greater than $N/2$, and the overall effect is that

$$\text{Re}(X_{N-k}) = \text{Re}(X_k) \qquad \text{Im}(X_{N-k}) = -\text{Im}(X_k)$$

In Table 9.2 it will be noticed that all the DFT results follow this pattern. The effect is intentional, and is simply due to the way the standard complex DFT is defined.

The indicated components at and below $k = N/2$ are correct, but the coefficients above $k = N/2$ are spurious, and must not be interpreted as Fourier coefficients at the frequencies indicated. They are there for good reason, however, and the following should be noted:

(1) The IDFT, in the form of Eq. (9.46), needs the 'spurious' values of X_k in order to reproduce the time series x_j correctly.

(2) As shown in Section 9.1.3, in the two-sided complex Fourier series, the components at negative frequencies are the complex conjugates of those at the corresponding positive frequencies, which is just what the 'spurious' components are, except that they appear at values of k that are N points to the right of their correct positions.

It can be seen that the standard DFT deliberately uses aliasing to make it reversible, but it is the cosine and sine multiplying functions that are aliased, not the input data, and nothing is lost in the process. Aliasing of *data*, as discussed in Section 9.3, is different, and a potentially serious problem.

Although the process described in Example 9.2 works, it can be seen that since a typical realistic value for N is about 1000, to find the complex coefficients for all N values of k by this means would require a very large number of multiplications, and in practice it is replaced by the fast Fourier transform (FFT) algorithm [9.2, 9.3], which produces exactly the same result, but much more efficiently. FFT codes usually require N to be a power of 2, and a typical practical value is likely to be 1024 ($=2^{10}$) rather than 1000.

9.2.2 Proprietary DFT Codes

Equations (9.45) and (9.46) are the forward and inverse DFTs, respectively, and this would appear to require two separate computer codes. However, proprietary software libraries usually provide only one, which is used both ways. A typical example is represented by the following equation:

$$ z_K = \frac{1}{\sqrt{N}} \sum_{J=0}^{N-1} z_J e^{-i\left(\frac{2\pi JK}{N}\right)} \tag{9.48} $$

This represents a general-purpose code, which can be used for transformation in both directions, and both z_J and z_K may be complex, although one of them will actually be real in ordinary vibration work. Equation (9.48) is correct as it stands for forward transformation, with J representing time and K representing frequency. The output would, of course, have to be multiplied by $1/\sqrt{N}$ to be equivalent to that of Eq. (9.45).

For inverse transformation, the same code is used, but in the following way:

(1) the roles of J and K are reversed, K now representing time and J representing frequency;

(2) the input file is changed to its complex conjugate values;

(3) a *forward* transformation is carried out;

(4) the output file is changed to its complex conjugates (if necessary).

The output would also have to be multiplied by $\sqrt{N}$ to be eqivalent to Eq. (9.46).

It is easily shown that this procedure works, as follows. A complex-to-complex inverse transformation is assumed.

If the inverse transformation corresponding to Eq. (9.48) existed, it would be

$$z_J = \frac{1}{\sqrt{N}} \sum_{K=0}^{N-1} z_K e^{i\left(\frac{2\pi JK}{N}\right)} \tag{9.49}$$

which is the same as Eq. (9.48) except that the exponent is positive.

Introducing the notation: $z_K = (A_K + iB_K)$, $z_J = (A_J + iB_J)$ and $\theta = 2\pi JK/N$, Equation (9.49) can be written as:

$$z_J = \frac{1}{\sqrt{N}} \sum_{K=0}^{N-1} (A_K + iB_K)(e^{i\theta}) = \frac{1}{\sqrt{N}} \sum_{K=0}^{N-1} (A_K + iB_K)(\cos\theta + i\sin\theta)$$

$$= \frac{1}{\sqrt{N}} \sum_{K=0}^{N-1} (A_K\cos\theta - B_K\sin\theta) + i(A_K\sin\theta + B_K\cos\theta) \tag{9.50}$$

However, since there is no inverse code, we use the forward code, Eq. (9.48), interchanging J and K, and input the data in the form of its complex conjugates, $z_J = (A_J - iB_J)$, instead of $z_J = (A_J + iB_J)$:

$$z_K = \frac{1}{\sqrt{N}} \sum_{J=0}^{N-1} (A_J - iB_J)(e^{-i\theta}) = \frac{1}{\sqrt{N}} \sum_{J=0}^{N-1} (A_J - iB_J)(\cos\theta - i\sin\theta)$$

$$= \frac{1}{\sqrt{N}} \sum_{J=0}^{N-1} (A_J\cos\theta - B_J\sin\theta) - i(A_J\sin\theta + B_J\cos\theta) \tag{9.51}$$

Comparing Eq. (9.50) with Eq. (9.51) shows that the *forward* code gives the correct answer for *inverse* transformation, provided that the complex conjugates of the input file are entered, and the resulting output file is also changed to its complex conjugates.

9.2.3 The fast Fourier transform

In 1965, Cooley and Tukey presented [9.2] an ingenious computer algorithm, known as the *fast Fourier transform*, or *FFT*, for calculating the DFT or IDFT many times faster than had been possible before. Digital Fourier analysis then became feasible for everyday work, replacing the analog methods used until then. It should be stressed that the FFT algorithm is just that, an algorithm, and should not be confused with the DFT. However, every program for evaluating the DFT will incorporate the FFT algorithm, or possibly a later development of it.

No attempt is made here to describe how the FFT algorithm works. The very clear explanation given by Newland [9.3] is recommended.

9.3 Aliasing

In Example 9.2, we encountered a harmless form of aliasing caused by the way the DFT is defined. Aliasing of actual records, due to analog to digital conversion, is a serious problem, however, and this is now discussed.

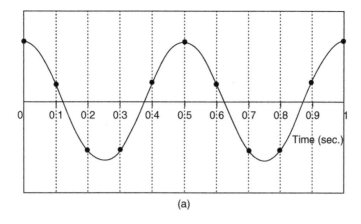

(a)

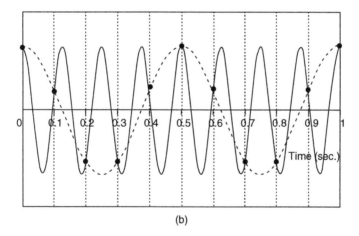

(b)

Fig. 9.7   (a) 2 Hz cosine wave sampled at 0.1 s intervals, (b) 8 Hz cosine wave sampled at 0.1 s intervals.

Figure 9.7(a) shows two cycles of a 2 Hz cosine wave with *sampling intervals*, Δ, of 0.1 s, or a *sampling rate*, $f_s = 1/\Delta$, of 10 samples per second (sps). Clearly, in this case, the wave could be reconstructed correctly knowing only the sampled values.

On the other hand, Fig. 9.7(b) shows eight cycles of an 8 Hz cosine wave, also sampled at 0.1 s intervals, and joining up the sampled values with a dotted line, we see that the only wave that could be reconstructed, given only the samples, is a 2 Hz wave, identical to that in Fig. 9.7(a). The 8 Hz wave is said to be *aliased*, and would be interpreted by any analysis program as having a frequency of 2 Hz, instead of the actual value of 8 Hz.

If the exercise shown in Fig. 9.7(b) is repeated with an 8 Hz *sine* wave rather than a cosine wave, it is found that the result is a 2 Hz sine wave, but reversed in sign compared with a genuine 2 Hz sine wave.

Restricting the argument, for the moment, to signal frequencies between 5 Hz and 10 Hz, in this example, or between $\frac{1}{2}f_s$ and f_s in general, we see that 7 Hz becomes 3 Hz; 9 Hz becomes 1 Hz; and so on, or, in general, any signal frequency $(f_s - f')$ becomes f', and is, in effect, 'reflected' or 'folded' about the frequency $\frac{1}{2}f_s$, known as the *Nyquist frequency*, f_N. The spurious component at f' will, of course, add to any genuine component already at that frequency, making the frequency analysis inaccurate. The process is irreversible, and the genuine information in the range 0 to f_N is corrupted forever.

This is not the whole story, since it is also possible for signal components at frequencies above f_s to be shifted into the range 0 to $\frac{1}{2}f_s$. This can be explained [9.3] by returning to the definition of the DFT, Eq. (9.45):

$$X_k = \frac{1}{N}\sum_{j=0}^{N-1} x_j e^{-i\left(\frac{2\pi jk}{N}\right)}$$

Suppose we replace k in Eq. (9.45) by $(k + N)$, i.e.:

$$X_{k+N} = \frac{1}{N}\sum_{j=0}^{N-1} x_j e^{-i\left(\frac{2\pi j}{N}\right)(k+N)} \tag{9.52}$$

Equation (9.52) can be written as:

$$X_{k+N} = \frac{1}{N}\sum_{j=0}^{N-1} x_j e^{-i\left(\frac{2\pi jk}{N}\right)} e^{-i2\pi j} \tag{9.53}$$

However, since $e^{-i2\pi j} = 1$ for any integer value of j:

$$X_{k+N} = X_k \tag{9.54}$$

The same argument holds for $(k + 2N)$, $(k + 3N)$ and so on, indefinitely, so:

$$X_k = X_{k+N} = X_{k+2N} = X_{k+3N} = \cdots \tag{9.55}$$

Therefore the spectrum, as indicated by the DFT, repeats exactly at intervals of N.

Converting now to frequencies, f, in hertz, since, from Eq. (9.43a), $f = k/T$ and the sampling frequency, $f_s = N/T$, we can see that $k = 0,\ N,\ 2N,\ 3N,\ \ldots$ correspond to $f = 0, f_s, 2f_s, 3f_s, \ldots$ respectively, and the spectrum therefore repeats at intervals of the sampling frequency, f_s. This means that, as shown in Fig. 9.8, signal components at any of the following frequencies

$$(f_s - f'),\quad (f_s + f'),\quad (2f_s - f'),\quad (3f_s + f'),\quad \cdots$$

would all appear aliased at frequency f', adding to, and distorting, any genuine component already there.

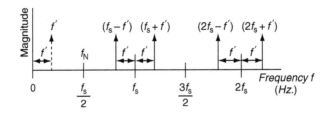

Fig. 9.8 Aliasing of high frequency components.

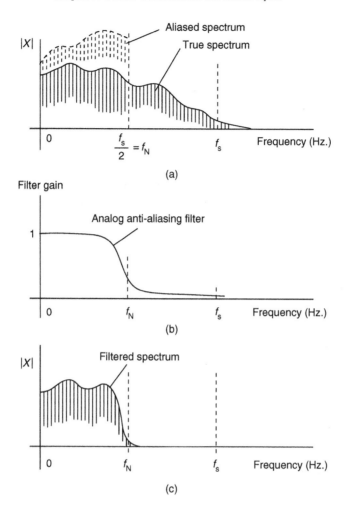

Fig. 9.9 Distortion of the spectrum by aliasing, and its prevention by an anti-aliasing filter.

This cannot be allowed to happen, and the only way to prevent it is to remove any signal components higher than the Nyquist frequency, f_N, or $\frac{1}{2}f_s$, *before* digitizing the data, by using an *anti-aliasing filter*. This must, of course, be an analog filter.

Figure 9.9 shows, at (a) a sketch representing the true spectrum, and the aliased spectrum caused by components above f_N folding back and adding to the part of the true spectrum between 0 and f_N. Figure 9.9 at (b) shows a low-pass analog filter intended to remove (or in practice severely attenuate), any signal components higher than f_N. The filtered spectrum is shown at (c) and after this treatment the analog time history is suitable for digitization at f_s samples per second.

The following practical points about analog anti-aliasing filters should be mentioned.
(1) With good anti-aliasing filter design, it is usually possible to leave signal components at frequencies up to about 0.8 f_N largely unchanged, at least in magnitude, while, at the same time, reducing components above f_N that would otherwise

cause aliasing to very low levels. This is why a typical DFT-based spectrum analyser, or spectral analysis program (to be discussed in Chapter 10), using, say, $N = 1024$ can, in practice, only produce about 400 output frequencies, rather than the 512 that could be expected theoretically.

(2) If the components in the original signal at frequencies above $\frac{1}{2}f_s$ are actually of interest, the only option is to increase the sampling rate so that they fall below the new, increased, value of f_N, or, in practice, below about $0.8\,f_N$.

(3) The phase distortion due to the analog filter is likely to be considerable, and can cause problems. Fortunately, in two common applications, the problem disappears.
  (a) When measuring (auto-) power spectra, phase is ignored anyway. This does not apply to cross-spectra, but note (b) below.
  (b) Cross-functions between channels will be unaffected, provided the filters have identical characteristics in each channel, since the phase shifts will then cancel out.

9.4 Response of systems to periodic vibration

In practice, periodic excitation is mostly associated with rotating machines, including engines of all kinds, propellers, helicopter rotors and machine tools.

Since a periodic force or other input, such as base motion, can be broken down into its Fourier components, if we can find the response of the system to these individually, then we can find the response to the complete periodic input, assuming linearity, by superposition, a relatively simple process.

The response of a single-DOF system to each individual Fourier component is just the steady-state response of the system to a simple harmonic (i.e. sinusoidal) input. Since all the Fourier components, at their various frequencies, act simultaneously, the total response is then the summation of all such responses. Only the steady-state response of the system is required, justified by the fact that since the periodic input has been applied for a long time, all transient effects will have died away.

For multi-DOF systems, normal mode superposition, as discussed in Chapter 6, will usually be used, each mode behaving in the same way as a single-DOF system.

In most practical cases of periodic excitation, the magnitude of the response depends largely upon whether or not a Fourier component coincides in frequency with a resonant peak. It is fairly rare for several Fourier components to coincide with resonance peaks at the same time, but it is possible.

In practice, the analysis can usually be considerably simplified, for example:

(1) For most periodic inputs, only the Fourier components at the fundamental frequency, and the first few integer multiples of that frequency, are of significant magnitude.

(2) The fundamental frequency is nearly always the rotation frequency of a major component of the machine. There may, of course, be several components rotating at different speeds.

(3) If the damping of the system is low, the response to components that do not coincide with a resonant frequency will often be negligible compared with those that do.

In Section 9.1, we saw that a periodic waveform can be expressed mathematically in several different ways, and any of these can be used to represent the forcing or other input function, and the steady-state response of the system, i.e. (1) as a combination of sines and cosines, (2) as magnitudes and phases, or (3) in complex form.

9.4.1 Response of a Single-DOF System to a Periodic Input Force

The following describes the calculation of the displacement response, $z(t)$, of a single-DOF system, as shown in Fig. 9.10, when a periodic force, $F(t)$, is applied. The input and response are both expressed in magnitude and phase form, in this case.

Suppose that the force time history is initially represented as a Fourier series in the manner of Eq. (9.3):

$$F(t) = a_0 + \sum_{n=1}^{\infty} [a_n \cos n\omega_0 t + b_n \sin n\omega_0 t] \quad n = 1, 2, 3, \dots \tag{9.56}$$

Using Eq. (9.15), Eq. (9.56) can be written in magnitude and phase form:

$$F(t) = a_0 + \sum_{n=1}^{\infty} [d_n \cos(n\omega_0 t - \psi_n)] \tag{9.57}$$

where from Eq. (9.12):

$$d_n = \sqrt{a_n^2 + b_n^2}$$

and from Eq. (9.16)

$$\psi_n = \tan^{-1}\left(\frac{b_n}{a_n}\right)$$

Now from Eq. (4.9), since the individual Fourier components of the input are just sinusoidal forces, the steady-state value of the modulus, $|z|_n$ of the displacement response to a Fourier component of magnitude d_n is

$$|z|_n = \frac{d_n}{k} \cdot \frac{1}{\sqrt{(1 - \Omega_n^2)^2 + (2\gamma\Omega_n)^2}} \tag{9.58}$$

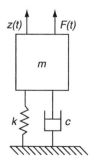

Fig. 9.10   Schematic diagram of a single-DOF system with applied force.

where k is the stiffness, γ the viscous damping coefficient expressed as a fraction of critical and Ω_n the excitation frequency divided by the undamped natural frequency of the single-DOF system. In this case the excitation frequency is the frequency of the Fourier component $d_n \cos(n\omega_0 t - \psi_n)$, which is $n\omega_0$ rad/s, and the undamped natural frequency is $\omega_u = \sqrt{k/m}$ rad/s. Thus

$$\Omega_n = \frac{n\omega_0}{\omega_u} \tag{9.59}$$

where, from Eq. (9.2), $\omega_0 = 2\pi/T$.

The phase angle, ϕ_n, by which the displacement vector lags the force component $d_n \cos(n\omega_0 t - \psi_n)$ is given by Eq. (4.10), which, in this case, becomes

$$\phi_n = \tan^{-1} \frac{2\gamma\Omega_n}{1 - \Omega_n^2} \tag{9.60}$$

Thus the time history, $z_n(t)$, of a single Fourier component of the displacement, for a single value of n, excluding $n = 0$, is

$$z_n(t) = \frac{d_n}{k} \cdot \frac{1}{\sqrt{\left(1 - \Omega_n^2\right)^2 + (2\gamma\Omega_n)^2}} \cos(n\omega_0 t - \psi_n - \phi_n) \tag{9.61}$$

The complete time history of the displacement, $z(t)$, is the summation of all such components for $n = 1, 2, 3, \ldots$ plus the mean displacement, a_0/k, i.e.,

$$z(t) = \frac{a_0}{k} + \sum_{n=1}^{\infty} \left[\frac{d_n}{k} \cdot \frac{1}{\sqrt{\left(1 - \Omega_n^2\right)^2 + (2\gamma\Omega_n)^2}} \cos(n\omega_0 t - \psi_n - \phi_n) \right], \quad n = 1, 2, 3, \ldots \tag{9.62}$$

Sometimes the time history of the response is not required, and it is sufficient to know the mean square or root mean square value. The mean square value, $\langle z^2(t) \rangle$, is

$$\langle z^2(t) \rangle = \left\langle \left\{ \frac{a_0}{k} + \sum_{n=1}^{\infty} \left[\frac{d_n}{k} A_n \cos(n\omega_0 t - \psi_n - \phi_n) \right] \right\}^2 \right\rangle \tag{9.63}$$

where $\langle \rangle$ is used to indicate 'the mean value of' Equation (9.63) can be simplified to:

$$\langle z^2(t) \rangle = \frac{a_0^2}{k^2} + \frac{1}{2k^2} \sum_{n=1}^{\infty} (d_n A_n)^2 \tag{9.64}$$

The RMS value is $\sqrt{\langle z^2(t) \rangle}$.

The mean displacement, a_0/k, affects the mean square and RMS values, and if these are being calculated primarily to give an indication of the vibration level, it is generally better to omit the mean level and deal with it separately.

The approach outlined here is easily adapted for other input and response quantities, for example the input could be base motion, defined as displacement, velocity or acceleration, and the response could also be in terms of any of these quantities, using any of the relationships developed in Chapter 4.

Example 9.3

The vertical motion of a machine tool can be represented schematically by Fig. 9.10. The mass, m, of 200 kg, is carried on elastic supports, so that the natural frequency for vertical motion is 30 Hz, and the viscous damping coefficient, γ, is 0.1 of critical. A mechanism applies a vertical, periodic force, $F(t)$, that can be approximated by a symmetrical square wave of period $T = 0.1$ s, and a magnitude of ± 3000 N.

Plot the vertical displacement time history, $z(t)$, of the machine.

Solution

The notation used is as defined in the preceding section.

It is first noted that since the natural frequency of the system is 30 Hz, it is likely to be excited by the third harmonic of the excitation, which is a square wave of $1/T = 10$ Hz, and that the response to the 10 Hz fundamental will be essentially static.

From Example 9.1 it can be seen that the square wave force input, in newtons, treated as an even function, can be represented by the Fourier series:

$$F(t) = 3000 \left(\frac{4}{\pi} \cos \omega_0 t - \frac{4}{3\pi} \cos 3\,\omega_0 t + \frac{4}{5\pi} \cos 5\,\omega_0 t - \frac{4}{7\pi} \cos 7\,\omega_0 t + \cdots \right) \qquad \text{(A)}$$

Using the fact that $- \cos \theta = \cos(\theta - \pi)$, Eq. (A) can be written as:

$$F(t) = d_1 \cos \omega_0 t + d_3 \cos(3\omega_0 t - \pi) + d_5 \cos 5\,\omega_0 t + d_7 \cos(7\omega_0 t - \pi) + \cdots \qquad \text{(B)}$$

or in the form of Eq. (9.57):

$$F(t) = a_0 + \sum_{n=1}^{\infty} [d_n \cos(n\omega_0 t - \psi_n)] \qquad \text{(C)}$$

where

$$\omega_0 = 2\pi/T$$
$$a_0 = 0, \qquad \text{(D)}$$
$$d_n = 0 \quad \text{for even values of } n.$$

Values of d_n and ψ_n for odd values of n are independent of T, and are as follows:

$$\begin{aligned} d_1 &= (3000 \times 4)/\pi &&= 3819 \text{ N} &&\psi_1 = 0 \\ d_3 &= (3000 \times 4)/(3\pi) &&= 1273 \text{ N} &&\psi_3 = \pi \\ d_5 &= (3000 \times 4)/(5\pi) &&= 763.9 \text{ N} &&\psi_5 = 0 \\ d_7 &= (3000 \times 4)/(7\pi) &&= 545.7 \text{ N} &&\psi_7 = \pi \end{aligned}$$

$$\vdots$$

The following numerical values are constant throughout:
$\omega_u = (2\pi \times 30) = 60\pi$ rad/s $=$ natural frequency of system $= 30$ Hz;
$m = 200$ kg $=$ mass of machine;
$k = m\omega_u^2 = 200(2\pi \times 30)^2 = 7.106 \times 10^6$ N/m $=$ stiffness of supports;
$\gamma = 0.1 =$ viscous damping coefficient.
The displacement response is given by Eq. (9.62), which can be written as:

$$z(t) = \frac{a_0}{k} + \sum_{n=1}^{\infty} \left[\frac{d_n}{k} A_n \cos(n\omega_0 t - \psi_n - \phi_n) \right] \qquad \text{(E)}$$

where

$$A_n = \frac{1}{\sqrt{\left(1 - \Omega_n^2\right)^2 + \left(2\gamma\Omega_n\right)^2}}$$ (F)

Table 9.3   Numerical Values for Example 9.3.

Frequency (Hz)	n	$n\omega_0$ (rad/s)	d_n (N)	ψ_n (rad)	Ω_n	A_n	ϕ_n (rad)
10	1	20π	3819	0	0.3333	1.1218	0.2213
30	3	60π	1273	π	1.0000	5.0000	1.5707
50	5	100π	763.9	0	1.6666	0.5528	3.029
70	7	140π	545.7	π	2.3333	0.2238	3.096
90	9	180π	424.4	0	3.0000	0.1246	3.116
110	11	220π	347.2	π	3.6666	0.0802	3.125
130	13	260π	293.8	0	4.3333	0.0561	3.130

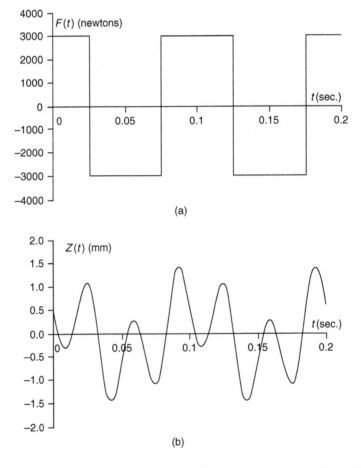

(a)

(b)

Fig. 9.11   (a) Input force time history, Example 9.3, (b) displacement response time history, Example 9.3.

$$\Omega_n = \frac{n\omega_0}{\omega_u} = \frac{2\pi n}{2\pi(30)T} = \frac{n}{30T} \tag{G}$$

and ϕ_n is given by Eq. (9.60):

$$\phi_n = \tan^{-1}\frac{2\gamma\Omega_n}{1 - \Omega_n^2} \tag{H}$$

Numerical values are shown in Table 9.3.

The displacement time history, $z(t)$, is now given by Eq. (E). It can be seen from Table 9.3 that the largest contribution to the response is at 30 Hz, where the third harmonic of the force waveform coincides with the natural frequency. There are also significant components at 10 and 50 Hz.

Figure 9.11(a) shows the input force waveform, and Figure 9.11(b) the displacement response waveform, including components up to 70 Hz.

References

9.1 Meirovich, L. (1986). *Elements of Vibration Analysis*. McGraw-Hill Inc.

9.2 Cooley, JW and Tukey, JW (1965). An algorithm for the machine calculation of complex Fourier Coefficients. *Mathematics of Computation*, 19, 297–301.

9.3 Newland, DE (1993). *Random Vibrations, Spectral and Wavelet Analysis*. Longman (London) and John Wiley (New York).

10 Random Vibration

Contents

Random vibration may be caused by the turbulent flow of gases or liquids, the passage of vehicles over rough surfaces, rough seas acting on ships and marine structures, and earthquakes. The aerospace field also provides many examples of random vibration, and these tend to fall into three groups: flight through atmospheric turbulence, 'separated' airflow over wings and other surfaces and 'mixing noise' from rocket and jet exhaust plumes.

So far, we have looked at the response of systems to deterministic inputs consisting of known functions of time, and there is, potentially at least, an exact answer, provided we have all the data. If the input is a random function of time, and has been recorded, then this approach is still possible, but the calculation would have to be stepwise, very lengthy, and the response would be specific to that particular input. However, if the average properties of the input do not change too much with time, much easier methods can be used. These require the introduction of new ways of describing the average properties of time histories. In this chapter, therefore, we first introduce two new concepts: amplitude probability and the power spectrum. These allow two of the most common tasks in day-to-day work to be tackled: measuring random vibration and calculating the response of linear systems to it. We can then fill in some of the theoretical gaps, introducing correlation functions and their relationship to power spectra and cross-power spectra; some applications of cross-power spectra; and the practical computation of basic functions, using the DFT. Finally, we look at the statistical accuracy of spectral estimates; and fatigue due to random vibration.

10.1 Stationarity, ergodicity, expected and average values

Mathematically, the conventional analysis of random waveforms requires them to be *stationary* and *ergodic*. Stationarity applies to a single waveform, and implies that its average properties are constant with time. Ergodicity applies to an ensemble of a

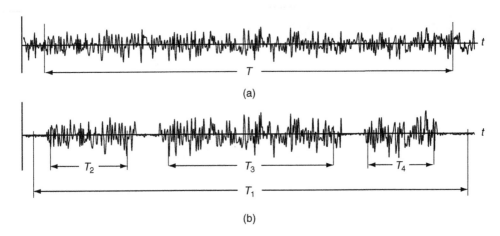

Fig. 10.1   Stationary and non-stationary waveforms.

large number of nominally similar waveforms, recorded in similar conditions. If these all have similar average properties, it suggests that it is reasonable to assume that the one or two records that we are able to analyse are typical of the process as a whole. Since there are no hard and fast criteria, in practice we must rely on common sense to ensure that the records we analyse are at least reasonably stationary. The concept of stationarity is illustrated by Fig. 10.1, which shows sketches of two random waveforms. That shown at (a) is obviously fairly stationary, and we would be justified in choosing the period T for analysis. On the other hand, the record shown at (b) is clearly non-stationary, and conventional analysis over the period T_1 would, at best, only give average properties over that period, and not of the periods T_2, T_3 and T_4, which are obviously more severe, and therefore more important in practice.

Techniques such as *wavelet analysis* [10.1] have been developed, specifically for dealing with non-stationary data, but these are beyond the scope of this book. However, even using conventional methods, there are steps that can be taken to mitigate the effects of non-stationarity. The most obvious approach would be to analyse the periods T_2, T_3 and T_4, in Fig. 10.1(b), separately, regarding each as stationary. Unfortunately, as we shall see, the use of short samples reduces the statistical accuracy of the results. We can sometimes, in effect, increase the length of the samples by combining the results of several sets of measurements. It may even be justifiable to modify the experimental procedure, in order to lengthen the samples. As an example, an aircraft maneuver producing severe buffeting may last for less than a second, which may not be long enough to measure the random response accurately. However, it is sometimes possible to prolong the maneuver artificially, enabling better measurements to be made. In this way, a better understanding of the response may be obtained; for example, the modes taking part may be identified. Of course, any assessment of the fatigue life of the structure should be obtained from the actual maneuvers.

The *expected value* of a random process, denoted, for example, by $E(x)$, is the value that would be obtained by averaging over a very long, theoretically infinite, time. In the case of a random time history, $x(t)$ say, the expected value, $E[x(t)]$, is the mean value $\langle x(t) \rangle$. Similarly, $E[x^2(t)]$ is the mean square value, $\langle x^2(t) \rangle$, and so on.

It is possible to find a few basic properties of a random waveform without considering how its amplitude and frequency components are distributed, by operating directly on a recorded time history. The *mean value*, the *mean square value*, the *RMS value*, the *variance* and the *standard deviation* can all be found in this way. Since these values can also be found, for example, from the probability density function, or in some cases, from the power spectrum, this provides a valuable cross check in practical work.

Assuming that a recorded time history is available, the mean value, $\langle x(t) \rangle$, is given by:

$$\langle x \rangle = \frac{1}{T} \int_0^T x \, dt \tag{10.1}$$

where $x(t)$ has been abbreviated to x and the notation $\langle x \rangle$ is used to indicate the mean value of x. The length of the sample, in seconds, is T.

In practice, the waveform will usually have been digitized at equal time intervals. The mean value is then simply the average of all the recorded discrete values, x_j, that is,

$$\langle x \rangle = \frac{1}{N} \sum_1^N x_j \tag{10.2}$$

where N is the total number of recorded values. Clearly, $N = T/\Delta$ where Δ is the sampling interval, in seconds.

The mean square value is

$$\langle x^2 \rangle = \frac{1}{T} \int_0^T x^2 \, dt \tag{10.3}$$

or for discrete, digitized values:

$$\langle x^2 \rangle = \frac{1}{N} \sum_{j=1}^N x_j^2 \tag{10.4}$$

The RMS value is, of course, $\sqrt{\langle x^2 \rangle}$.

The mean square and RMS values contain the effect of the mean value, if any. It is usually desirable to separate out the mean or steady level before squaring, since in many practical applications, such as fatigue life assessment, the steady stress will probably be treated differently from the oscillatory stress. The mean square value about the mean, i.e. taking the mean as zero, is known as the *variance*, usually denoted by σ^2, and this is given by:

$$\sigma_x^2 = \frac{1}{T} \int_0^T [x - \langle x \rangle]^2 \, dt \tag{10.5}$$

Equation (10.5) can be expressed in terms of the mean and mean square values separately, as follows. First, multiplying out:

$$\sigma_x^2 = \frac{1}{T} \int_0^T \left(x^2 - 2x\langle x \rangle + \langle x \rangle^2 \right) dt = \frac{1}{T} \int_0^T x^2 \, dt - 2\langle x \rangle \int_0^T x \, dt + \langle x \rangle^2$$

But noting that $\langle x \rangle^2$ is a constant, that from Eq. (10.1), $\frac{1}{T}\int_0^T x\,dt = \langle x \rangle$; and from Eq. (10.3) that $1/T\int_0^T x^2\,dt = \langle x^2 \rangle$ then Eq. (10.5) can be written as:

$$\sigma_x^2 = \langle x^2 \rangle - \langle x \rangle^2 \tag{10.6}$$

where $\langle x^2 \rangle$ is the mean square value and $\langle x \rangle^2$ is the mean value squared, a simpler expression than might have been expected.

The discrete, or digital, version is

$$\sigma_x^2 = \frac{1}{N}\sum_{j=1}^{N}\left(x_j - \langle x \rangle\right)^2 = \langle x^2 \rangle - \langle x \rangle^2 \tag{10.7}$$

The square root of the variance, equal to σ_x, is known as the *standard deviation*.

In a zero-mean waveform, the mean square is equal to the variance and the RMS is equal to the standard deviation. These terms are therefore often interchanged, and some care is necessary to check that this is permissible.

10.2 Amplitude probability distribution and density functions

Figure 10.2(a) shows a random waveform, $x(t)$, from which a sample of length, T, has been selected for analysis. A small part of the waveform is shown enlarged in Fig. 10.2(b). The probability, $P(\underline{x})$, that the waveform has a value less than $\underline{x}$, a particular value of x, is given by counting up the total time that it spends below $\underline{x}$, which will be the sum of periods such as $t_1, t_2, t_3, \ldots$ expressed as a fraction of the total time T, thus,

$$P(\underline{x}) = \frac{1}{T}(t_1 + t_2 + t_3 + \cdots) \tag{10.8}$$

If this is repeated with $\underline{x}$ set to a number of different values of x, a plot of $P(x)$ versus x can be obtained. $P(x)$ is known as the *cumulative amplitude probability distribution*. Although many different shapes are possible, all must have the following features:

(1) The value of $P(x)$ when $x = -\infty$, that is, $P(-\infty)$, must always be zero, since there is no chance that the waveform lies below $-\infty$. It may, of course, be zero up to a value of x greater (less negative) than $-\infty$.

(2) The value of $P(x)$ can never decrease as x becomes more positive.

(3) The value of $P(x)$ when $x = \infty$ must always be 1, since it is certain that all of the waveform is smaller than ∞. It is, however, possible for $P(x)$ to reach 1 at a value of x less than ∞.

A possible cumulative probability distribution is shown in Fig. 10.2(c). The probability that the original waveform has an amplitude between x and $(x + \delta x)$ is $P(x + \delta x) - P(x)$, and if δx is small, this probability can be seen to be equal to $(d[P(x)]/dx)\delta x$. The quantity $d[P(x)]/dx$, which is the slope of the graph of $P(x)$ versus x, is known as the *amplitude probability density function, $p(x)$*, that is,

$$p(x) = \frac{d[P(x)]}{dx} \tag{10.9}$$

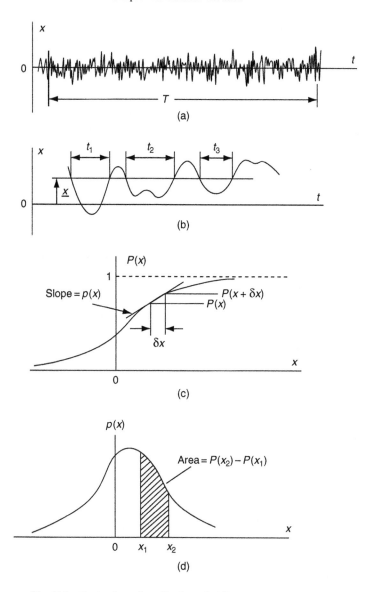

Fig. 10.2   Derivation of amplitude probability and density functions.

Figure 10.2(d) is a sketch of $p(x)$ corresponding to the plot of $P(x)$ shown in Fig. 10.2(c). Since $p(x)$ is the slope of the graph of $P(x)$ versus x, it follows that $P(x)$ is the integral of $p(x)$ from $-\infty$ to x, i.e.,

$$P(x) = \int_{-\infty}^{x} p(x)\mathrm{d}x \qquad (10.10)$$

The probability that any point in the waveform lies between $x = x_1$ and $x = x_2$ is therefore:

$$P(x_2) - P(x_1) = \int_{-\infty}^{x_2} p(x)\mathrm{d}x - \int_{-\infty}^{x_1} p(x)\mathrm{d}x = \int_{x_1}^{x_2} p(x)\mathrm{d}x \qquad (10.11)$$

which is simply the area under the $p(x)$ curve between x_1 and x_2, as shown shaded in Fig. 10.2(d). The *total* area under the probability density curve, $p(x)$, plotted against x, must always be unity, since it is certain that any point in the waveform has a value between $\pm\infty$ (or its minimum and maximum values, if finite). Thus it is always true that

$$\int_{-\infty}^{\infty} p(x)\,\mathrm{d}x = 1 \qquad (10.12)$$

The averages derived in Section 10.1 directly from the time history, the mean value; the mean square value; the root mean square value; the variance and the standard deviation, can also be derived from the probability density function. In practice, this is now a simpler process, since the operation of counting the times spent in the various amplitude bands has already been done to produce $p(x)$. The mean value, $\langle x \rangle$, for example, can be derived from $p(x)$ as follows. In Fig. 10.3, the probability of the waveform being in the band from x to $x + \delta x$ is represented by the area of the strip of width δx, which is $p(x)\,\delta x$. The contribution of this strip to the mean value is $x \cdot p(x)\,\delta x$, and the mean value, $\langle x \rangle$, must be given by the summation of all such strips, so

$$\langle x \rangle = \sum_{-\infty}^{\infty} x \cdot p(x)\delta x \qquad (10.13)$$

If $p(x)$ is known (or is approximated) analytically, then Eq. (10.13) can be written as an integral:

$$\langle x \rangle = \int_{-\infty}^{\infty} x \cdot p(x)\mathrm{d}x \qquad (10.14)$$

The mean square value can be found in a similar way, using either

$$\langle x^2 \rangle = \sum_{-\infty}^{\infty} x^2 \cdot p(x)\delta x \qquad (10.15)$$

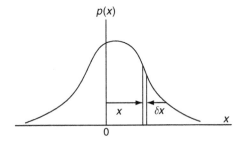

Fig. 10.3   Derivation of the mean value from the probability density function.

for discrete values or

$$\langle x^2 \rangle = \int_{-\infty}^{\infty} x^2 \cdot p(x) dx \qquad (10.16)$$

in the case of analytical representation.

The RMS value is, in either case $\sqrt{\langle x^2 \rangle}$.

For analytical description, the variance is given by:

$$\sigma_x^2 = \int_{-\infty}^{\infty} (x - \langle x \rangle)^2 p(x) dx \qquad (10.17)$$

The discrete form is given by substituting a summation for the integral and δx for dx. Equation (10.17) can be simplified as follows:

$$\sigma_x^2 = \int_{-\infty}^{\infty} \left[(x - \langle x \rangle)^2 p(x) \right] dx = \int_{-\infty}^{\infty} \left[(x^2 - 2x\langle x \rangle + x^2) p(x) \right] dx$$

$$= \int_{-\infty}^{\infty} x^2 p(x) dx - 2\langle x \rangle \int_{-\infty}^{\infty} x p(x) dx + \langle x \rangle^2 \int_{-\infty}^{\infty} p(x) dx = \langle x^2 \rangle - 2\langle x \rangle^2 + \langle x \rangle^2$$

$$\sigma_x^2 = \langle x^2 \rangle - \langle x \rangle^2 \qquad (10.18)$$

where Eqs (10.16), (10.14) and (10.12) have been used. Thus, the variance is the mean square value minus the mean value squared, agreeing with Eq. (10.6). The standard deviation, σ_x is, of course, the square root of the variance, σ_x^2.

In practical, everyday, work, the time histories to be analysed into cumulative probability distributions or probability density functions will usually have been digitized as part of the recording process, and determining the amplitude probability density function, $p(x)$, and the cumulative probability distribution function, $P(x)$ become sorting and counting operations. Assuming that the waveform to be analysed has been sampled at equal intervals of time, say Δ, it is a simple computer operation to sort the digitized amplitudes into bands, and in principle:

$$P(x_i) = \frac{n_i \Delta}{T} = \frac{n_i}{N} \qquad (10.19)$$

where $P(x_i)$ is the cumulative probability distribution value corresponding to amplitude level x_i; T the time duration of the sample; n_i the number of data points having an amplitude less (more negative) than x_i and $N = T/\Delta$ the total number of data points in the sample. The probability density, $p(x_i)$, can then be found by differencing adjacent values of $P(x_i)$. Alternatively, the probability density values can be found first, by counting the number of data points between adjacent levels, and the cumulative distribution by progressively summing the results.

Example 10.1

The waveform shown in Fig. 10.4 at (a) increases and decreases alternately at a constant rate between the two extremes $\pm X$, the rates being random from cycle to cycle. Derive the probability density and cumulative distribution functions.

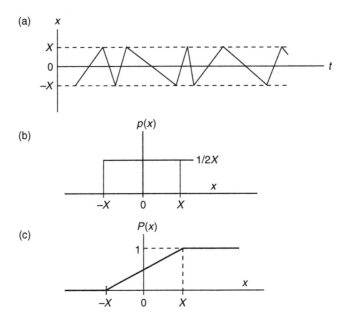

Fig. 10.4   Random waveform discussed in Example 10.1.

Solution

The probability of any point in the waveform being outside the range $\pm X$ is zero, and the probability density $p(x)$ is also zero. Within the range $\pm X$, the probability of any point being between x and $x + \delta x$ is proportional to the time spent between these levels, so $p(x)$ must be constant. The probability density function must therefore be rectangular, as shown in Fig. 10.4(b). Since the probability of the waveform having an amplitude in the range $\pm X$ is 1, that is,

$$\int_{-X}^{X} p(x)\mathrm{d}x = 1$$

the area of the rectangle in Fig. 10.4(b) is 1, and since its width is $2X$, its height must be $1/2X$, as shown.

The cumulative amplitude distribution, $P(x)$, is the integral of the density function with respect to x, and since its value must be 0 at $x = -X$ and 1 at $x = X$, it must be as shown in Fig. 10.4(c).

10.2.1 The Gaussian or Normal Distribution

The majority of probability distributions occurring naturally are of this form, the shape of the probability density curve being the well-known bell shape. The fact that this curve occurs almost universally in natural phenomena can be explained by the *central limit theorem*, which states that if a random time history is made up from the sum of a large number of small, unrelated components, then it will tend to have a Gaussian amplitude distribution. This tendency can be demonstrated in a simple way by the dice-throwing exercise described in the following example.

Example 10.2

Derive the expected probability density functions for the score when (a) a single die is thrown; (b) two dice are thrown, and the results added and (c) three dice are thrown and the results added. (d) Show by plotting the probabilities that the results tend towards a Gaussian distribution as the number of dice thrown increases.

Solution

Part (a)

If a single die is thrown, the probability of any score from 1 to 6 is obviously 1/6; thus we have the discrete uniform distribution shown in Table 10.1.

Part (b)

If two dice are thrown each time, they can fall in $6 \times 6 = 36$ ways, each with probability 1/36. However, there are only ten different total scores (the integers 2–12), because the same score can be obtained in a number of different ways. As an example, a total score of 4 is obtained from the three possible throws: $(1+3)$, $(3+1)$ and $(2+2)$, so the probability of scoring 4 is 3/36. The probability of each possible score is as shown in Table 10.2.

Part (c)

If three dice are thrown each time, and again the resulting numbers are added, there are $6 \times 6 \times 6 = 216$ possible outcomes, but only 16 different scores: for example, a score of 7 or 14 can both be obtained in 15 ways, so the probability of these scores is 15/216. All the probabilities for three dice are as shown in Table 10.3.

Part (d)

The discrete probability functions for one, two and three dice thrown are plotted in Figs 10.5(a), (b) and (c). Figure 10.5(a), for one die, obviously shows a uniform

Table 10.1

Score x	1	2	3	4	5	6
Probability $p(x)$	1/6	1/6	1/6	1/6	1/6	1/6

Table 10.2

Score x	2	3	4	5	6	7	8	9	10	11	12
Probability $p(x)$	1/36	2/36	3/36	4/36	5/36	6/36	5/36	4/36	3/36	2/36	1/36

Table 10.3

Score x	3 or18	4 or 17	5 or 16	6 or 15	7 or 14	8 or 13	9 or 12	10 or 11
Probability $p(x)$	1/216	3/216	6/216	10/216	15/216	21/216	25/216	27/216

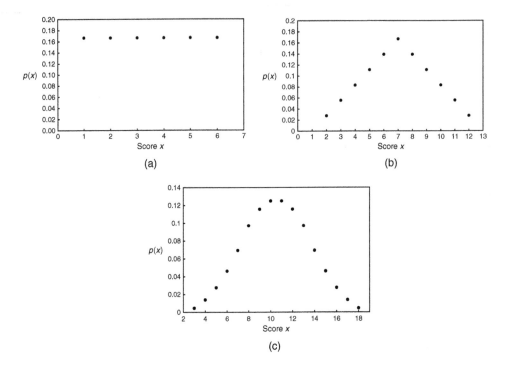

Fig. 10.5   (a) One die thrown, (b) two dice thrown, (c) three dice thrown.

distribution. However, in Fig. 10.5(b), for two dice, it is apparent that the curve is already approaching the Gaussian shape. For three dice, Fig. 10.5(c) is remarkably close to a Gaussian probability density curve.

It should be noted that the probabilities plotted in Figs 10.5(a), (b) and (c) are expected values that would be realized in a very large number, theoretically an infinite number, of throws.

It can be shown that the basic form of the probability density function for a zero-mean Gaussian process is

$$p(x) = ae^{-bx^2} \tag{10.20}$$

where a and b are constants. However, since these constants are almost always chosen to satisfy the two equations, $\int_{-\infty}^{\infty} p(x)\mathrm{d}x = 1$ and $\sigma^2 = \int_{-\infty}^{\infty} x^2 p(\mathrm{d}x)$, where σ is the standard deviation, Eq. (10.20) usually appears in the following form:

$$p(x) = \frac{1}{\sigma\sqrt{2\pi}} e^{-\frac{x^2}{2\sigma^2}} \tag{10.21}$$

If the mean value, m, say, is non-zero, then Eq. (10.21) becomes

$$p(x) = \frac{1}{\sigma\sqrt{2\pi}} e^{-\frac{(x-m)^2}{2\sigma^2}} \tag{10.22}$$

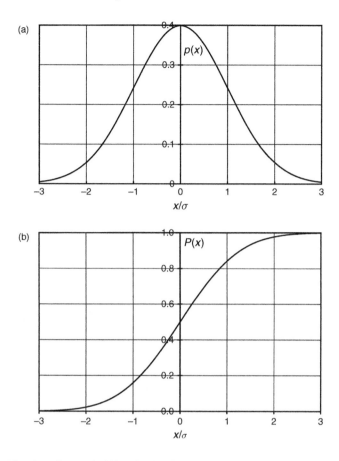

Fig. 10.6 (a) The Gaussian probability density function; (b) the cumulative Gaussian probability distribution function.

It can be seen that $p(x)$ is completely specified when the mean value, m, and the standard deviation, σ, are known. and that the function extends from $x = -\infty$ to ∞, although values of $p(x)$ for x/σ less than –3, or greater than 3, are small.

The probability density, $p(x)$, as given by Eq. (10.21), for zero mean value, is plotted as Fig. 10.6(a). From Eq. (10.10), the cumulative probability distribution, $P(x)$, is always the integral of $p(x)$ with respect to x. In the case of the Gaussian probability density, in spite of the apparent simplicity of the expression, it cannot be integrated easily, and numerical values from tables, spreadsheet programs, etc., must be used. Figure 10.6(b) is a plot of $P(x)$ versus x/σ for values of the latter from –3 to +3, and some numerical values of $P(x)$ are given in Table 10.4. These values of $P(x)$ are cumulative probabilities from $-\infty$ to x/σ. Some tables give different quantities, such as the cumulative value between the limits $\pm\lambda\sigma$, where λ is a constant, and care is necessary in interpreting exactly what values are given by any particular table or program.

The value of $P(x)$ at $x/\sigma = -3.1$ is approximately 0.001, and the value at $x/\sigma = +3.1$ is 0.999, so there is only a 1 in 1000 chance that the waveform will lie outside each

Table 10.4
Probability, $P(x)$, of a Gaussian random waveform lying between $-\infty$ and x.

x/σ	$P(x)$
$-\infty$	0
-3.1	0.0010
-3.0	0.0013
-2.5	0.0062
-2.0	0.0228
-1.5	0.0668
-1.0	0.1587
-0.5	0.3085
0	0.5000
0.5	0.6915
1.0	0.8413
1.5	0.9332
2.0	0.9772
2.5	0.9938
3.0	0.9987
3.1	0.9990
∞	1

(not both) of these limits. This fact is often used when setting the working range of transducers, recorders, exciters, etc. With a truly Gaussian distribution, which theoretically extends to $\pm \infty$, some signal information is always lost. However, ensuring that the system will not overload up to $\pm 3\sigma$, or preferably a little over, will keep the signal loss small.

The following properties of the Gaussian probability distribution are important.

(1) If the input waveform to any linear system has a Gaussian distribution, then so does the output. As an example, if broad band random excitation is applied to a linear single-DOF system with low damping, the output will be narrow band, and look quite different from the input. Nevertheless, its amplitude probability density will still be Gaussian.

(2) Time derivatives of a Gaussian waveform are also Gaussian. So, for example, if a displacement time history has a Gaussian amplitude probability distribution, then so do the corresponding velocity and acceleration waveforms. This obviously also works for time integrals.

(3) The sum of two or more Gaussian signals is also Gaussian.

Example 10.3

The output signal from an accelerometer in a certain test is expected to have a Gaussian, or normal, probability density, with zero mean and an RMS value of 1 V. The recording device, to which the accelerometer is connected, has a switchable maximum recording range of ± 1 V, ± 2 V or ± 3 V. For what percentage of the total recording time could the signal be expected to be correctly recorded for each of the three switch settings?

Solution

Since the signal is assumed to have a zero mean value, the standard deviation, σ, is equal to the RMS value in this case, and is also 1 V.

In Table 10.5, x is the instantaneous value of the voltage, so the ± 1 V recorder setting corresponds to values of x/σ between -1 and $+1$. The corresponding values of $P(x)$ from Table 10.4 are 0.1587 and 0.8413, so the probability of the signal lying between these limits, and therefore of being correctly recorded, is 0.8413–0.1587 = 0.6826 or 68.26%. Alternatively, it can be said that the signal would be correctly recorded for that percentage of the time.

Table 10.5

Recorder range	x/σ ($\sigma 1$)	$P(x)$ Values	Fraction correctly recorded	%Time
± 1 V	± 1	$P(-1) = 0.1587$ $P(+1) = 0.8413$	0.6826	68.3
± 2 V	± 2	$P(-2) = 0.0228$ $P(+2) = 0.9772$	0.9544	95.4
± 3 V	± 3	$P(-3) = 0.0013$ $P(+3) = 0.9987$	0.9974	99.7

This calculation is easily repeated for the other two ranges, and Table 10.5 gives the fraction of the signal likely to be correctly recorded, in terms of probability, and as a percentage of the recording time, for all three.

10.3 The power spectrum

In Section 10.2, we discussed how some of the properties of a random waveform can be expressed in terms of its amplitude probability distribution. However, if we wish to predict or measure the response of systems to random inputs, we must know how the power is spread over the frequency range, and this information is conveniently represented by the *power spectrum*. Strictly, the rigorous derivation [10.1] of the power spectrum is as the Fourier transform of the autocorrelation function, which we shall discuss later, and it is double-sided, i.e. it extends to negative frequencies. We shall look at this more mathematical approach later, but for most practical work, the following derivation from the Fourier series, leading to a single-sided definition of the power spectrum, is adequate.

10.3.1 Power Spectrum of a Periodic Waveform

Although the main application of the power spectrum is in dealing with random vibration, it can also be used with periodic waveforms. Let us look at this simpler application first. From Eq. (9.3), we know that a *periodic* time history, $x(t)$, say, can be expressed as a Fourier series:

$$x(t) = a_0 + \sum_{n=1}^{\infty} [a_n \cos n\omega_0 t + b_n \sin n\omega_0 t] \quad n = 1, 2, 3, \ldots, \infty \tag{10.23}$$

where a_0 is the mean level, a_n and b_n are the amplitudes of the cosine and sine components that together represent the waveform and ω_0 is the fundamental frequency in rad./s. From Eq. (9.2), $\omega_0 = 2\pi/T$, where T is the period of the waveform, and this can be substituted into Eq. (10.23). If we also remove the mean level, a_0, and deal with it separately, Eq. (10.23) can be written as:

$$x(t) = \sum_{n=1}^{\infty} \left[a_n \cos n\frac{2\pi}{T} t + b_n \sin n\frac{2\pi}{T} t \right] \quad n = 1, 2, 3, \ldots, \infty \tag{10.24}$$

The 'power' (actually the mean square value) associated with any single frequency component, at $f_{(n)}$, say, where $f_{(n)} = n/T$, in Hz, for $n = 1, 2, 3, \ldots$, is seen to be $\frac{1}{2}(a_n^2 + b_n^2)$. The *power spectrum* of $x(t)$ consists of a plot of *discrete values* of this power versus the discrete frequencies, $f_{(n)}$. The notation $f_{(n)}$ is used to indicate discrete frequencies of Fourier components, in Hz, associated with the integer n, to avoid confusion with f_n, which is used to indicate an undamped natural frequency in hertz.

The use of the term 'power' merely indicates that the quantity involved is squared, by analogy with the power in an electric circuit, which is *proportional to*, but not necessarily equal to, the voltage or current squared.

Example 10.4

Plot the power spectrum of the ± 1 V square wave discussed in Example 9.1, if the period of the wave is 1 s.

Solution

In Example 9.1, depending upon whether the square wave was treated as an even or an odd function, the Fourier coefficients were given by either a cosine series or a sine series. It was shown that this affects only the phases of the coefficients, not their magnitudes, and since the power spectrum does not depend on phase, either can be used. So taking the sine series, Eq. (H) in Example 9.1 gave

$$x(t) = \frac{4}{\pi} \left[\sin\left(\frac{2\pi}{T}\right) t + \frac{1}{3}\sin 3\left(\frac{2\pi}{T}\right) t + \frac{1}{5}\sin 5\left(\frac{2\pi}{T}\right) t + \frac{1}{7}\sin 7\left(\frac{2\pi}{T}\right) t + \cdots \right] \tag{A}$$

Table 10.6 lists the frequency, $f_{(n)} = n/T$, with $T = 1$; the Fourier components a_n and b_n and the power, $\frac{1}{2}(a_n^2 + b_n^2)$, at each value of n.

Table 10.6

n	1	2	3	4	5	6	7	8	9
$f_{(n)}$ (Hz)	1	2	3	4	5	6	7	8	9
a_n (V)	0	0	0	0	0	0	0	0	0
b_n (V)	$4/\pi$	0	$4/(3\pi)$	0	$4/(5\pi)$	0	$4/(7\pi)$	0	$4/(9\pi)$
$\frac{1}{2}(a_n^2 + b_n^2)$ (V)^2	0.8105	0	0.0900	0	0.0324	0	0.0165	0	0.0100

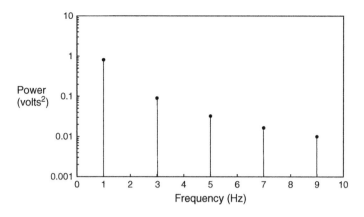

Fig. 10.7

The total 'power' in the waveform, which is 1 V^2, should be given by summing the power in all the harmonics. For the first nine terms shown in Table 10.6 this is 0.96 V^2, the remaining 0.04 V^2 therefore being accounted for by higher frequency terms.

Figure 10.7 is a plot of the discrete power spectrum of the waveform up to 10 Hz.

It can be seen from Example 10.4 that the power spectrum of a zero-mean periodic function consists of discrete amounts of power concentrated at the fundamental frequency and its harmonics, which are spaced at intervals of $1/T$ Hz. If T is small, then these frequency intervals may be quite large, for example as in Fig. 10.7. In cases like this it can be seen that the power spectrum should remain discrete, and should *not* be expressed in the form of power spectral density, discussed below, which could lead to serious errors.

10.3.2 The Power Spectrum of a Random Waveform

If we now wish to find the frequency content of a random waveform, we can use the same method as we did for the periodic waveform considered above, by assuming that the sample analysed is one period of a very long periodic function. With this assumption, the Fourier coefficients, a_n and b_n, as defined by Eq. (10.24), can be found, typically by the use of the DFT, and will lie at the discrete frequencies $1/T, 2/T, 3/T, \ldots$, corresponding to $n = 1, 2, 3, \ldots$.

Figure 10.8 illustrates the derivation of a power spectrum in this way. The random time history $x(t)$ is shown at (a), and a portion of length T seconds has been selected for analysis. This is now regarded as one period of a periodic waveform, and its Fourier coefficients, a_n and b_n, defined by Eq. (10.24), are shown at (b)(i) and (b)(ii), respectively. Individually, these are random, but this is largely due to phase, and if phase is eliminated by calculating the mean square value, or 'power', $\frac{1}{2}(a_n^2 + b_n^2)$, in each pair of coefficients, at each frequency, the result shown at (c) is obtained. This is the *discrete power spectrum* of the original time history. The length of each line

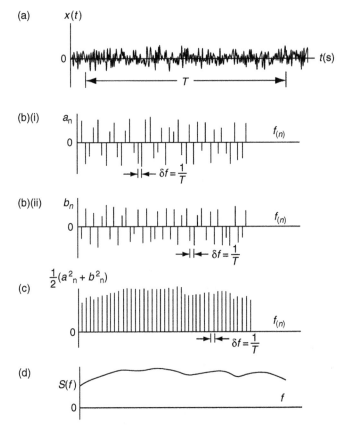

Fig. 10.8   Derivation of a power spectrum.

represents the power in the Fourier components at that particular frequency, $f_{(n)}$, and its units are the square of the physical quantity represented, for example $(V)^2$, $(mm)^2$, g^2, etc.

Power spectral density

Power spectra are sometimes left in this discrete form, and it is usually the appropriate format for periodic data, as discussed above. In the case of random data, however, the assumed period, T, is usually much longer, and the spacing of the spectral lines, $1/T$ Hz, is much closer. It is then usual to express the power as a continuous *power spectral density* (PSD) curve. The PSD, $S(f_{(n)})$, at any discrete frequency, $f_{(n)}$, is given by dividing the power at that frequency by the bandwidth over which it is considered to be distributed, say δf. This can be assumed to be equal to the interval between adjacent Fourier components, which is $1/T$ Hz. Thus $\delta f = 1/T$, and

$$S(f_{(n)}) = \frac{\frac{1}{2}(a_n^2 + b_n^2)}{\delta f} = \frac{T}{2}(a_n^2 + b_n^2) \qquad (10.25)$$

It should be pointed out that the derivation of Eq. (10.25) used above is an intuitive one, rather than mathematically rigorous. Strictly, in a more rigorous derivation [10.1], use is made of the fact that the Fourier transform of Eq. (10.24) consists of a series of Dirac delta functions in the frequency domain, spaced at intervals of $1/T$ Hz. However, the effect is the same, and for practical work Eq. (10.25) is an adequate definition of the PSD.

Power spectral density has the units (physical quantity)^2 per unit of frequency, for example (V)^2/Hz, g^2/Hz and so on, and is independent of the frequency intervals at which it is computed, a major advantage, making it easy to compare PSD plots from different sources, or obtained by different methods.

The values of $S(f_{(n)})$ are still, strictly, discrete quantities, plotted at frequencies, $f_{(n)}$, which are at the centers of each band, of width δf or $1/T$. If the frequencies are closely spaced, however, the values of $S(f_{(n)})$ can be considered to form a continuous power spectrum, $S(f)$, plotted against the continuous frequency f. The final plot, in Fig. 10.8, at (d) is therefore labeled in this way.

It is important to note that $S(f)$, as derived above, is single-sided, and does not extend to negative frequencies. This is usual in practical work, but it should be pointed out that in theoretical derivations, it is necessary to introduce a double-sided version, where the power is spread over the frequency range $\pm \infty$. This will be referred to as $S(f_{\pm})$ as a reminder. At any frequency, f, then $S[f_{\pm}] = \frac{1}{2}S(f)$. It should also be mentioned that PSD plots can also be presented using rad/s as the unit of frequency. However, the standard method for practical work is single-sided, with frequencies in Hz.

Mean square value or variance from the PSD

The total power (or mean square value) in the original waveform is equal to the sum of the small amounts of power in each Fourier component. The power in any one component is $\frac{1}{2}(a_n^2 + b_n^2)$, but since $\delta f = 1/T$, this can be written as $(T/2)(a_n^2 + b_n^2)\delta f$. From Eq. (10.25):

$$\frac{T}{2}(a_n^2 + b_n^2)\delta f = S(f_{(n)})\delta f \qquad (10.26)$$

The power in the sum of all the Fourier components, equal to the mean square value of the original waveform, $\langle x^2(t) \rangle$, is given by summation:

$$\langle x^2(t) \rangle = \sum_{n=1}^{\infty} \frac{T}{2}(a_n^2 + b_n^2)\delta f = \sum_{n=1}^{\infty} S(f_{(n)})\delta f \qquad (10.27)$$

Since $x(t)$ was defined by Eq. (10.24) as having zero mean value, the mean square value $\langle x^2(t) \rangle$ is equal to the variance, σ^2, in this case, and therefore:

$$\langle x^2(t) \rangle = \sigma^2 = \sum_{n=1}^{\infty} S(f_{(n)})\delta f \qquad (10.28)$$

and the mean square value or the variance is the area under the PSD plot.

The *measured* PSD of a random time history will never be an exact analytical function, and finding the area under its plot will, in practice, always be a summation rather than an integration. However, PSD functions can be analytic, or idealized as such, for example in vibration test specifications, and it is then reasonable to say that in the limit, as $\delta f \to 0$, the discrete spectral density, $S(f_{(n)})$, becomes the continuous function, $S(f)$, and Eq. (10.28) becomes the integral:

$$\sigma^2 = \int_0^\infty S(f)\mathrm{d}f \tag{10.29}$$

The same arguments can be used to show that the mean square value of a waveform between any two frequencies, f_1 and f_2, is

$$\sigma^2_{f_1,f_2} = \int_{f_1}^{f_2} S(f)\mathrm{d}f \tag{10.30}$$

Power spectra of time derivatives and integrals

A power spectrum can be presented in terms of the time derivative or the time integral of the original quantity. If, for example, we have computed the PSD function of a random *displacement* history, $x(t)$, then the PSD function in terms of *velocity*, $\dot{x}(t)$, or *acceleration*, $\ddot{x}(t)$, is easily found. Similarly, the displacement PSD can be found if either the velocity or acceleration PSD is known.

We saw above that the Fourier series of a periodic waveform can be written in the form of Eq. (10.24):

$$x(t) = \sum_{n=1}^\infty \left[a_n \cos n \frac{2\pi}{T} t + b_n \sin n \frac{2\pi}{T} t \right] \quad n = 1, 2, 3, \ldots, \infty \tag{10.31}$$

where $x(t)$ is now specifically a displacement time history. Since $f_{(n)} = n/T$, Eq. (10.31) can be written as:

$$x(t) = \sum_{n=1}^\infty \left(a_n \cos 2\pi f_{(n)} t + b_n \sin 2\pi f_{(n)} t \right) \tag{10.32}$$

Differentiating Eq. (10.32) twice with respect to t, we have

$$\dot{x}(t) = \sum_{n=1}^\infty 2\pi f_{(n)} \left[a_n \left(-\sin 2\pi f_{(n)} t \right) + b_n \left(\cos 2\pi f_{(n)} t \right) \right] \tag{10.33}$$

and

$$\ddot{x}(t) = \sum_{n=1}^\infty \left(2\pi f_{(n)} \right)^2 \left[a_n \left(-\cos 2\pi f_{(n)} t \right) + b_n \left(-\sin 2\pi f_{(n)} t \right) \right] \tag{10.34}$$

The power associated with the discrete frequency $f_{(n)}$ can be expressed in the following different ways:

In terms of *displacement,* from Eq. (10.32):   power $= \frac{1}{2}(a_n^2 + b_n^2)$;

In terms of *velocity,* from Eq. (10.33):   power $= \left(2\pi f_{(n)} \right)^2 \frac{1}{2}(a_n^2 + b_n^2)$;

In terms of *acceleration*, from Eq. (10.34), $\text{power} = \left(2\pi f_{(n)}\right)^4 \frac{1}{2}\left(a_n^2 + b_n^2\right)$.

Defining the displacement, velocity and acceleration discrete PSD values at frequency $f_{(n)}$ as $S_x(f_{(n)})$, $S_{\dot{x}}(f_{(n)})$ and $S_{\ddot{x}}(f_{(n)})$, respectively, they are given by dividing the power by the frequency bandwidth over which they are considered to be spread, i.e. δf or $1/T$ Hz. Thus,

$$S_x\left(f_{(n)}\right) = \frac{T\left(a_n^2 + b_n^2\right)}{2} \tag{10.35}$$

$$S_{\dot{x}}\left(f_{(n)}\right) = \left(2\pi f_{(n)}\right)^2 \frac{T\left(a_n^2 + b_n^2\right)}{2} = \left(2\pi f_{(n)}\right)^2 S_x\left(f_{(n)}\right) \tag{10.36}$$

$$S_{\ddot{x}}\left(f_{(n)}\right) = \left(2\pi f_{(n)}\right)^4 \frac{T\left(a_n^2 + b_n^2\right)}{2} = \left(2\pi f_{(n)}\right)^4 S_x\left(f_{(n)}\right) \tag{10.37}$$

In the limit, as the frequency intervals become infinitesimal, the discrete spectral density functions become the continuous functions, $S_x(f)$, $S_{\dot{x}}(f)$ and $S_{\ddot{x}}(f)$, and we have the following relationships:

$$S_{\dot{x}}(f) = (2\pi f)^2 S_x(f) \tag{10.38a}$$

$$S_{\ddot{x}}(f) = (2\pi f)^4 S_x(f) \tag{10.38b}$$

$$S_x(f) = \frac{1}{(2\pi f)^2} S_{\dot{x}}(f) \tag{10.39a}$$

$$S_x(f) = \frac{1}{(2\pi f)^4} S_{\ddot{x}}(f) \tag{10.39b}$$

Example 10.5

A vibration test table is set up to produce a constant PSD level of 0.02 g^2/Hz over the frequency range 10–1000 Hz. Find
(a) The RMS acceleration level of the table.
(b) The RMS displacement of the table, and the total movement if the displacement is Gaussian and limited to ± 3 times the RMS value.

Solution

Let S_g = acceleration PSD level in g^2/Hz;
$\qquad S_{\ddot{x}}$ = acceleration PSD level in m/s^2/Hz;
$\qquad S_x$ = displacement PSD level in m^2/Hz;
$\qquad \sigma_g$ = RMS level in g units ($1g = 9.81$ m/s^2);
$\qquad \sigma_x$ = RMS level in meters.

Part (a):
From Eq. (10.30),

$$\sigma_g^2 = \int_{f_1}^{f_2} S_g df \tag{A}$$

$$S_g = 0.02 g^2/Hz \quad \text{for} \quad 10 < f < 1000$$
$$S_g = 0 \quad \quad \quad \text{for} \quad f < 10 \quad \text{and} \quad f > 1000$$

From Eq. (A),

$$\sigma_g^2 = \int_{10}^{1000} 0.02 \, df = [0.02f]_{10}^{1000} = 0.02(1000 - 10) = 19.8 \, g^2.$$

RMS acceleration of table $= \sigma_g = \sqrt{19.8} = 4.45 \, g$

Part (b):
In this case, since acceleration is to be converted to displacement, consistent units must be used, and the acceleration must be expressed in m/s². Since $1g = 9.81$ m/s²: $S_{\ddot{x}} = (9.81^2 \, S_g) = (9.81^2 \times 0.02) = 1.925$ m/s²/Hz, in the range 10–1000 Hz, and zero elsewhere. Then from Eq. (10.39b):

$$S_x = \frac{1}{(2\pi f)^4} S_{\ddot{x}} = \frac{1.925}{(2\pi)^4} f^{-4} = (1.235 \times 10^{-3}) f^{-4} \tag{B}$$

From Eq. (10.30) and Eq. (B),

$$\sigma_x^2 = \int_{f_1}^{f_2} S_x \, df = (1.235 \times 10^{-3}) \int_{10}^{1000} f^{-4} df = (1.235 \times 10^{-3}) \left[-\frac{1}{4} f^{-3} \right]_{10}^{1000}$$

$$= (0.3088 \times 10^{-6}) m^2 \quad \text{and} \quad \sigma_x = 0.556 \times 10^{-3} m = 0.556 \, mm.$$

If the displacement has a Gaussian amplitude distribution limited to $\pm 3\sigma_x$ the total movement is 6×0.556 mm $= 3.34$ mm.

10.4 Response of a system to a single random input

Many practical problems involving the response of a system to random excitation can be solved by considering a single random input function. As will be seen, the process is then relatively straightforward, requiring only two ingredients: (1) the power spectrum of the input and (2) the frequency response function (FRF) of the system. We have discussed the power spectrum, and its derivation; now we must review the FRF.

10.4.1 The Frequency Response Function

The FRF, usually denoted by $H(\omega)$ or $H(f)$, depending on whether it is expressed in terms of rad/s or Hz, respectively, is simply the ratio of the steady-state response of a system to an applied sinusoidal input, which can be a force, an imposed displacement, or almost any other quantity. The FRF is usually expressed in complex form, but can also be expressed in magnitude and phase form. When calculating an FRF, the sinusoidal input is considered to have started at $-\infty$, and any transient response to have died away. The response is therefore also sinusoidal, but generally different in magnitude and phase from the input.

The FRFs of several forms of single-DOF system were introduced in Section 4.1.3. As has been seen, FRFs of single-DOF systems are easily calculated, or an FRF can, of course, be measured, by applying a sinusoidal input, and measuring the response.

The FRFs of multi-DOF systems are usually calculated by the normal mode summation method, introduced in Chapter 6. This allows the response of a multi-DOF system to be expressed as the summation of several single-DOF systems, one for each normal mode.

The FRF of a system is often referred to as its 'transfer function' but this term should strictly be reserved for the response of the system expressed in Laplace notation. The two terms are, however, closely related, and we saw in Chapter 3 that if the Laplace operator, s, in a system transfer function, is replaced by $i\omega$, the transfer function becomes the FRF, $H(\omega)$, in complex notation.

10.4.2 *Response Power Spectrum in Terms of the Input Power Spectrum*

Let the time history, $x(t)$, of length T, of the input to a system, be represented by the Fourier series:

$$x(t) = \sum_{n=1}^{\infty} \left(a_n \cos\ 2\pi f_{(n)}\, t + b_n \sin\ 2\pi f_{(n)} t \right) \tag{10.40}$$

The frequency interval, δf, between adjacent Fourier components is $1/T$. The power in a single component, at frequency f_n, is $\frac{1}{2}(a_n^2 + b_n^2)$, and the corresponding PSD is

$$S_x\left(f_{(n)}\right) = \frac{\frac{1}{2}\left(a_n^2 + b_n^2\right)}{\delta f} = \frac{T}{2}\left(a_n^2 + b_n^2\right) \tag{10.41}$$

If we let the response or output time history of the system be $y(t)$, then it can be represented by a Fourier series in the same way as Eq. (10.40), except that both the coefficients a_n and b_n will be multiplied by the modulus of the FRF of the system, $\left|H(f_{(n)})\right|$, at each component frequency $f_{(n)}$. The phase of each component will also be shifted by ϕ_n, the phase angle of the FRF at frequency $f_{(n)}$. If we are only interested in the power in the output, however, the phase shift has no effect, and it is easily shown that the output power at frequency $f_{(n)}$ is $\left|H(f_{(n)})\right|^2 \frac{1}{2}\left(a_n^2 + b_n^2\right)$, and the corresponding PSD is

$$S_y\left(f_{(n)}\right) = \left|H(f_{(n)})\right|^2 \frac{\frac{1}{2}\left(a_n^2 + b_n^2\right)}{\delta f} = \left|H(f_{(n)})\right|^2 \frac{T}{2}\left(a_n^2 + b_n^2\right) \tag{10.42}$$

Thus from Eqs (10.41) and (10.42):

$$S_y\left(f_{(n)}\right) = \left|H(f_{(n)})\right|^2 S_x\left(f_{(n)}\right) \tag{10.43}$$

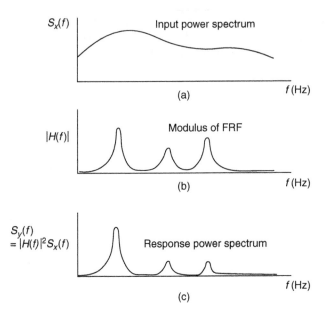

Fig. 10.9   Response PSD from input PSD and system FRF.

In the limit, as the frequency intervals become infinitesimal, and the discrete frequencies, $f_{(n)}$, merge to become the continuous frequency, f, we have one of the most useful results in practical random vibration analysis:

$$S_y(f) = |H(f)|^2 S_x(f) \tag{10.44}$$

where $S_x(f)$ is the power spectral density of the input or forcing function; $S_y(f)$ the power spectral density of the output or response and $|H(f)|$ the modulus of the steady-state frequency response function relating input and output. So in the case of a single random input to a system, the response can be calculated without requiring the phase characteristics of the system.

Figure 10.9 is a sketch showing a typical implementation of Eq. (10.44). The input power spectrum $S_x(f)$ is shown at (a). This could be a PSD plot derived from a random waveform, representing the force, displacement or other input to the system. The modulus of the frequency response function, $|H(f)|$, for the system, is shown at (b). In this case the system clearly has three significant vibration modes. Figure 10.9(c) is the PSD of the response $S_y(f)$ at the single output point for which $|H(f)|$ was defined. The response at other points, for the same input, could easily be found by defining alternate FRFs.

10.4.3 Response of a Single-DOF System to a Broadband Random Input

Many day-to-day problems in random vibration involve only the response of single-DOF system to a single random input. If the damping of the system is fairly low, and the input spectrum, in terms of PSD, is fairly flat in the region of the resonant frequency, several simplifications are possible, leading to useful simple expressions for the mean square and RMS value of the response.

Single-DOF system with random applied force excitation

Consider first the single-DOF system shown in Fig. 10.10. A random external force F is applied, and the displacement of the mass is z. From Eq. (4.1), the equation of motion is $m\ddot{z} + c\dot{z} + kz = F$, and the corresponding FRF, in complex form, with the force vector, $\underline{F}$, as the input, and the displacement vector, $\underline{z}$, as the response, is given by Eq. (4.21):

$$\frac{\underline{z}}{\underline{F}} = H(f) = \frac{(1 - \Omega^2) - i(2\gamma\Omega)}{m\omega_n^2\left[(1 - \Omega^2)^2 + (2\gamma\Omega)^2\right]} \tag{10.45}$$

where

$\omega_n = 2\pi f_n$, $\Omega = \omega/\omega_n = f/f_n$, ω is the excitation frequency in rad/s, f the excitation frequency in Hz, ω_n the undamped natural frequency in rad/s, f_n the natural frequency in Hz and $\gamma = c/2m\omega_n$ is the non-dimensional viscous damping coefficient.

In this case we require only the modulus, $|H(f)|$, which is

$$|H(f)| = \frac{1}{m\omega_n^2\sqrt{(1 - \Omega^2)^2 + (2\gamma\Omega)^2}} = \frac{1}{m(2\pi f_n)^2\sqrt{(1 - \Omega^2)^2 + (2\gamma\Omega)^2}} \tag{10.46}$$

Defining the input force spectral density as $S_F(f)$, and the response displacement spectral density as $S_z(f)$, we have from Eq. (10.44):

$$S_z(f) = |H(f)|^2 S_F(f) \tag{10.47}$$

Substituting Eq. (10.46) into Eq. (10.47):

$$S_z(f) = \frac{S_F(f)}{m^2(2\pi f_n)^4\left[(1 - \Omega^2)^2 + (2\gamma\Omega)^2\right]} \tag{10.48}$$

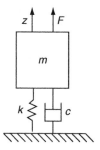

Fig. 10.10 Single-DOF system with applied external force.

Equation (10.48) is the general expression giving the displacement power spectral density, $S_z(f)$, in terms of the applied force power spectral density, $S_F(f)$, of a single-DOF system as shown in Fig. 10.10, with undamped natural frequency, f_n, in Hz, and non-dimensional damping coefficient, γ. The mean square displacement response, $\langle z^2 \rangle$, equal to the variance of the displacement, σ_z^2, if the input force, and hence the displacement, has zero mean value, can be found from Eq. (10.29), which applied to this case is

$$\langle z^2 \rangle = \sigma_z^2 = \int_0^\infty S_z(f)\,\mathrm{d}f \tag{10.49}$$

Substituting Eq. (10.48) into Eq. (10.49) gives

$$\langle z^2 \rangle = \sigma_z^2 = \frac{1}{m^2(2\pi f_n)^4} \int_0^\infty \frac{S_F(f)}{(1-\Omega^2)^2 + (2\gamma\Omega)^2}\,\mathrm{d}f \tag{10.50}$$

Equation (10.50) is a general expression for the displacement mean square or variance. Exact evaluation requires $S_F(f)$ to be known, but a useful result can be derived by setting it to a constant value, S_F, at all frequencies. With this assumption:

$$\sigma_z^2 = \frac{S_F}{m^2(2\pi f_n)^4} \int_0^\infty \frac{1}{(1-\Omega^2)^2 + (2\gamma\Omega)^2}\,\mathrm{d}f$$

Noting that $f = f_n\Omega$ and $\mathrm{d}f = f_n\mathrm{d}\Omega$:

$$\sigma_z^2 = \frac{S_F f_n}{m^2(2\pi f_n)^4} \int_0^\infty \frac{1}{(1-\Omega^2)^2 + (2\gamma\Omega)^2}\,\mathrm{d}\Omega \tag{10.51}$$

The integral is a standard form, and equal to $\pi/4\gamma$, so Eq. (10.51) can be written as:

$$\sigma_z^2 = \frac{S_F}{64\pi^3 m^2 f_n^3 \gamma} \quad\text{or}\quad \sigma_z = \frac{1}{8m}\left(\frac{S_F}{\pi^3 f_n^3 \gamma}\right)^{\frac{1}{2}} \tag{10.52}$$

Since $k = m(2\pi f_n)^2$, Eq. (10.52) can also be written as:

$$\sigma_z^2 = \frac{\pi f_n S_F}{4k^2 \gamma} \quad\text{or}\quad \sigma_z = \frac{1}{2k}\left(\frac{\pi f_n S_F}{\gamma}\right)^{\frac{1}{2}} \tag{10.53}$$

Equation (10.52) was derived by assuming that the input force spectral density, S_F, remains constant up to infinite frequency, and it would appear to give a gross over-estimate of the displacement response in practical cases. However, nearly all the response occurs in a fairly narrow band around the natural frequency, f_n, and provided the excitation is reasonably constant over this band, the actual error is surprisingly small. If the half-power bandwidth of the single-DOF system is defined as $2\gamma f_n$ Hz, then it can be shown by numerical integration that for values of γ up to 0.1 of critical, even if the excitation bandwidth extends for only five times the half-power bandwidth each side of the natural frequency, f_n, Eq. (10.53) over-estimates the *RMS* displacement response by only about 3%. The error increases to about 7% if the excitation extends for only 2.5 times the half-power bandwidth each side of the natural frequency. If the accuracy of Eq. (10.53) is doubted, in particular cases, perhaps because the input PSD function is not flat around the resonant frequency, it is quite easy to integrate Eq. (10.51) numerically, provided the shape of S_F is known.

Example 10.6

A proposed environmental vibration test on a radar pod, attached to the wing of a high performance aircraft, involves applying a single lateral random force to the pod by means of an exciter and rigid connecting rod, as shown in Fig. 10.11 at (a). The pod, considered rigid, is attached to the wing by a flexible pylon, and the combination has a lateral mode of vibration at $f_n = 15\,\mathrm{Hz}$, with a non-dimensional viscous damping coefficient equal to $\gamma = 0.05$ of critical. Its lateral vibration behavior at the exciter attachment point can be represented by the single-DOF lumped equivalent system shown in Fig. 10.11 at (b), where the equivalent mass m is 100 kg, and the natural frequency and damping coefficient are as given above. The applied force time history is $F(t)$ and the resulting displacement history is $y(t)$. The force power spectral density, defined at the exciter attachment rod, is held constant at $4000\,\mathrm{N^2/Hz}$ in the range 7.5–50 Hz, and is zero outside these limits. The force supplied by the exciter has a Gaussian probability density, with zero mean, but is limited at source to 3 times the RMS value in each direction.
Find
(a) the RMS (or standard deviation) value, σ_F, of the force required from the exciter, and its maximum and minimum values.
(b) the RMS (or standard deviation) displacement, σ_y, of the point on the pod where the exciter rod is attached, and the total travel of the exciter rod.

Solution

Part (a)
From Eq. (10.30), the mean square value of a random time history (equal to the variance if the mean value is zero), between the frequency limits f_1 and f_2, is given by integrating the PSD function between these limits. Applied to this case:

$$\langle F^2(t) \rangle = \sigma_F^2 = \int_{f_1}^{f_2} S_F(f)\mathrm{d}f \qquad (A)$$

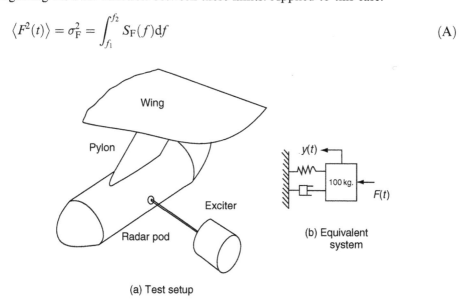

(a) Test setup

Fig. 10.11   Vibration test on a radar pod discussed in Example 10.6.

If $S_F(f)$ has the constant value $S_F = 4000$ N^2/Hz between 7.5 and 50 Hz:

$$\sigma_F^2 = \int_{7.5}^{50} 4000 \, df = [4000f]_{7.5}^{50} = 170 \times 10^3 \, \text{N}^2 \quad \text{and} \quad \sigma_F = 412 \, \text{N}$$

The RMS force required from the exciter is 412 N, and since the force is limited to ± 3 times the RMS value in this case, the maximum and minimum values are ± 1236 N.

Part (b)
The natural frequency of the system is 15 Hz, and the half-power bandwidth of the single mode is $2\gamma f_n = (2 \times 0.05 \times 15) = 1.5$ Hz. The excitation PSD is constant in the band from 7.5 Hz to 50 Hz, and thus extends from 5 times the half-power bandwidth below the natural frequency, to over 23 times above it. The error in the RMS value due to using Eq. (10.52) will therefore be negligible, at less than 3%. The RMS displacement at the exciter attachment point (also equal to the standard deviation, since the mean value is zero) is therefore given by Eq. (10.52):

$$\sigma_y = \frac{1}{8m} \left(\frac{S_F}{\pi^3 f_n^3 \gamma} \right)^{\frac{1}{2}} \tag{B}$$

where $m = 100$ kg; $S_F = 4000$ N^2/Hz; $f_n = 15$ Hz; $\gamma = 0.05$
The RMS value, or standard deviation, of the point on the pod where the exciter is attached is
$\sigma_y = 1.093 \times 10^{-3}$ m $= 1.093$ mm
Since the input force was assumed to have a Gaussian probability density, and the system is assumed linear, then the displacement response will also be Gaussian. Assuming that this is limited, like the input force, to 3 times the standard deviation in each direction, then the maximum exciter travel is $6\sigma_y$ or $6 \times 1.093 = 6.59$ mm peak to peak.

Single-DOF system with random base motion
Another single-DOF configuration with many practical applications is shown schematically in Fig. 10.12. The frequency response functions were discussed in Chapter 4. The input is the base motion, and for random vibration this will usually

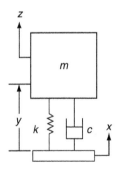

Fig. 10.12   Single-DOF system with base excitation.

be defined in terms of displacement, velocity or acceleration power spectral density. Given the modulus of the FRF between the base motion and the required response, the PSD of the response, and hence its mean square and RMS values, can be calculated in much the same way as for the applied force case discussed above. The response parameter may be the displacement y of the mass relative to the base, the corresponding relative velocity, $\dot{y}$, the absolute displacement, z, of the mass, one of its time derivatives, $\dot{z}$ or $\ddot{z}$, and so on. Some of the FRFs derived in Chapter 4 are as follows. They are given here only in modulus form, since we do not need phase.

From Eq. (4.26), we have

$$\frac{|y|}{|x|} = \frac{\Omega^2}{\sqrt{(1-\Omega^2)^2 + (2\gamma\Omega)^2}} \tag{10.54}$$

where, as before, $\Omega = \omega/\omega_n = f/f_n$, and, ω is the excitation frequency in rad/s $= 2\pi f$, ω_n the undamped natural frequency in rad/s, f the excitation frequency in Hz, f_n the natural frequency in Hz and $\gamma = c/2m\omega_n$ the non-dimensional viscous damping coefficient.

The moduli $|\dot{y}|/|\dot{x}|$ and $|\ddot{y}|/|\ddot{x}|$ are the same as $|y|/|x|$, and from Eq. (4.32):

$$\frac{|y|}{|\dot{x}|} = \frac{1}{\omega}\left(\frac{|y|}{|x|}\right), \quad \frac{|y|}{|\ddot{x}|} = \frac{1}{\omega^2}\left(\frac{|y|}{|x|}\right), \quad \frac{|\dot{y}|}{|x|} = \omega.\left(\frac{|y|}{|x|}\right), \quad \frac{|\ddot{y}|}{|x|} = \omega^2\left(\frac{|y|}{|x|}\right)$$

A frequently required FRF is the relative displacement, y, between the mass and the base, per unit base acceleration, $\ddot{x}$. From Eq. (10.54) its modulus is

$$\left|H_{y\ddot{x}}(\omega)\right| = \frac{|y|}{|\ddot{x}|} = \frac{1}{\omega^2}\left(\frac{|y|}{|x|}\right) = \frac{\Omega^2}{\omega^2\sqrt{(1-\Omega^2)^2 + (2\gamma\Omega)^2}} = \frac{1}{\omega_n^2\sqrt{(1-\Omega^2)^2 + (2\gamma\Omega)^2}} \tag{10.55}$$

or

$$\left|H_{y\ddot{x}}(f)\right| = \frac{1}{(2\pi f_n)^2\sqrt{(1-\Omega^2)^2 + (2\gamma\Omega)^2}} \tag{10.56}$$

If the absolute motion of the mass rather than its value relative to the base is required, Eq. (4.35) gives

$$\frac{|z|}{|x|} = \frac{|\dot{z}|}{|\dot{x}|} = \frac{|\ddot{z}|}{|\ddot{x}|} = \sqrt{\frac{1 + (2\gamma\Omega)^2}{(1-\Omega^2)^2 + (2\gamma\Omega)^2}} \tag{10.57}$$

which is known as the transmissibility.

A simple expression for the RMS or standard deviation of the displacement, y, of the spring, k, in Fig. 10.12, can be found, given the acceleration spectral density of the base, $S_{\ddot{x}}(f)$, as follows. First, the PSD function of the displacement, $S_y(f)$, is given by Eq. (10.44), which in this case becomes

$$S_y(f) = \left|H_{y\ddot{x}}(f)\right|^2 S_{\ddot{x}}(f) \tag{10.58}$$

where $|H_{y\ddot{x}}(f)|^2$ is given by Eq. (10.56). Substituting Eq. (10.56) into Eq. (10.58):

$$S_y(f) = \frac{S_{\ddot{x}}(f)}{(2\pi f_n)^4\left((1-\Omega^2)^2+(2\gamma\Omega)^2\right)} \tag{10.59}$$

The mean square value, and, if the input is assumed to have zero mean, the variance, of y, equal to σ_y^2, is given by Eq. (10.29), which becomes in this case:

$$\sigma_y^2 = \int_0^\infty S_y(f)\mathrm{d}f \tag{10.60}$$

or using Eq. (10.59):

$$\sigma_y^2 = \frac{1}{(2\pi f_n)^4}\int_0^\infty \frac{S_{\ddot{x}}(f)}{\left((1-\Omega^2)^2+(2\gamma\Omega)^2\right)}\mathrm{d}f \tag{10.61}$$

Equation (10.61) is the general expression for the variance of y, and is of the same form as Eq. (10.50) except for the constant factor $1/m^2$. If $S_{\ddot{x}}(f)$ is constant at the value $S_{\ddot{x}}$ around the natural frequency, f_n, then Eq. (10.61) can be integrated in the same way as Eq. (10.51) to give the approximations:

$$\sigma_y^2 = \frac{S_{\ddot{x}}}{64\pi^3 f_n^3\gamma} \quad \text{or} \quad \sigma_y = \frac{1}{8}\left(\frac{S_{\ddot{x}}}{\pi^3 f_n^3\gamma}\right)^{\frac{1}{2}} \tag{10.62}$$

where the mean square or variance of the spring displacement, σ_y^2, and its RMS value or standard deviation, σ_y, are seen to depend only upon the spectral density of the base acceleration, $S_{\ddot{x}}$, the natural frequency, f_n, in Hz and the damping coefficient γ.

Example 10.7

A hydraulic pipe installation, exactly as described in Example 8.1, is to be used in a location where the vibration input, specified at the supports, is random, and defined in terms of acceleration power spectral density. If the input level is defined as $2.0\mathrm{g}^2/\mathrm{Hz}$ from 10 to 1000 Hz, find
(a) The RMS displacement at the center of the pipe.
(b) The RMS stress at the center of the pipe.
   Note: lbf in. units are used, as in the original Example 8.1.

Solution

Part (a):
It was shown in Example 8.1 that the vibration behavior of the hydraulic pipe, considered as a simply supported uniform beam excited by acceleration at the supports, as shown in Fig. 10.13 at (a), can be represented by the equivalent system shown in Fig. 10.13 at (b). Here, $\ddot{x}_B$ represents the acceleration at the pipe attachments, and y_c represents the maximum displacement of the pipe, relative to the attachments,

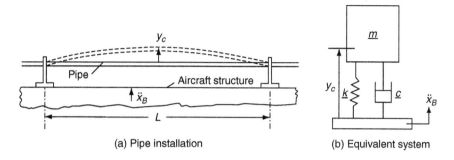

(a) Pipe installation   (b) Equivalent system

Fig. 10.13   Hydraulic pipe installation discussed in Example 10.7.

which occurs at the semi-span position. The equivalent mass, stiffness and damper were shown to be given by:

   Equivalent mass, $\underline{m} = \frac{1}{2}\mu L$, referred to the center of the span;
   Equivalent stiffness, $\underline{k} = \frac{1}{2}EI(\pi^4/L^3)$, also referred to the center of the span;
   Equivalent damper, $\underline{c} = 2\gamma\sqrt{km}$, where $\gamma = 0.02$ in this case.

Other data for the pipe, from Example 8.1, were

D is the outer diameter of pipe $= 0.3125$ in., d the inner diameter of pipe $= 0.2625$ in., E the Young's modulus $= 30 \times 10^6$ lbf/in^2., I the second moment of area of cross-section $= \frac{\pi}{64}\left(D^4 - d^4\right) = 0.2350 \times 10^{-3}$ in^4., μ the mass of pipe, plus contained fluid, per inch $= 0.0214 \times 10^{-3}$ lb in^{-1} s^2 per in. and L the center to center spacing of pipe supports $= 17$ in.

Figure 10.13 can be seen to be identical to Fig. 10.12, and therefore Eq. (10.62) can be used directly to find the RMS displacement, σ_{y_c} at the center of the pipe, i.e.,

$$\sigma_{y_c} = \frac{1}{8}\left(\frac{S_{\ddot{x}_B}}{\pi^3 f_n^3 \gamma}\right)^{\frac{1}{2}} \qquad \text{(A)}$$

where $S_{\ddot{x}_B}$ is the base acceleration spectral density in (in./s^2)^2/Hz, f_n the natural frequency in Hz and γ the damping coefficient $= 0.02$.

The base PSD is 2.0g^2/Hz, so $S_{\ddot{x}_B} = \left(386^2 \times 2.0\right) = 298\ 000$ (in./s^2)^2/Hz.

The natural frequency, f_n, from Example 8.1, is 98.6 Hz. Substituting these numerical values into Eq. (A) gives $\sigma_{y_c} = 0.0885$ in. RMS.

Part (b):
From Eq. (R) in Example 8.1, the relationship between the amplitude of the stress, $|s_c|_{MAX}$, at the center of the pipe, and the displacement amplitude at the center of the pipe, $|y_c|_{MAX}$, is

$$|s_c|_{MAX} = \frac{\pi^2 DE}{2L^2}|y_c|_{MAX} \qquad \text{(B)}$$

The same factor applies to the corresponding RMS values, so

$$\sigma_{sc} = \frac{\pi^2 DE}{2L^2}\sigma_{y_c} \qquad \text{(C)}$$

where σ_{s_c} is the RMS stress and σ_{y_c} is the RMS displacement. Inserting the numerical values $D = 0.3125$ in.; $E = 30 \times 10^6$ lbf/in.²; $L = 17$ in. and $\sigma_{y_c} = 0.0885$ in. gives the RMS stress at the center of the pipe:

$$\sigma_{s_c} = 14\ 170\,\text{lbf/in.}^2$$

10.4.4 Response of a Multi-DOF System to a Single Broad-band Random Input

Provided a multi-DOF system is excited by only a single random input, calculation of the response can still be based on Eq. (10.44):

$$S_y(f) = |H(f)|^2 S_x(f) \tag{10.44}$$

where $S_x(f)$ is the power spectral density of the single input or forcing function; $S_y(f)$ the power spectral density of the output or response and $|H(f)|$ the modulus of the steady-state frequency response function, relating input and output. The process of evaluating Eq. (10.44) was sketched in Fig. 10.9.

Having found the response spectrum, $S_y(f)$, the mean square and RMS levels of the response can be found from Eq. (10.29), i.e.,

$$\sigma_y^2 = \int_0^\infty S_y(f)\mathrm{d}f \tag{10.63}$$

The procedure for a multi-DOF system is thus seen to be precisely the same as for the single-DOF case, except that the modulus of the FRF, $|H(f)|$, contains the effect of several modes acting together, instead of just one mode. Deriving $|H(f)|$ is much easier if, as is almost always the case in practice, the system is represented by a set of normal modes, allowing the normal mode summation method to be used.

The following example uses the simple 2-DOF system introduced in Example 6.5 in which the equations of motion in normal mode coordinates were derived. The method can be generalized to any number of modes, but is, of course, limited to a single input.

Example 10.8

The system used in Example 6.5 is shown again in Fig. 10.14. The beam is assumed massless, and uniform, with $EI = 2 \times 10^6$ N m² . Two discrete masses $m_1 = 10$ kg and $m_2 = 8$ kg are attached, as shown. The total length of the beam, $L = 4$ m. The viscous

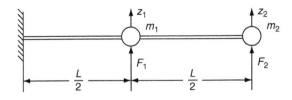

Fig. 10.14   2-DOF system discussed in Example 10.8.

damping coefficient is assumed to be 0.05 for both modes. F_1 is a random force with constant power spectral density of 100 N^2/Hz, from 10 to 120 Hz, and F_2 is zero.

Find the RMS value of the displacement, z_1.

Solution

From Example 6.5, the equations of motion in orthonormal form (i.e. with the mass matrix defined as a unit diagonal) are

$$\begin{bmatrix} 1 & 0 \\ 0 & 1 \end{bmatrix}\begin{Bmatrix} \ddot{q}_1 \\ \ddot{q}_2 \end{Bmatrix} + \begin{bmatrix} 2\gamma_1\omega_1 & 0 \\ 0 & 2\gamma_2\omega_2 \end{bmatrix}\begin{Bmatrix} \dot{q}_1 \\ \dot{q}_2 \end{Bmatrix} + \begin{bmatrix} \omega_1^2 & 0 \\ 0 & \omega_2^2 \end{bmatrix}\begin{Bmatrix} q_1 \\ q_2 \end{Bmatrix} = \begin{Bmatrix} Q_1 \\ Q_2 \end{Bmatrix} \tag{A}$$

where the undamped natural frequencies, in rad/s, are $\omega_1 = 102.02$ and $\omega_2 = 621.30$, and in this case $\gamma_1 = \gamma_2 = 0.05$.

Also from Example 6.5, the relationship between the actual displacements,$\{z\}$, and the generalized displacements, $\{q\}$, is

$$\{z\} = [X]\{q\} \quad \text{or} \quad \begin{Bmatrix} z_1 \\ z_2 \end{Bmatrix} = \begin{bmatrix} 0.1071 & -0.2975 \\ 0.3326 & 0.1198 \end{bmatrix}\begin{Bmatrix} q_1 \\ q_2 \end{Bmatrix} \tag{B}$$

and the relationship between the actual external forces, $[F]$, and the generalized forces, $[Q]$, is

$$\begin{Bmatrix} Q_1 \\ Q_2 \end{Bmatrix} = \begin{bmatrix} 0.1071 & 0.3326 \\ -0.2975 & 0.1198 \end{bmatrix}\begin{Bmatrix} F_1 \\ F_2 \end{Bmatrix} \tag{C}$$

Since Eq. (A) is expressed in normal modes, with all matrices diagonal, it can now be treated as two uncoupled, single-DOF equations:

$$\ddot{q}_1 + 2\gamma_1\omega_1\dot{q}_1 + \omega_1^2 q_1 = Q_1 \tag{D}$$

and

$$\ddot{q}_2 + 2\gamma_2\omega_2\dot{q}_2 + \omega_2^2 q_2 = Q_2 \tag{E}$$

From Chapter 4, the complex receptance expressions, or FRFs ,$H_{11}(f)$ and $H_{22}(f)$, corresponding to Eqs (D) and (E), respectively, are as follows, noting that the generalized mass is unity in both cases, and that (f) has been omitted throughout for clarity:

$$H_{11} = \frac{q_1}{Q_1} = \frac{(1 - \Omega_1^2) - i(2\gamma_1\Omega_1)}{(2\pi f_1)^2\left[(1 - \Omega_1^2)^2 + (2\gamma_1\Omega_1)^2\right]} = R_{11} + iI_{11} \tag{G}$$

and

$$H_{22} = \frac{q_2}{Q_2} = \frac{(1 - \Omega_2^2) - i(2\gamma_2\Omega_2)}{(2\pi f_2)^2\left[(1 - \Omega_2^2)^2 + (2\gamma_2\Omega_2)^2\right]} = R_{22} + iI_{22} \tag{H}$$

where R_{11} and I_{11} are the real and imaginary parts of H_{11} and R_{22} and I_{22} are the real and imaginary parts of H_{22}. Also, $\Omega_1 = f/f_1$ and $\Omega_2 = f/f_2$, where f is the excitation frequency in Hz and $f_1 = \omega_1/2\pi = 16.236$ Hz is the natural frequency of mode 1 and $f_2 = \omega_2/2\pi = 98.883$ Hz is the natural frequency of mode 2.

Equations (B) and (C) apply equally well when z_1, z_2, q_1, q_2, F_1, F_2, Q_1, Q_2 are replaced by the complex vectors $\underline{z}_1$, $\underline{z}_2$, $\underline{q}_1$, $\underline{q}_2$, $\underline{F}_1$, $\underline{F}_2$, $\underline{Q}_1$, $\underline{Q}_2$. Therefore, from Eq. (B):

$$\underline{z}_1 = 0.1071\underline{q}_1 - 0.2975\underline{q}_2 \tag{I}$$

From Eqs (G) and (H):

$$\underline{q}_1 = (R_{11} + iI_{11})\underline{Q}_1 \quad \text{and} \quad \underline{q}_2 = (R_{22} + iI_{22})\underline{Q}_2 \tag{J}$$

and from Eq. (C), since $F_2 = \underline{F}_2 = 0$, in this case:

$$\underline{Q}_1 = 0.1071\underline{F}_1 \quad \text{and} \quad \underline{Q}_2 = -0.2975\underline{F}_1 \tag{K}$$

From Eqs (I), (J) and (K), the complex receptance between the applied force F_1 and the displacement z_1 is

$$H_{z_1 F_1}(f) = \frac{\underline{z}_1}{\underline{F}_1} = 0.1071^2(R_{11} + iI_{11}) + 0.2975^2(R_{22} + iI_{22}) \tag{L}$$

For random response we require only the modulus, which is

$$|H_{z_1 F_1}|(f) = \frac{|\underline{z}_1|}{|\underline{F}_1|} = \sqrt{(0.1071^2 R_{11} + 0.2975^2 R_{22})^2 + (0.1071^2 I_{11} + 0.2975^2 I_{22})^2} \tag{M}$$

Using a standard spreadsheet program, this was plotted from $f = 10\text{--}120$ Hz, and appears as Fig. 10.15.

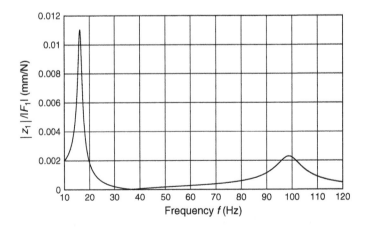

Fig. 10.15   Modulus of the FRF of the 2-DOF system, Example 10.8.

From Eq. (10.44), we can now find the displacement PSD, S_{z_1}, of z_1, given that the force PSD of F_1 is constant at 100 N^2/Hz, from 10–120 Hz. In the present notation, Eq. (10.44) can be written as:

$$S_{z_1}(f) = |H_{z_1 F_1}(f)|^2 S_{F_1}(f) \qquad (N)$$

where $|H_{z_1 F_1}(f)|$ is given by Eq. (M), and is plotted as Fig. 10.15, and $S_{F_1}(f) = 100$ N^2/Hz from $f = 10$–120 Hz.

The mean square, and hence the RMS value, of the displacement z_1, can now be found from Eq. (10.63):

$$\sigma_{z_1}^2 = \int_0^\infty S_{z_1}(f) \mathrm{d}f$$

However, the system is excited only between the frequency limits 10–120 Hz, and the response will be zero outside these limits, so the mean square displacement is given by:

$$\sigma_{z_1}^2 = \int_{10}^{120} S_{z_1}(f) \mathrm{d}f \qquad (O)$$

This was integrated numerically, using a spreadsheet program, giving

$$\sigma_{z_1} = 0.191 \,\text{mm RMS}$$

10.5 Correlation functions and cross-power spectral density functions

In this chapter, so far we have seen that many problems in random vibration can be solved by the use of the power spectral density function alone, together with an understanding of probability density and distribution functions. There are some problems, however, for which this will not be sufficient. Before looking at these, we must first introduce a few more topics, such as *correlation functions*, and how they are related to spectral density functions, including *cross-power spectral density functions*.

10.5.1 Statistical Correlation

Before considering autocorrelation and cross-correlation functions, we should briefly consider *correlation*, a basic concept in statistics. The *correlation coefficient* is often applied to pairs of properties, such as the height and weight of men in a given population. The idea can be extended to include sampled values of two time histories, $x(t)$ and $y(t)$, say. The correlation coefficient for simultaneous pairs of samples of x and y, taken from $x(t)$ and $y(t)$, where the mean values are zero, is defined as:

$$\rho_{xy} = \left(\frac{\langle xy \rangle}{\sigma_x \sigma_y} \right) = \frac{\sigma_{xy}}{\sigma_x \sigma_y} \qquad (10.64)$$

where σ_x and σ_y are the standard deviations of x and y, respectively and $\sigma_{xy} = \langle xy \rangle$ is known as the *covariance*. It can be seen from Eq. (10.64), incidentally, why ρ_{xy} is sometimes called the *normalized covariance*.

The correlation coefficient, ρ_{xy}, is dimensionless, and must lie in the range ± 1. A value of $+1$ indicates complete positive correlation, and all values of y would then be related to the corresponding values of x by $y = Ax$, where A is a positive constant. A value of -1 indicates complete negative correlation, with A as a negative constant. If $\rho_{xy} = 0$, of course, there is no statistical correlation at all between the x and y values. This does not necessarily mean that there is no relationship between them: for example, $\sin \omega t$ and $\cos \omega t$ are uncorrelated in a statistical sense, yet there is a strong relationship between them.

The correlation coefficient takes no account of the time order of the pairs of sampled values, and is therefore of limited use in vibration work. The auto- and cross-correlation functions, however, discussed next, do make use of the relative timing between samples.

10.5.2 The Autocorrelation Function

The autocorrelation function (ACF) can be derived from a single waveform, $x(t)$ say, as shown in Fig. 10.16(a). It is defined as:

$$R_x(\tau) = E[x(t)x(t+\tau)] = \langle x(t)x(t+\tau) \rangle \tag{10.65}$$

It is the mean of all the products of $x(t)$ and $x(t+\tau)$, where $x(t+\tau)$ is simply the value of x at a time, $t + \tau$, which is τ later than time t.

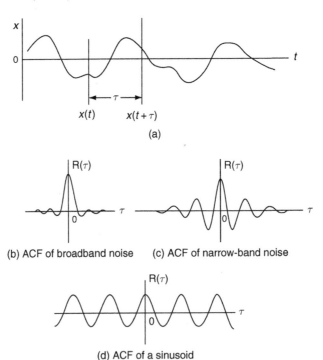

(a)

(b) ACF of broadband noise   (c) ACF of narrow-band noise

(d) ACF of a sinusoid

Fig. 10.16   The autocorrelation function.

The ACF can be evaluated digitally, directly from Eq. (10.65), for a suitable range of values of τ, by multiplying and averaging. However, this is a relatively inefficient process computationally, and, as will be seen later in this chapter, the ACF is more usually computed as the inverse Fourier transform of the power spectrum. This is much quicker, thanks to the efficiency of the FFT algorithm.

In analog days, the ACF was, in fact, derived directly from Eq. (10.65). The values of $x(t)$ and $x(t + \tau)$ were represented by voltages derived from a magnetic tape loop with two replay heads, one of which could be moved relative to the other to vary the time lag τ. Analog multiplication was then used to produce the product $x(t)x(t + \tau)$, which was then integrated to find the average. References to this now obsolete method can still be found in the literature.

Some of the main properties of $R_x(\tau)$, the ACF of $x(t)$, can now be deduced. It will be assumed, for the present, that any mean value in $x(t)$ has been removed.

(1) Its value is closely related to the correlation coefficient, ρ, discussed above. If we regard $x(t + \tau)$ as a separate waveform from $x(t)$, and call it $y(t)$, then Eq. (10.64) can be used to find the correlation coefficient, ρ_{xy}, between $x(t)$ and $y(t) = x(t + \tau)$, for any fixed value of τ. From Eq. (10.64):

$$\rho_{xy} = \left(\frac{\langle x(t)y(t) \rangle}{\sigma_x \sigma_y} \right) \tag{10.66}$$

but since we have assumed that $y(t) = x(t + \tau)$, and if $x(t)$ is stationary, then $\sigma_x = \sigma_{x(t)} = \sigma_{x(t+\tau)} = \sigma_y$, and Eq. (10.66) can be written as:

$$\rho_{x(t)x(t+\tau)} = \frac{\langle x(t)x(t + \tau) \rangle}{\sigma_x^2} = \frac{R_x(\tau)}{\sigma_x^2} \tag{10.67}$$

Thus the correlation coefficient between $x(t)$ and $x(t + \tau)$ is equal to the value of the ACF at the same value of τ, divided by the variance. The ACF can therefore be used to find the expected correlation coefficient between a point in a waveform and its future or past values.

(2) The value of the ACF at $\tau = 0$ is equal to the mean square value, which is equal to the variance, if the mean value of $x(t)$ is zero. This can be seen from Eq. (10.65), since at $\tau = 0$, $R_x = \langle x^2(t) \rangle = \sigma_x^2$. It can never exceed this for any other value of τ, since $\langle x(t)x(t + \tau) \rangle$ can never be greater than $\langle x^2(t) \rangle$.

(3) It is an even function, i.e. $R_x(\tau) = R_x(-\tau)$.

(4) For a broad-band random function it dies away quickly for positive and negative values of τ, from its maximum value at $\tau = 0$, as shown in Fig. 10.16 at (b).

(5) For a narrow-band random input it consists of a decaying cosine function, as shown in Fig. 10.16 at (c). The decay is exponential if the narrow-band response is due to the response of a single-DOF system. In the case of the response of a multi-DOF system to broad-band random excitation, the ACF of the response consists of a number of superimposed, exponentially decaying cosine functions, each of which, using suitable analysis, can give the natural frequency and damping coefficient of one of the modes of the system directly. This method (which can be extended to include cross-correlation functions, introduced below) was formerly used in such applications as flight flutter analysis. Even today it is sometimes preferred to the more usual frequency domain method based on power spectra.

(6) In the case of any sinusoidal wave, the ACF is a cosine function, as shown in Fig. 10.16 at (d). This gives a useful method for identifying a mixture of random noise and periodic functions, since the random components will die out for large values of τ, whereas the periodic components will not.

(7) If the mean value of $x(t)$ is not removed before calculating the ACF, it is easily seen that this adds a constant, μ_x^2, to the ACF at all values of τ, where μ_x is the mean value of $x(t)$.

10.5.3 The Cross-Correlation Function

There are two possible cross-correlation functions (CCFs) between two waveforms $x(t)$ and $y(t)$ and they are defined as:

$$R_{xy}(\tau) = E[x(t)y(t+\tau)] = \langle x(t)y(t+\tau) \rangle \tag{10.68}$$

and

$$R_{yx}(\tau) = E[y(t)x(t+\tau)] = \langle y(t)x(t+\tau) \rangle \tag{10.69}$$

Equation (10.68), giving $R_{xy}(\tau)$, is illustrated in Fig. 10.17 at (a). Each value of $x(t)$ is multiplied by $y(t+\tau)$, which is the value of $y(t)$ an amount of time τ later, and the average of all the products is calculated.

As shown in Fig. 10.17 at (b), $R_{yx}(\tau)$ is obtained in the same way, except that in this case $x(t)$ is delayed with respect to $y(t)$, becoming $x(t+\tau)$. In general, $R_{xy}(\tau)$ and $R_{yx}(\tau)$ are not the same, but if the original waveforms are stationary, it can be seen that $R_{xy}(\tau) = R_{yx}(-\tau)$ and $R_{yx}(\tau) = R_{xy}(-\tau)$.

The main properties of the CCF are

(1) Assuming that the mean values of $x(t)$ and $y(t)$, if any, have been removed, the value of a CCF at a particular value of τ is related to the corresponding correlation coefficient as follows:

$$\rho_{x(t)y(t+\tau)} = \frac{R_{xy}(\tau)}{\sigma_x \sigma_y} \quad \text{and} \quad \rho_{y(t)x(t+\tau)} = \frac{R_{yx}(\tau)}{\sigma_x \sigma_y} \tag{10.70}$$

where $\rho_{x(t)y(t+\tau)}$ is the correlation coefficient between $x(t)$ and $y(t+\tau)$, and $\rho_{y(t)x(t+\tau)}$ is the correlation coefficient between $y(t)$ and $x(t+\tau)$.

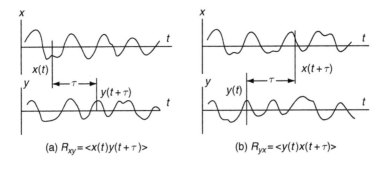

(a) $R_{xy} = \langle x(t)y(t+\tau) \rangle$ (b) $R_{yx} = \langle y(t)x(t+\tau) \rangle$

Fig. 10.17   Definition of the cross-correlation function.

(2) The CCF is not, generally, an even function of τ, and its maximum value does not necessarily occur at $\tau = 0$.

(3) For zero mean values, the CCF must always lie in the range $\pm \sigma_x \sigma_y$.

(4) If the mean levels of $x(t)$ and $y(t)$ are not removed, $\mu_x \mu_y$ will be added to the CCF at all values of τ, where μ_x is the mean value of $x(t)$ and μ_y is the mean value of $y(t)$.

10.5.4 *Relationships Between Correlation Functions and Power Spectral Density Functions*

Although correlation functions and spectral density functions are apparently quite different concepts, it can be shown that the power spectrum is the Fourier transform of the ACF, and the *cross-power spectrum*, which we have not yet introduced, is similarly the Fourier transform of the CCF.

The Wiener–Khintchine relationships

These relationships, discovered independently by Wiener and Khintchine in the 1940s, show that the power spectral density function is the Fourier transform of the ACF, and that the ACF is the inverse Fourier transform of the PSD function, where both functions are derived from the same underlying waveform, $x(t)$. They can be proved as follows [10.2], using Fourier transforms.

From Chapter 9, we have the Fourier transform pair

$$x(t) = \int_{-\infty}^{\infty} X(f) \, e^{i2\pi ft} df \tag{9.32}$$

and

$$X(f) = \int_{-\infty}^{\infty} x(t) \, e^{-i2\pi ft} dt \tag{9.33}$$

From Eq. (9.32) we can write

$$x(t + \tau) = \int_{-\infty}^{\infty} X(f) \, e^{i2\pi f(t+\tau)} df \tag{10.71}$$

The basic definition of the ACF, $R_x(\tau)$ is, from Eq. (10.65):

$$R_x(\tau) = \langle x(t)x(t+\tau) \rangle = \lim_{T \to \infty} \frac{1}{T} \int_{-\infty}^{\infty} x(t)x(t+\tau) dt$$

Substituting for $x(t + \tau)$ from Eq. (10.71):

$$R_x(\tau) = \lim_{T \to \infty} \frac{1}{T} \int_{-\infty}^{\infty} x(t) \int_{-\infty}^{\infty} X(f) \, e^{i2\pi ft} e^{i2\pi f\tau} df \, dt$$

$$= \int_{-\infty}^{\infty} \lim_{T \to \infty} \frac{1}{T} \left[\int_{-\infty}^{\infty} x(t) e^{i2\pi ft} dt \right] X(f) e^{i2\pi f\tau} df \tag{10.72}$$

But from Eq. (9.33), $X(f) = \int_{-\infty}^{\infty} x(t) e^{-i2\pi ft} dt$ and therefore:

$$X^*(f) = \int_{-\infty}^{\infty} x(t) e^{i2\pi ft} dt \qquad (10.73)$$

where $X^*(f)$ is the complex conjugate of $X(f)$. Substituting Eq. (10.73) into Eq. (10.72):

$$R_x(\tau) = \int_{-\infty}^{\infty} \left[\lim_{T \to \infty} \frac{1}{T} X^*(f) X(f) \right] e^{i2\pi f\tau} df \qquad (10.74)$$

The quantity $\left[\lim_{T \to \infty} \frac{1}{T} X^*(f) X(f) \right]$ is the power spectral density, $S_x(f_\pm)$, and so Eq. (10.74) can be written as:

$$R_x(\tau) = \int_{-\infty}^{\infty} S_x(f_\pm) e^{i2\pi f\tau} df \qquad (10.75)$$

showing that that the ACF is the inverse Fourier transform of the power spectral density function. By analogy with Eqs (9.32) and (9.33) it can be seen that:

$$S_x(f_\pm) = \int_{-\infty}^{\infty} R_x(\tau) e^{-i2\pi f\tau} d\tau \qquad (10.76)$$

showing that the power spectral density function is the Fourier transform of the ACF.

In Eqs (10.75) and (10.76) the spectral density has been written as $S_x(f_\pm)$ to indicate that it is double-sided, and not the same as the single-sided spectral density, $S_x(f)$, derived in Section 10.3.2, and used generally in practical work. Since $S_x(f_\pm)$ is an even function, however, the power is the same at corresponding positive and negative frequencies, and $S_x(f) = 2S_x(f_\pm)$.

As a check, the mean square value of the underlying time history, $\langle x^2(t) \rangle$, should be given by the value of $R_x(\tau)$ at $\tau = 0$, and also by the area under the PSD curve, however it is defined. Substituting $\tau = 0$ into Eq. (10.75) gives

$$R_x(0) = \int_{-\infty}^{\infty} S_x(f_\pm) df = \langle x^2(t) \rangle \quad \text{as required}$$

Relationships between cross-correlation functions and cross-power spectra

The Wiener–Khintchine equations can be extended to show [10.2] that the *cross-power spectrum* is the Fourier transform of the CCF:

$$S_{xy}(f_\pm) = \int_{-\infty}^{\infty} R_{xy}(\tau) e^{-i2\pi f\tau} d\tau \qquad (10.77)$$

and that the CCF is the inverse Fourier transform of the cross-power spectrum:

$$R_{xy}(\tau) = \int_{-\infty}^{\infty} S_{xy}(f_\pm) e^{i2\pi f\tau} df \qquad (10.78)$$

Equation (10.77) is the usual definition of the cross-power spectrum $S_{xy}(f_\pm)$, giving the cross-power between two waveforms, say $x(t)$ and $y(t)$. The power spectrum, $S_x(f_\pm)$, can be regarded as a special case of the cross-spectrum when $y(t) = x(t)$.

The cross-power spectrum is, in general, complex, and not an even function. Its value at a positive frequency, f_+ is the complex conjugate of its value at a corresponding negative frequency, f_-, i.e.,

$$S_{xy}^*(f_+) = S_{xy}(f_-) \tag{10.79}$$

which is sometimes known as 'conjugate even'. Like the CCF, the cross-power spectrum can be defined the other way round, for example $S_{yx}(f_\pm)$ and $R_{yx}(\tau)$ are related by:

$$S_{yx}(f_\pm) = \int_{-\infty}^{\infty} R_{yx}(\tau) \, e^{-i2\pi f\tau} d\tau \tag{10.80}$$

and

$$R_{yx}(\tau) = \int_{-\infty}^{\infty} S_{yx}(f_\pm) \, e^{i2\pi f\tau} df \tag{10.81}$$

10.6 The Response of structures to random inputs

We have seen from the preceding section that the auto-power spectrum and the cross-power spectrum are the Fourier transforms, respectively, of the ACF and CCF. It can also be shown that the frequency response function, $H(\omega)$, is the Fourier transform of the impulse response $h(t)$. This means that, in theory, there are always two ways to calculate the response of a structure to a random input, or inputs, or to calculate the characteristics of a structure when the input and response are given:

(a) In the time domain, using correlation functions to describe the input and response, and the impulse response to represent the structure, or:

(b) In the frequency domain, using power spectra to describe the input and output, and the frequency response function to represent the structure.

In practice, the first method is much more difficult to use than the second, and most practical work is carried out using power spectra and FRFs.

10.6.1 The Response of a Structure to Multiple Random Inputs

When a system, assumed linear, is excited by several random forces, which may be correlated, the auto-power and cross-power spectral density functions provide a convenient method for finding its response, provided all the individual frequency response functions between excitation points and response points are known.

The following example describes the calculation of the response at a single point on a structure, when the loading can be represented by two concentrated, random, forces at two other points. The method can easily be generalized to any number of excitation and response locations.

A typical practical application would be the calculation of the response of an aircraft component, such as a wing or fin, from measurements of fluctuating pressures, at a number of locations, on the component concerned. The measured pressures can be converted to an array of discrete forces by multiplying each by the area over which it can be considered to act.

Example 10.9

The aircraft wing shown in Fig. 10.18 is excited by two random forces, $P(t)$ and $Q(t)$, assumed to have zero mean values. Show how the variance of the displacement at the wing tip can be found in the following cases:
(a) When the forces $P(t)$ and $Q(t)$ are correlated, say, because they are derived from turbulence in the same airstream.
(b) When the forces are uncorrelated, because they are derived from entirely different sources, such as different jet engines.

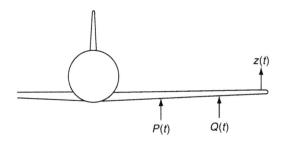

Fig. 10.18   Two random forces applied to an aircraft structure.

Solution

The following standard result relates the displacement power spectrum at the wing tip to the power and cross-power spectra of the applied forces:

$$S_z = H^*_{zP}H_{zP}S_P + H^*_{zP}H_{zQ}S_{PQ} + H^*_{zQ}H_{zP}S_{QP} + H^*_{zQ}H_{zQ}S_Q \qquad \text{(A)}$$

where all the quantities are functions of frequency, f, and are defined as follows:
S_z = power spectral density of the wing tip displacement $z(t)$,
S_P = power spectral density of the force $P(t)$,
S_Q = power spectral density of the force $Q(t)$,
S_{PQ} = complex cross-power spectral density between force $P(t)$ and force $Q(t)$,
S_{QP} = complex cross-power spectral density between force $Q(t)$ and force $P(t)$,
H_{zP} = complex frequency response function giving $z(t)$ per unit $P(t)$,
H_{zQ} = complex frequency response function giving $z(t)$ per unit $Q(t)$.

Case (a):
When there is correlation between the applied forces $P(t)$ and $Q(t)$, Eq. (A) applies in full, and all the quantities on the right side of Eq. (A) must be known.

The variance of the wing tip displacement, σ_z^2, is then given by Eq. (10.30):

$$\sigma_z^2 = \int_{f_1}^{f_2} S_z(f)\mathrm{d}f \tag{B}$$

that is by integrating (or summing) the displacement PSD at the wing tip over an appropriate frequency range, f_1 to f_2.

Case (b)
If the forces $P(t)$ and $Q(t)$ are completely uncorrelated, the cross-power spectra, S_{PQ} and S_{QP} are zero, and Eq. (A) becomes

$$S_z = H_{zP}^* H_{zP} S_P + H_{zQ}^* H_{zQ} S_Q = |H_{zP}|^2 S_P + |H_{zQ}|^2 S_Q \tag{C}$$

Thus, the response spectrum can be found by simply adding the response power spectra due to each force acting separately. The variance of the displacement at the wing tip is given by Eq. (B), as before.

10.6.2 *Measuring the Dynamic Properties of a Structure*

Power and cross-power spectra can also be used to find the dynamic properties of a structure, by analysing the input and response time histories. There are two main methods:
(1) In the first method, a *known* input, usually a force, is applied to the structure, and one or more responses are measured. The known input, which can be random, pseudo-random or deterministic, for example a sinusoidal frequency sweep, or even a blow from an instrumented hammer, is recorded, together with the responses. By suitable processing of the resulting data, it is possible to find the FRFs between the applied force and each response, and hence define the dynamic properties of the structure. This is possible, up to a point, even when the structure is excited by unknown random forces at the same time, provided they are uncorrelated with the known input. The device used to provide the input force depends upon the application. When the method is used to test an aircraft in flight, for example, the forcing may be provided through an existing control surface, such as an aileron, by a special moving vane, or other excitation device, installed for the purpose. The method can also be used in modal testing, as discussed in Chapter 13, as an alternative to sinusoidal testing, in which case a standard exciter is used.
(2) The second method relies upon 'natural' random excitation, applied to the structure by its environment, such as aerodynamic turbulence or jet noise. This input cannot usually be measured, so only response measurements are possible in this case. FRFs cannot be found, but it is still possible to find the natural frequencies, damping coefficients and approximate mode shapes of the structure, provided all the required modes are excited. The fact that no special excitation equipment is required by this method is an attractive feature, with the possibility of cost savings. It can give very good results when the excitation level is adequate, and the spectra can be averaged over long periods.

The first of the two methods described above is now illustrated by the following example.

Example 10.10

The aircraft shown in Fig. 10.19, of which only one side is shown, is fitted with a device which applies a known force time history, $F(t)$. This force is recorded, together with the responses $z_1(t)$, $z_2(t)$, $z_3(t)$, etc., which are derived from accelerometers.

The aircraft is assumed to be in flight, and subjected to unknown random forces due to turbulent airflow, in addition to the known force $F(t)$.

Show how the FRFs between the applied force $F(t)$ and each response location can be found, enabling the dynamic properties of the aircraft wing to be determined.

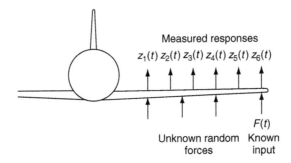

Fig. 10.19   Measuring FRFs of an aircraft in flight using a known input force.

Solution

Consider first one typical response, $z(t)$, and let its Fourier transform be $Z(f)$. Also let the Fourier transform of the input force, $F(t)$, be $F(f)$. Then the FRF, $H_{zF}(f)$, between the force $F(t)$ and the response $z(t)$ is given by:

$$H_{zF}(f) = \frac{Z(f)}{F(f)} \tag{A}$$

Multiplying numerator and denominator by $F^*(f)$, the complex conjugate of $F(f)$, we have

$$H_{zF}(f) = \frac{Z(f)F^*(f)}{F(f)F^*(f)} = \frac{S_{zF}(f)}{S_F(f)} \tag{B}$$

Thus, the FRF, $H_{zF}(f)$, between the applied force and the response is given by dividing the cross-power spectrum between input and response, $S_{zF}(f)$, by the input power spectrum $S_F(f)$. The unknown forcing due to the turbulent flow is, of course, present, and the structure will respond to it, but since it is uncorrelated with $F(t)$, it will not affect the measurement of $S_{zF}(f)$, if the latter is obtained by averaging over a sufficiently long period of time.

The FRF, $H_{zF}(f)$, obtained from this procedure is, in theory, exactly the same as the FRF that would have been obtained if $F(t)$ had been a sinusoidal force, applied at the same series of frequencies, and contains all the information needed to find the main dynamic properties of the structure, such as its natural frequencies and damping coefficients.

If the exercise described above is carried out with the single force, $F(t)$, but several responses, $z_1(t)$, $z_2(t)$, $z_3(t)$, etc., Eq. (B) can be used to find all the FRFs between the force and the response points, enabling approximations to the normal mode shapes to be plotted also.

If a special forcing device cannot be provided, but there is sufficient 'natural' random excitation to excite the structure, the natural frequencies, damping coefficients and approximate mode shapes of a structure can still be found, using the second method described above, as in the following example.

Example 10.11

The aircraft shown in Fig. 10.20 is fitted with transducers enabling the responses $z_1(t)$, $z_2(t)$, $z_3(t)$, etc. to be recorded, but there is no special provision for a known applied force. The aircraft is, however, excited by random forces due to turbulent airflow, or from the jet engines.

Show how the ambient random excitation can be utilized to find the natural frequencies, damping coefficients and approximate mode shapes of the wing.

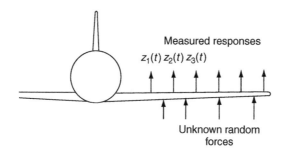

Fig. 10.20

Solution

Consider a typical response, $z(t)$, of the wing, and let the Fourier transform of $z(t)$ be $Z(f)$. Although we do not know the input force producing this response, let us call it $F(t)$, and let its Fourier transform be $F(f)$.

We can then write

$$|H_{zF}(f)|^2 = \frac{Z(f)Z^*(f)}{F(f)F^*(f)} = \frac{S_z(f)}{S_F(f)} \tag{A}$$

where $|H_{zF}(f)|$ is the modulus of the FRF relating $z(t)$ and $F(t)$, $S_z(f)$ the auto power spectrum of $z(t)$, and is known and $S_F(f)$ the auto power spectrum of $F(t)$, and is unknown.

Although the input power spectrum $S_F(f)$ is unknown, if we can make the assumption that it is reasonably flat in a frequency band surrounding each resonance of the structure, then it can be taken as a constant for any given mode. Thus we know the *shape* of $|H_{zF}(f)|$ in that frequency band, but not its overall size. Nevertheless, this is sufficient to enable the natural frequency and damping coefficient of each mode to be found.

If several responses, $z_1(t)$, $z_2(t)$, $z_3(t)$, etc., are measured, at suitable locations, it is possible to find not only the frequencies and damping coefficients of the normal modes, but also the approximate mode shapes, in the following way.

One of the responses, $z_1(t)$, say, is designated as the 'master' or 'reference' response, and the following spectra are calculated:

The auto-spectrum of the master response, $z_1(t)$:

$$S_1(f) = Z_1(f)Z_1^*(f) \tag{B}$$

The cross-spectrum between $z_1(t)$ and $z_2(t)$:

$$S_{12}(f) = Z_1(f)Z_2^*(f) \tag{C}$$

The cross-spectrum between $z_1(t)$ and $z_3(t)$:

$$S_{13}(f) = Z_1(f)_1 Z_3^*(f) \tag{D}$$

and similarly the cross-spectra between $z_1(t)$ and $z_4(t)$, $z_1(t)$ and $z_5(t)$ and so on, where $Z_1(f)$, $Z_2(f)$, etc. are the Fourier transforms of $z_1(t)$, $z_2(t)$, etc.

If all the spectra are now plotted, for example in magnitude and phase form, the resonant frequencies will be identifiable as peaks, and the approximate mode shapes can be found by comparing the magnitudes and phases of the cross spectra with the auto power spectrum at the master location, at each resonant frequency.

10.7 Computing power spectra and correlation functions using the discrete Fourier transform

The operations described next are normally carried out in one of two ways:

(1) A general-purpose computer, with appropriate software, may be used. In this case anti-aliasing filtering of the analog data, as described in Chapter 9, must first be carried out, followed by analog to digital (A to D) conversion. These preliminary steps may well be performed remotely, for example in an air or space vehicle, and the digital data stored on tape or disk, for later analysis, or transmitted to a ground station by a radio link.

(2) Alternatively, all these functions can be combined in a single box known as a *spectrum analyzer*. These usually accept two, four or more simultaneous channels of analog data directly, and are very easy to use, as the integrated design enables appropriate anti-aliasing filters and A to D conversion to be switched in automatically.

In either case, the digital operations required to transform random time histories into useful functions, such as power spectra, enabling practical random vibration problems to be understood and solved, will usually be as described in the following sections. The DFT, as introduced in Chapter 9, and implemented by the FFT algorithm, is used almost exclusively in this work.

We should now review some essential results from Chapter 9 before proceeding further.

(1) The Fourier series representing a periodic waveform, $x(t)$, is defined, by Eq. (9.3), as:

$$x(t) = a_0 + \sum_{n=1}^{\infty} [a_n \cos n\omega_0 t + b_n \sin n\omega_0 t] \tag{9.3}$$

where a_0 is the mean value, and a_n and b_n are defined by Eqs (9.8) and (9.10):

$$a_n = \frac{2}{T} \int_{-T/2}^{T/2} x(t) \cos n\omega_0 t \cdot dt \quad (n = 1, 2, 3, \ldots, \infty) \tag{9.8}$$

$$b_n = \frac{2}{T} \int_{-T/2}^{T/2} x(t) \sin n\omega_0 t \cdot dt \quad (n = 1, 2, 3, \ldots, \infty) \tag{9.10}$$

(2) For a waveform of length T, represented by N samples taken at equal time intervals of $\Delta = T/N$, Eqs (9.3), (9.8) and (9.10) can be represented approximately by the DFT equations, in complex form, Eqs (9.45) and (9.46):

$$X_k = \frac{1}{N} \sum_{j=0}^{N-1} x_j e^{-i\left(\frac{2\pi jk}{N}\right)} \tag{9.45}$$

$$x_j = \sum_{k=0}^{N-1} X_k e^{i\left(\frac{2\pi jk}{N}\right)} \tag{9.46}$$

(3) The following relationships between the Fourier series equations, Eqs (9.3), (9.8), and (9.10), and the DFT equations, Eqs (9.45) and (9.46), were established:

(a) The integers n and k are essentially the same, except that there are only N values of k, whereas n extends to ∞, in theory.
(b) Time t and the integer j are related by Eq. (9.42b): $t = jT/N$, for j = 0, 1, 2, \ldots, $(N–1)$.
(c) Frequency, f, in hertz, and the integer k are related by Eq. (9.43a): $f = k/T$, for k =0, 1, 2, \ldots, $(N–1)$;
(d) The complex DFT coefficients, X_k, and the real Fourier coefficients, a_n and b_n, where they exist, are related by:

$$X_k = \frac{a_n}{2} - i\left(\frac{b_n}{2}\right) \quad (n = k) \tag{9.47a}$$

from which:

$$X_k^* X_k = \frac{1}{4}(a_n^2 + b_n^2) \quad (n = k) \tag{10.82}$$

where X_k^* is the complex conjugate of X_k.

10.7.1 Computing Spectral Density Functions

Auto- and cross-power spectra are the most frequently used functions in everyday vibration work.

Computing auto-PSD functions

From Eq. (10.25), the power spectral density, $S_x(f_{(n)})$, at a discrete frequency, $f_{(n)}$, the frequency of the Fourier coefficients, a_n and b_n, is given by:

$$S_x(f_{(n)}) = \frac{\frac{1}{2}(a_n^2 + b_n^2)}{\delta f} = \frac{T}{2}(a_n^2 + b_n^2) \tag{10.25}$$

where $\delta f = 1/T$ is the width of the small frequency band over which the power $\frac{1}{2}(a_n^2 + b_n^2)$ is considered to be spread. Thus, from Eqs (10.25) and (10.82), noting that $n = k$:

$$S_x(f_{(n)}) = S_x(f_k) = 2TX_k^*X_k \tag{10.83}$$

where $S_x(f_k)$ is the *single-sided* power spectral density, in Hz, at frequency f_k.

From Eq. (9.43a), $f_k = k/T$, and when $k = N/2$, then $f_k = N/2T$, which will be recognized from Chapter 9 as the Nyquist frequency, f_N. Values of X_k above this frequency will be spurious, and cannot be used to form valid spectra. As discussed in Section 9.3, however, practical limitations in the design of anti-aliasing filters usually restrict the useful output to frequencies below about $0.8 f_N$.

Example 10.12

It is desired to monitor the vibration level of a test structure by measuring auto power spectral density functions, at frequencies up to 1000 Hz, using accelerometers. The analog outputs from the accelerometers are sampled digitally at $f_s = 2500$ samples/s, after passing through analog anti-aliasing filters, designed so that the response is nearly flat up to 1000 Hz, but then falls sharply, so that it is negligible at the Nyquist frequency, f_N, which is $f_s/2 = 1250$ Hz. The samples are stored in a computer containing a DFT program, which evaluates Eq. (9.45):

$$X_k = \frac{1}{N}\sum_{j=0}^{N-1} x_j \, e^{-i\left(\frac{2\pi jk}{N}\right)} \tag{A}$$

The value of N is fixed at 2048. The digital samples from the accelerometers are represented by x_j ($j = 0, 1, 2, 3, \ldots, 2047$). The program outputs 2048 complex values of X_k ($k = 0, 1, 2, 3, \ldots, 2047$).

Derive expressions enabling acceleration PSD plots, without refinements such as data windows, overlap or sequential averaging, to be produced by the computer.

Solution

The output of a single accelerometer is considered. Since the sampling rate is $f_s = 2500$ samples/s, and the computer takes the data in a batch of $N = 2048$ values, the time duration of the sample is T where

$$T = N/f_s = 2048/2500 = 0.8192\,\text{s}$$

The frequency spacing of the 2048 values of X_k is $1/T = f_s/N = 2500/2048 = 1.2207$ Hz. Values of X_k above the Nyquist frequency, 1250 Hz, will be spurious, and the useful frequency range will be further limited by the anti-aliasing filter cut-off at 1000 Hz. So, of the 2048 values of X_k output by the DFT program, only those below 1000 Hz will be useful, that is (1000/1.2207) or about 819 discrete frequencies, spaced at intervals of 1.2207 Hz.

From Eq. (10.82), $X_k^* X_k = \frac{1}{4}(a_n^2 + b_n^2)$, the power at each frequency, $\frac{1}{2}(a_n^2 + b_n^2)$, is given by:

$$\frac{1}{2}(a_n^2 + b_n^2) = 2X_k^* X_k \quad (n = k) \tag{B}$$

The PSD at each frequency, f_k, will be the power at that frequency, $\frac{1}{2}(a_n^2 + b_n^2)$, divided by the frequency interval, $\delta f = 1/T$, i.e.,

$$\frac{\frac{1}{2}(a_n^2 + b_n^2)}{\delta f} = \frac{\frac{1}{2}(a_n^2 + b_n^2)}{1/T} = \frac{2X_k X_k^*}{1/T} = \frac{2X_k X_k^*}{1.2207} \tag{C}$$

If the original samples, x_j, are scaled in g units, the PSD values will be in g^2/Hz. The values of frequency will be

$$f_k = k/T \quad (k = 0, 1, 2, 3, \cdots, 819) \tag{D}$$

that is from 0 to 1000 Hz (approximately) in steps of 1.2207 Hz.

Computing cross-PSD functions

If there is a second random waveform, $y(t)$, in addition to $x(t)$, of the same length, T, and sampled at the same instants, for the same number of samples, N, then applying Eq. (9.45) again, this time to values of y, gives

$$Y_k = \frac{1}{N} \sum_{j=0}^{N-1} y_j \, e^{-i\left(\frac{2\pi jk}{N}\right)} \tag{10.84}$$

where Y_k is the DFT of the discrete time samples, y_j.

We already have, from Eq. (9.45),

$$X_k = \frac{1}{N} \sum_{j=0}^{N-1} x_j \, e^{-i\left(\frac{2\pi jk}{N}\right)} \tag{9.45}$$

and from Eq. (10.83),

$$S_x\left(f_{(n)}\right) = S_x(f_k) = 2TX_k^*X_k$$

We can now compute another PSD function, $S_y(f_k)$, and two cross-power spectral density functions, $S_{xy}(f_k)$ and $S_{yx}(f_k)$. The complete set of four spectra derivable from the two time series, $x(t)$ and $y(t)$, are thus:

$$S_x(f_k) = 2TX_k^*X_k \tag{10.85a}$$

$$S_y(f_k) = 2TY_k^*Y_k \tag{10.85b}$$

$$S_{xy}(f_k) = 2TX_k^*Y_k \tag{10.85c}$$

$$S_{yx}(f_k) = 2TY_k^*X_k \tag{10.85d}$$

These spectra are single-sided, i.e. expressed in terms of positive frequencies only, in Hz, as is usual in practical work.

10.7.2 Computing Correlation Functions

The need to compute correlation functions still arises occasionally. Although they are functions of time, thanks to the power of the fast Fourier transform (FFT), it is much quicker to compute them from the corresponding spectra, than directly from the original time samples.

There is no need to form the spectra from Eqs (10.85a–10.85d)), unless they are actually required as output. Assuming that the complete set of four functions from two input time histories, i.e. two ACFs and two CCFs, is required, the procedure is
(1) Starting with the discrete time samples, x_j and y_j, the DFT, Eq. (9.45), is used to compute X_k and Y_k. Then $X_k^*X_k$, $Y_k^*Y_k$, $X_k^*Y_k$ and $Y_k^*X_k$ are formed. They are kept at the full length, N, including the 'spurious' values.
(2) The IDFT, Eq. (9.46), is then used to find the correlation functions from:

$$R_x(j) = \sum_{k=0}^{N-1} X_k^*X_k e^{i\left(\frac{2\pi jk}{N}\right)} \tag{10.86a}$$

$$R_y(j) = \sum_{k=0}^{N-1} Y_k^*Y_k e^{i\left(\frac{2\pi jk}{N}\right)} \tag{10.86b}$$

$$R_{xy}(j) = \sum_{k=0}^{N-1} X_k^*Y_k e^{i\left(\frac{2\pi jk}{N}\right)} \tag{10.86c}$$

$$R_{yx}(j) = \sum_{k=0}^{N-1} Y_k^*X_k e^{i\left(\frac{2\pi jk}{N}\right)} \tag{10.86d}$$

A factor of 2 is not required because the IDFT uses the 'spurious' Fourier coefficients to represent coefficients at negative frequencies. Thus, while the 'spurious' values generated by the DFT must be removed before finally outputting a result, in internal operations within the computer they are retained and used.

The following example, a DFT consisting of only ten points, is used in a simple 'hand' calculation to illustrate the preceding points.

Example 10.13

Extend the 10-point DFT calculation of Example 9.2, to illustrate the computation of:
(a) the PSD function of $x(t)$;
(b) the ACF of $x(t)$.

Solution

Part (a):
Table 10.7 repeats the values of j, t and x_j from Table 9.1 of Example 9.2.
The value of x_j^2 is not strictly required, but has been added to enable the mean square value to be calculated directly for subsequent checks. For later reference, the mean square value of the input is seen to be 0.710.
Table 10.8 repeats the values of k, f, $\mathrm{Re}(X_k)$ and $\mathrm{Im}(X_k)$ from Table 9.2, of Example 9.2. The values of $X_k^* X_k$ are calculated, and in the last column the PSD values are given by Eq. (10.85a):

$$S_x(f_k) = 2TX_k^* X_k \tag{A}$$

with $T = 0.5$. The 'spurious' values above $k = 5$ cannot be used for calculating the PSD, but $X_k^* X_k$ is retained at full length, N, for subsequent use.
As a check, the mean square value of the input time function should be given by summing the power in the PSD function, as given in the last column. The width of each band, δf, is 2 Hz, so the mean square value is $0.355\ \delta f = 0.710$, agreeing with the mean square value from Table 10.7. Note that the sum of the $X_k^* X_k$ values is also 0.710.

Table 10.7

j	$t(s)$	x_j	x_j^2
0	0	0.1000	0.0100
1	0.05	0.3165	0.1002
2	0.10	0.1958	0.0384
3	0.15	1.7063	2.9114
4	0.20	0.8591	0.7381
5	0.25	−0.1000	0.0100
6	0.30	−0.7973	0.6357
7	0.35	−1.5445	2.3854
8	0.40	−0.3576	0.1279
9	0.45	−0.3783	0.1431
			$\sum = 7.10$
			$\langle x_j^2 \rangle = 0.710$

Table 10.8

k	f (Hz)	Re(X_k)	Im(X_k)	$X_k^* X_k$	PSD = $2TX_k^* X_k$ (T=0.5)
0	0	0	0	0	0
1	2	0	−0.5	0.2500	0.2500
2	4	0	0.25	0.0625	0.0625
3	6	0.05	0	0.0025	0.0025
4	8	0	−0.2	0.0400	0.0400
5	10	0	0	0	0
6	12	0	0.2	0.0400	−
7	14	0.05	0	0.0025	−
8	16	0	−0.25	0.0625	−
9	18	0	0.5	0.2500	−
				$\sum = 0.710$	$\sum = 0.355$
					$0.355\, \delta f = 0.710$

Part (b)

The ACF is now given by applying the IDFT in the form of Eq. (10.86a):

$$R_x(j) = \sum_{k=0}^{N-1} X_k^* X_k e^{i\left(\frac{2\pi jk}{N}\right)} \tag{B}$$

which for the purposes of this example is written as:

$$R_x(j) = \sum_{k=0}^{N-1} X_k^* X_k \left[\cos\left(\frac{2\pi jk}{N}\right) + i \sin\left(\frac{2\pi jk}{N}\right) \right] \tag{C}$$

Equation(C) is evaluated for the single value $j = 3$ in Table 10.9. It is seen that the real part of $R_x(j = 3)$ is −0.2268. The imaginary part is zero, as must be the case. Repeating Table 10.9 for all the other values of j gives Table 10.10.

Table 10.9

k	$X_k^* X_k$	$X_k^* X_k \cos\left(\frac{2\pi jk}{N}\right) j = 3\ N = 10$	$X_k^* X_k \sin\left(\frac{2\pi jk}{N}\right) j = 3\ N = 10$
0	0	0	0
1	0.2500	−0.07725	0.23776
2	0.0625	−0.05056	−0.03674
3	0.0025	0.00202	−0.00147
4	0.0400	0.01236	0.03804
5	0	0.00000	0.00000
6	0.0400	0.01236	−0.03804
7	0.0025	0.00202	0.00147
8	0.0625	−0.05056	0.03674
9	0.2500	−0.07725	−0.23776
		$\sum = -0.2268 = R_x(j = 3)$	$\sum = 0$

Table 10.10

j	τ	R_x
0	0	0.7100
1	0.05	0.3768
2	0.10	0.0740
3	0.15	−0.2268
4	0.20	−0.4290
5	0.25	−0.3000
6	0.30	−0.4290
7	0.35	−0.2268
8	0.40	0.0740
9	0.45	0.3768

The values of R_x corresponding to values of j larger than 5 are spurious, and are not used in the final output. Thus R_x is output for positive values of τ from 0 to 0.25 s.

10.7.3 Leakage and Data Windows

When a time history of length T seconds is sampled and converted into Fourier components, say by using the DFT, the assumption is made that it is periodic, and can be represented by sine and cosine waves at the harmonic frequencies $1/T, 2/T, 3/T, \ldots$ and so on. If the waveform actually is periodic, and the period is known, T can sometimes be chosen so that it is exactly equal to the period. This can be achieved in the analysis of data from some rotating machines, perhaps by using a synchronizing pulse, but generally, and always with random data, this is not possible, and there will be frequencies in the waveform that do not correspond to any of the frequencies in the assumed harmonic series. These will produce output at adjacent frequencies in the harmonic series, a phenomenon known as *leakage*. To illustrate this, if the complete sinusoid shown in Fig. 10.21 at (a) was present in the signal, it would appear, correctly, in the resulting power spectrum, as shown at (b), as a single line. However, the incomplete sinusoid shown at (c) does not coincide with any single harmonic frequency, and the result sketched at (d) is obtained. It can be shown, as might be expected, that the effect is largely due to the discontinuities at the ends of the sample, and it can be reduced by applying a *data window*, as shown at (e), to the whole sample. The sampled values are multiplied by the window shape before the DFT is carried out. The tapering is often, but not always, of the cosine form shown.

Tapering the ends of the data in this way represents a loss of data, which can be recovered, to some extent, by overlapping consecutive samples, if *sequential averaging*, as discussed later, is used. The effect of the window on the measured spectrum must, of course, be allowed for. This form of data window is not appropriate for transient data, for which rectangular or exponential windows are mostly used.

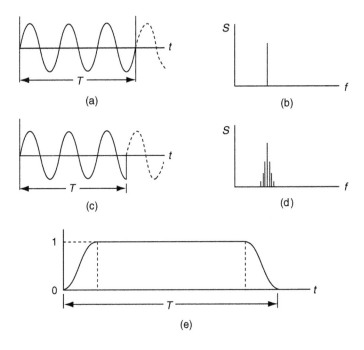

Figure 10.21

10.7.4 Accuracy of Spectral Estimates from Random Data

When a spectrum of any kind is derived from stationary random data, and the length of the time sample is restricted by practical considerations, such as for how long the response is available, we cannot expect the results to be perfect. Fortunately, there is a simple formula giving the approximate statistical variation of a PSD estimate in relation to its expected value in the long term, and this is now discussed.

The problem of estimating how a spectral estimate of a random signal in a narrow frequency band is likely to vary was first investigated in analog days, and the fundamental relationship is still expressed in terms of spectral analysis by analog methods. In the model usually considered, a stationary, broad-band signal, with Gaussian amplitude probability, is considered to be input to an old-time analog spectrum analyser. Briefly, this device works as follows: the original random signal, $x(t)$, say, is input to a narrow band filter, which passes only the power in a band centered on a specific frequency, f_0, and of bandwidth B. The output from the filter is squared and integrated for the duration of the signal, T, thus giving the mean square value of that part of the original signal within the bandwidth B, assuming that the gain of the filter is unity within its pass band, and zero elsewhere. An estimate of the PSD of the original signal at frequency f_0 is then given by dividing that mean square value of the output by the bandwidth of the filter, B. If this is repeated for many values of f_0, *with the same input*, a PSD plot of the original signal $x(t)$ can be built up.

Now consider the device to be used in a slightly different way. Instead of changing the center frequency of the filter sequentially, and repeating the same input, as above, suppose the center frequency to be kept constant, and *different* samples from the stationary input, $x(t)$, each of length T, to be input, one after the other. In this case we obtain many different estimates of the spectral density, at a single frequency, and it is the statistical properties of these estimates, in terms of mean and standard deviation, that we need to know. In fact, it can be shown theoretically [10.1] that, approximately:

$$\frac{\sigma}{\mu} \approx \frac{1}{\sqrt{BT}} \tag{10.87}$$

where the PSD estimates are assumed to have a Gaussian distribution, and

σ is the standard deviation of many estimates of spectral density in a bandwidth B, centered on any given frequency, μ the mean value of the estimates of spectral density for many different samples, regarded as the 'correct' value of the PSD, B the bandwidth over which the PSD is defined and T the time duration of each sample of random data.

Although strictly only valid for the analog process described above, it has been shown [10.1] that Eq. (10.87) also works for digital analysis, if the bandwidth, B, is interpreted as an equivalent bandwidth, B_e, implied by the digital method, which for practical purposes is equal to the spacing between the spectral lines, i.e.,

$$B \approx B_e \approx \frac{1}{T} \tag{10.88}$$

When the standard DFT is used to compute a power spectrum (without smoothing or sequential averaging), Eq. (10.88) always applies, and Eq. (10.87) then gives

$$\frac{\sigma}{\mu} \approx \frac{1}{\sqrt{T/T}} = 1 \tag{10.89}$$

Thus a single DFT will always give a PSD plot in which the standard deviation of each point about the 'true' value is approximately the same as the true value itself, a fairly poor result. However, there are two basic ways in which the accuracy can be improved. These are *smoothing* and *sequential averaging*.

Smoothing the spectrum

The basic DFT may give a frequency resolution finer than is actually needed, and the PSD plot can then be smoothed. The following example illustrates the idea.

Example 10.14

It is desired to use a DFT program to measure the PSD of random pressure fluctuations generated by a jet engine, in the frequency band from 25 to 800 Hz. It is known from experience that in this case the spectrum will be fairly flat, so that it will be sufficient to plot it at 25 Hz intervals. A data sample 1 second long, sampled at

2048 samples per second, is available. An anti-aliasing filter with a substantially flat response up to 800 Hz, and negligible response at the Nyquist frequency, 1024 Hz, is used before sampling. Show how the scatter can be reduced by smoothing.

Solution

Using a DFT with $N=2048$ points, sampled at 2048 samples/s, then $T=1$ and $B=1/T=1$. From Eq. (10.87):

$$\frac{\sigma}{\mu} \approx \frac{1}{\sqrt{BT}} = 1 \tag{A}$$

However, since it is acceptable, in this case, to present the power spectrum at intervals of 25 Hz, it may be smoothed by averaging in groups of twenty five. Now, in effect, $B=25$, and $T=1$, as before, so,

$$\frac{\sigma}{\mu} \approx \frac{1}{\sqrt{BT}} = \frac{1}{\sqrt{25}} = 0.2 \tag{B}$$

Thus the scatter that would be apparent if the PSD were plotted at 1 Hz intervals is considerably reduced.

It is important to note that the PSD values must be formed before smoothing takes place. The Fourier transform of random data is itself random, and would average to zero.

Sequential averaging

The product BT can effectively be increased in another way. This consists of breaking the input data down into shorter sections, computing the auto- or cross-spectrum of each, and averaging the results. This technique is used in all digital spectrum analysers and equivalent software packages. It is usual to overlap the sequential samples, to compensate for the loss of data due to the use of a data window. There is virtually no limit to the number of averages that can be used, permitting very accurate results, if a sufficiently long-time history is available.

In this method, the effective value of B is unchanged, and is the same as for each individual DFT. However, the effective value of T is that of each individual transform multiplied by the number of averages, and the product BT can be as large as the available data will permit.

10.8 Fatigue due to random vibration

When a structure is excited by a broad-band Gaussian random input, and the response is also broad band, there is no easy way, in general, to estimate its fatigue life. However, if the response can be considered to be confined to a narrow frequency band, as in a large proportion of practical cases, an approximate prediction of the fatigue life becomes feasible. It should be recognized that such estimates are far from reliable, and testing remains essential in critical cases. The method outlined here relies upon the fact that a narrow band Gaussian random waveform has a well-defined peak distribution, the Rayleigh distribution, enabling the Palmgren–Miner hypothesis to be used, provided an S–N diagram for the material is available. These concepts are first introduced, and then combined to develop a practical method.

10.8.1 The Rayleigh Distribution

If the response of a linear system to a random input is confined to a narrow band of frequencies, it has the appearance shown in Fig. 10.22. It is roughly sinusoidal, but with a randomly varying envelope. Although the amplitude distribution remains Gaussian, the distribution of the peaks can be shown to have a *Rayleigh probability density* given by:

$$p(a) = \frac{a}{\sigma_x^2} e^{\frac{-a^2}{2\sigma_x^2}}$$

(10.90)

where: a is a peak value of the response, which can be a displacement, stress, etc.; $p(a)$ is the probability density of the peaks, i.e. the probability of a peak lying between a and $a + da$; and σ_x is the standard deviation of $x(t)$, the narrow band time history, *not of the peak values*.

The probability $p(a)$ is shown plotted against a/σ_x in Fig. 10.23. The Rayleigh distribution is that of the positive peaks. The 'negative peaks', or 'troughs', are assumed to have the same distribution, but inverted. The area of the plot is unity, since this must represent the probability of all the peaks exceeding zero, which is 1, or certain.

It can be shown [10.2], that the mean value of the (positive) peaks, $\langle a \rangle$, is given by:

$$\langle a \rangle = \sqrt{\frac{\pi}{2}} \sigma_x$$

(10.91)

and that the mean square value of the peaks is

$$\langle a^2 \rangle = 2\sigma_x^2$$

(10.92)

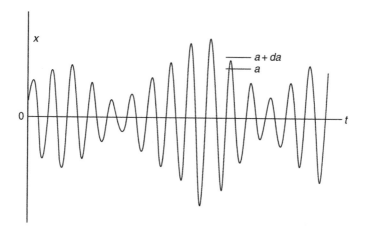

Fig. 10.22   Narrow-band random response.

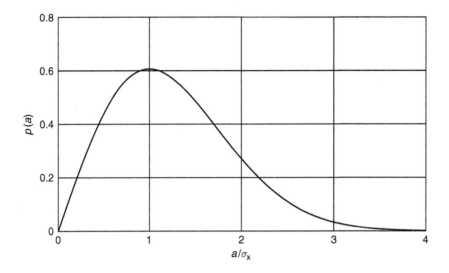

Fig. 10.23   Rayleigh probability density function.

Thus the variance of the peaks, σ_a^2, is

$$\sigma_a^2 = \langle a^2 \rangle - (\langle a \rangle)^2 = \left(\frac{4-\pi}{2}\right)\sigma_x^2 \tag{10.93}$$

10.8.2 The S–N Diagram

The fatigue characteristics of a given material are characterized by its *S–N* diagram, an example of which is shown in Fig. 10.24. In the form shown here, each of the dots represents a separate fatigue test, in which a sample of material is tested to failure by

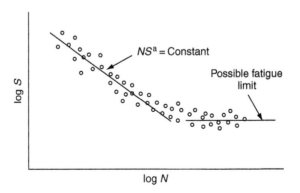

Fig. 10.24   Typical S–N diagram.

applying an alternating stress of magnitude S. This is repeated for many different values of S, and the number of cycles to failure, N, is plotted against S. For many materials it is found that, up to some value of N, the plot of $\log S$ against $\log N$ approximates to a straight line, implying that

$$NS^\alpha = \text{constant} \qquad (10.94)$$

where α is a constant, often in the range 4–8. This property of the curve is sometimes used to justify the practice of shortening vibration endurance tests by increasing the test level.

For large values of N, it is sometimes, but not always, found that the curve flattens out, and there is some value of S below which failure would never occur; however, many stress cycles are applied. This is known as the fatigue limit.

The Palmgren–Miner hypothesis

This hypothesis, which is entirely empirical, allows the Rayleigh peak distribution, and the S–N diagram, to be combined to calculate the fatigue life of a structure under narrow-band random loading.

The basic S–N diagram is defined by the fact that if *constant amplitude* stress reversals of magnitude $\pm S_i$ are applied, then failure occurs after N_i cycles. In the Palmgren–Miner hypothesis, this is assumed to represent an amount of damage, D, of 1. If, instead, a smaller number of cycles, n_i, are applied, the material does not fail, but is assumed to suffer an amount of damage, $D_i = n_i/N_i$, equal to the fraction of its life used up. This can be repeated at different stress levels, and failure eventually occurs when the total damage, from all the loadings, is equal to 1, i.e.,

$$D = \sum_i D_i = \sum_i \frac{n_i}{N_i} = 1 \qquad (10.95)$$

It is also assumed that the order in which the stress cycles are applied is unimportant, so the small amounts of damage, D_i, could be due to stress cycles in the Rayleigh distribution, where they occur in random order.

Fatigue life

If Eq. (10.90) is written in terms of peak stress, S, then:

$$p(S) = \frac{S}{\sigma_S^2} e^{\frac{-S^2}{2\sigma_S^2}} \qquad (10.96)$$

where σ_S is the RMS value of the stress.

If f_0 is the number of zero-crossings, with positive slope, of the stress, per unit time, the number of peaks between the stress levels S and $S + dS$, in time T, is $f_0 T p(S) dS$. The proportion of the damage in time T is $f_0 T p(S) dS/N(S)$. The total damage in time T is therefore D_T, say, where:

$$D_T = f_0 T \int_0^\infty \frac{p(S)}{N(S)} \, dS \qquad (10.97)$$

Since, for failure, the total damage $D = 1$, the time to failure, T_F is then:

$$T_F = \frac{T}{D_T} = \frac{1}{f_0 \int_0^\infty \frac{p(S)}{N(S)} dS} \tag{10.98}$$

where f_0 is defined above as the number of positive-going zero crossings in unit time, which can also be interpreted as the center frequency of the narrow band, or the natural frequency of the single-DOF system, if this is the reason for the narrow band, $p(S)$ the Rayleigh distribution, in terms of peak stress, S and $N(S)$ is essentially the S–N curve for the material.

In principle, the fatigue life can be calculated from Eq. (10.98), assuming that the standard deviation of the stress, σ_S, at the worst location, can be found; that the stress response is reasonably narrow band; and that the S–N curve is available. It is usual to apply a safety factor between 3 and 5 to the number of cycles, or time, to failure, as given by Eq. (10.98).

The accuracy of predictions of this kind can often be improved, particularly in the case of acoustic fatigue, by obtaining the S–N data for the material in a different way. If the test specimens consist, for example, of small cantilevers, and are excited in bending by a random input, their response in the region of resonance will automatically have the desired Rayleigh peak distribution, and a plot of cycles (and hence time) to failure versus the standard deviation of the stress can be obtained directly, rather than by the integration shown in Eq. (10.98).

References

10.1 Newland, DE (1993). *Random Vibrations, Spectral and Wavelet Analysis*. Longman (London) and John Wiley (New York).
10.2 Thomson, WT (1988). *Theory of Vibration with Applications*. Prentice Hall.

11 Vibration Reduction

Contents

There are many examples in engineering where vibration levels can become so high that they threaten the structural integrity of the machine or structure concerned, the reliability of essential electronic equipment, or the health and efficiency of those exposed to it. In some cases it can be said that a high level of vibration is inherent in the design of a machine or vehicle, and measures to reduce it or isolate it are almost always needed, for example:

(a) Reciprocating engines inevitably generate periodic forces that must be isolated from the vehicle interior in nearly all cases, and crankshaft torsional vibrations often have to be reduced, simply to prolong the life of the component.

(b) Helicopter rotors can rarely be balanced well enough to avoid unacceptable periodic vibration in fuselage areas, and vibration reduction techniques are practically always required.

(c) Combat aircraft often generate broad-band random vibration of sufficient severity to make vibration isolation necessary.

In a case of severe vibration response, if the source cannot be removed, the obvious first move is to try to increase the damping of the mode or modes affected. This is often not possible, and *vibration isolation* and *vibration absorption* may then be considered.

Vibration isolation uses the properties of a single-DOF system, either to isolate sensitive equipment from a severe vibration environment or to isolate an environment, such as the interior of a vehicle, from vibration forces developed within, say, the engine. Vibration isolation can be used with any form of vibration input. When used to reduce the effect of transients it is usually known as *shock isolation*.

Vibration absorption can be provided by several different devices. All use an auxiliary mass (or rotational inertia), and they can be roughly classified as follows:

(1) A *dynamic absorber* uses an auxiliary mass connected to the main system by a spring, with negligible damping. In its basic form it will absorb energy at one frequency only, but it does this very efficiently. A variant of the dynamic absorber, the *centrifugal dynamic absorber* uses the pendulum principle to self-tune the absorber to a multiple of the rotation speed of a component of a machine.

(2) A *damped absorber* or *auxiliary mass damper* is also connected to the main system by a spring, but unlike the dynamic absorber it also uses a damper of appreciable value in parallel with the spring.

(3) An absorber consisting of a rotational inertia with damping, but no spring, is known as a *Lanchester damper*. The term *Houdaille damper* is often used when the damping is viscous.

In principle, vibration can be canceled, by applying equal but opposite excitation, in a generalized sense, and such active systems have been developed for helicopters. The forces required to cancel the vibration can be applied by dedicated actuators, or through the hydraulic actuators controlling the swash plate. In the latter case the technique is known as *higher harmonic control* (HHC). Discussion of these active systems is beyond the scope of this text.

11.1 Vibration isolation

The principle of vibration isolation can be used in two basic ways:
(1) To protect sensitive items, such as delicate electronic components, from high environmental vibration levels.
(2) To reduce the magnitude of the oscillatory forces transmitted to the supports, when inertial or other forces are developed in a machine.

11.1.1 Isolation from High Environmental Vibration

This aspect of vibration isolation is illustrated by Fig. 11.1, which shows a mass m, connected to the base by a spring, k, and damper c. The base represents a location where there is a high level of vibration, and the mass represents a shelf or rack containing delicate equipment, which is to be protected from the vibration environment.

It was shown by Eq. (4.35) that if $|x|$ is the magnitude of a sinusoidal displacement of the base, and $|z|$ is the resulting steady-state displacement magnitude of the mass, m, then:

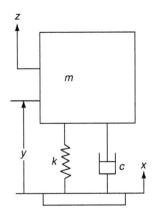

Fig. 11.1   Schematic model illustrating vibration isolation.

$$\frac{|z|}{|x|} = \sqrt{\frac{1 + (2\gamma\Omega)^2}{(1 - \Omega^2)^2 + (2\gamma\Omega)^2}}$$ (4.35)

where $\Omega = \omega/\omega_n$; ω is the excitation frequency, in rad/s; $\omega_n = \sqrt{k/m}$ the undamped natural frequency of the system, in rad/s and γ the non-dimensional viscous damping coefficient, equal to $c/2m\omega_n$.

The ratio $|z|/|x|$ is known as the displacement transmissibility, T, and it is easily seen from Chapter 4 that

$$T = \frac{|z|}{|x|} = \frac{|\dot{z}|}{|\dot{x}|} = \frac{|\ddot{z}|}{|\ddot{x}|}$$ (11.1)

so Eq. (4.35) also applies when the base and response magnitudes are *both* expressed as velocities or accelerations. Equation (4.35) was plotted for a wide range of values of the damping coefficient, γ, in Fig. 4.8. This is repeated here in Fig. 11.2 for a range of values of γ likely to be used in practical isolation systems.

Vibration attenuation is obtained for input frequencies greater than $\sqrt{2}$ times the natural frequency of the isolation system, and this is true for all values of damping. For input frequencies below $\sqrt{2}$ times the natural frequency, and especially those close to the natural frequency, the vibration level of the mass is actually increased compared with that of the base. The main objective in the design of a vibration isolation system is therefore to secure the necessary reduction at high frequencies without increasing the low frequency response too much.

When the input, x, at the base, is periodic, each harmonic component in the input is simply multiplied by the transmissibility, at that frequency, to obtain the magnitude of the corresponding component of the response, z. When the input is broad-band random, it is usually expressed in the form of a PSD function, and the response PSD is found by multiplying the input PSD by the square of the transmissibility. The following example illustrates a typical application with random vibration input.

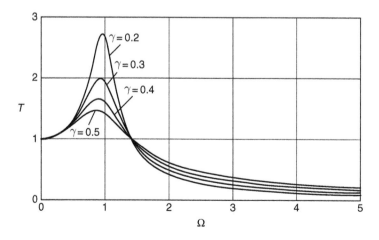

Fig. 11.2   The transmissibility, T, of a single-DOF system versus the non-dimensional frequency ratio, Ω.

Example 11.1

Figure 11.3 shows a proposed vibration-isolated shelf, to be fitted at a location in a high performance aircraft, where the environmental vibration is broad-band random. The measured PSD levels, in the vertical direction, are all enveloped by the values shown in Table 11.1, which can therefore be regarded as an upper bound of the input PSD level.

The shelf, and the equipment mounted on it, are assumed rigid. It is supported by four vibration isolators, each of which can be considered to consist of a linear spring and viscous damper in parallel.

For vertical behavior only, if the system is assumed balanced, i.e. the center of gravity coincides with the stiffness center, it can be represented schematically by Fig. 11.1. The undamped natural frequency is 20 Hz, and the assembled system has $Q = 2.5$, equivalent to a viscous damping coefficient, $\gamma = 0.2$.

(a) Calculate and plot the maximum vertical acceleration PSD level to be expected on the shelf for the frequency range 10–1000 Hz.

(b) Calculate and compare
  (i) the RMS vertical acceleration of the aircraft structure to which the shelf is attached;
  (ii) the RMS vertical acceleration of the shelf and supported equipment.

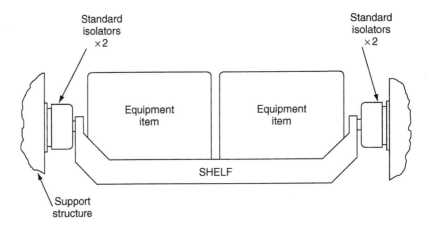

Fig. 11.3   Vibration-isolated shelf (Example 11.1).

Table 11.1

Frequency range (Hz)	Acceleration PSD (g^2/Hz)
10–50	0.01
50–1000	0.10

Solution

Part (a):
From Eq. (10.44), the response PSD function of any linear system with a single input is given by multiplying the input PSD function by the square of the frequency response function between input and output, so in this case:

$$S_{\ddot{z}} = T^2 \, S_{\ddot{x}} \tag{A}$$

where $S_{\ddot{z}}$ is the response or shelf PSD in g^2/Hz; $S_{\ddot{x}}$ the input or base PSD in g^2/Hz, given by Table 11.1 and T the transmissibility ratio for acceleration, which from Eqs (4.35) and (11.1) is given by:

$$T = \frac{|\ddot{z}|}{|\ddot{x}|} = \frac{|z|}{|x|} = \sqrt{\frac{1 + (2\gamma\Omega)^2}{(1 - \Omega^2)^2 + (2\gamma\Omega)^2}} \tag{B}$$

where $\Omega = f/f_n = f/20$ and $\gamma = 0.2$.
The transmissibility, T, is plotted in Fig. 11.4(a) for $f = 10$–1000 Hz.
The shelf PSD, $S_{\ddot{z}}$, is now given by Eq. (A), as follows.
For the range 10–50 Hz:

$$S_{\ddot{z}} = T^2 S_{\ddot{x}} = T^2 \times 0.01$$

For the range 50–1000 Hz:

$$S_{\ddot{z}} = T^2 S_{\ddot{x}} = T^2 \times 0.10$$

The shelf response PSD is plotted, together with the input PSD, for comparison, in Fig. 11.4(b). The large reduction of the shelf response PSD compared with the input PSD at higher frequencies can be seen, as can the increase around the natural frequency of the isolating system at 20 Hz.

Part (b):
From Eq. (10.30), the mean square value (and the variance, if the mean value is zero) of any random waveform is equal to the area under the PSD function.

$$\sigma^2 = \int_{f_1}^{f_2} S(f) \mathrm{d}f \tag{C}$$

(i) Applying Eq. (C) to the input PSD, $S_{\ddot{x}}$, as defined by Table 11.1:

$$\sigma_{\ddot{x}}^2 = 0.01(50-10) + 0.10(1000-50) = 95.4 \, g^2$$

where $\sigma_{\ddot{x}}^2$ is the mean square value of the input, in g units.
The RMS value of the input waveform is thus $\sigma_{\ddot{x}} = 9.77 \, g$.
(ii) The response PSD, $S_{\ddot{z}}$, was integrated numerically, giving $\sigma_{\ddot{z}}^2 = 0.970 g^2$, where $\sigma_{\ddot{z}}^2$ is the mean square value of the shelf response, in g units.
The RMS value of the shelf response is thus $\sigma_{\ddot{z}} = 0.985g$.
The isolators therefore reduce the overall RMS acceleration of the shelf compared with the aircraft structure by nearly a factor of ten in this case.

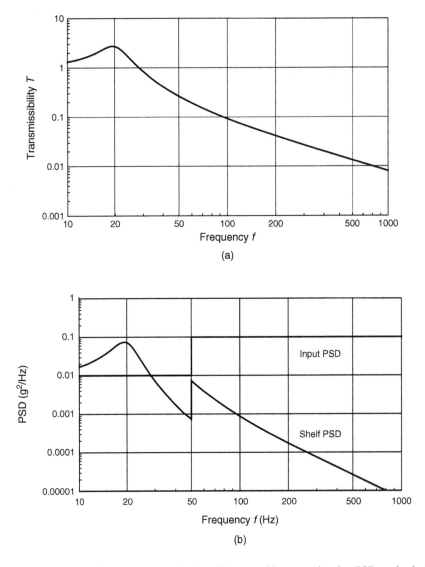

Fig. 11.4   (a) Transmissibility plot in Example 11.1, (b) assumed input acceleration PSD, and calculated shelf acceleration PSD, in Example 11.1.

The practical design of vibration isolation systems

All vibration isolated systems have six rigid modes of vibration, and it is only possible to treat these separately when the center of gravity coincides exactly with the stiffness center of the system, and, of course, the system is assumed linear. The rigid modes of the system then consist of three translations and three rotations. This situation should always be aimed for in practical cases, in order to avoid cross-coupling between translation and rotation modes, which leads to reduced efficiency of isolation and unequal dynamic loading on different isolators.

When a balanced system cannot be achieved, it is always worth carrying out a 6-DOF analysis, enabling natural frequencies, rigid mode shapes, and isolator loads and displacements to be estimated. Although isolators are often deliberately designed to have non-linear load-displacement characteristics, a linear model usually gives adequate results if the stiffnesses are well chosen to represent the working range in a particular application. An important aim of modeling should be to establish acceptable limits for the possible variation of the center of gravity (CG) position of the isolated system as a whole. Variations of the CG position, perhaps due to changes in the equipment carried, can have a surprisingly large effect on the relative dynamic loading of the individual isolators, leading to early failure in some cases.

A wide range of isolators is available commercially from specialist suppliers, who also offer a design service for particular applications. The majority of simple mounts use elastomers, i.e. natural or synthetic rubber, which can be formulated to give the required combination of stiffness and damping, but metal springs with friction damping are also used. Most isolators allow movement in three perpendicular directions, and are often designed to have similar characteristics in each, since usually the same natural frequency is required for translation in each direction.

An isolation system must, of course, be designed to take into account loads other than vibration loads. In an aircraft system, these may be transients due to landing, braking, turning, etc., and especially, steady accelerations due to maneuvers. The isolators should not reach their full travel, and this sets a lower limit to the natural frequency that can be used in practice, as shown by the following simple analysis, which assumes linear spring characteristics.

Consider, for example, the static 'down load', i.e. the normal force, Z, acting on the isolation system in a 'maximum g' turn. This is,

$$Z = ngm \tag{11.2}$$

where n is the normal force load factor acting on the aircraft at the location occupied by the isolation system; g the acceleration due to gravity, i.e. 9.81 m/s^2 and m the mass of the isolation system.

However, the force $Z = ky_s$ where k is the total stiffness of the isolation system in the normal direction, and y_s is the static displacement of the isolator springs. Since $k = m(2\pi f_n)^2$ where f_n is the undamped natural frequency of the isolation system, in Hz, we can write

$$Z = ngm = ky_s = m(2\pi f_n)^2 y_s$$

or

$$f_n = \frac{1}{2\pi}\sqrt{\frac{ng}{y_s}} \tag{11.3}$$

Thus, if y_s is the maximum displacement permitted by the isolators, the lowest possible natural frequency of the system, without allowance for additional displacement due to vibration, is as given by Eq. (11.3).

As an example, consider a combat aircraft where the maximum normal load factor is 9, and the maximum allowable displacement of the system from the unloaded condition is 10 mm = 0.01 m. Then the lowest possible natural frequency, if the isolator springs are not to run out of travel is

$$f_n = \frac{1}{2\pi}\sqrt{\frac{ng}{y_s}} = \frac{1}{2\pi}\sqrt{\frac{9 \times 9.81}{0.01}} = 15\,\text{Hz (approx.)}$$

In practice, an allowance would have to be made for the expected vibration amplitude, allowing at least 3σ, where σ is the standard deviation of the vibration displacement of the isolators, for a Gaussian random input.

Isolators having stiffening characteristics with increasing load would probably be used in such cases, somewhat modifying the simple calculation above.

11.1.2 Reducing the Transmission of Vibration Forces

The opposite problem to that considered above arises when forces developed within a machine are to be isolated from its surroundings. In principle, the forces can take any form, but most practical examples seem to involve periodic forces. The most familiar example is surely the reciprocating engine of a road vehicle. The inertia forces produced by the pistons and cranks cannot always be balanced perfectly, and the torque from the drive shaft is never perfectly smoothed by the flywheel, and these would transmit considerable vibration to the vehicle if flexible engine mounts were not used. Other cases where this form of vibration isolation may be required include propeller driven aircraft, pumps, compressors and machine tools.

Most practical cases can be dealt with by using the steady-state frequency response function relating the force applied to the machine to the force transmitted to its supports. For a single-DOF system, this function was derived in Chapter 4, as Eq. (4.42):

$$\frac{|F_T|}{|F|} = \sqrt{\frac{1 + (2\gamma\Omega)^2}{(1 - \Omega^2)^2 + (2\gamma\Omega)^2}} \tag{4.42}$$

where $|F|$ is the magnitude of a sinusoidal force applied to the mass and $|F_T|$ the magnitude of the sinusoidal force applied by the spring and damper to the supporting base, which is assumed rigid; and Ω and γ are defined as for Eq. (4.35).

As pointed out in Chapter 4, the right side of Eq. (4.42) is identical to that of Eq. (4.35), so the force transmissibity for a single-DOF system is of the same form as that of the displacement transmissibility shown in Fig. 11.2. Thus, force attenuation is only obtained for frequencies higher than $\sqrt{2}$ times the natural frequency, and forces close to the natural frequency are magnified.

Example 4.5 discussed the reduction of out of balance forces in a washing machine by the use of damped spring supports.

11.2 The dynamic absorber

In Fig. 11.5, suppose that initially m_2 and k_2 are not present, and the single-DOF system consisting of the mass m_1 and the spring k_1 is being excited at or near its

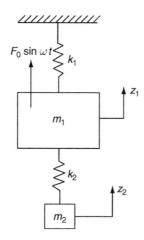

Fig. 11.5   Schematic diagram illustrating a dynamic absorber.

resonant frequency by the force $F_0 \sin \omega t$. This can be an actual sinusoidal force, as shown, or an equivalent force due to base motion.

If the resulting steady-state displacement, z_1, of the mass m_1 is considered excessive, a *dynamic absorber* consisting of the auxiliary system, m_2 and k_2, with nominally zero damping, is added, with the intention of reducing the response, z_1. We can investigate if this works by writing the equations of motion for the complete system. These were derived in Chapter 6, i.e.,

$$\begin{bmatrix} m_1 & 0 \\ 0 & m_2 \end{bmatrix} \begin{Bmatrix} \ddot{z}_1 \\ \ddot{z}_2 \end{Bmatrix} + \begin{bmatrix} (k_1 + k_2) & -k_2 \\ -k_2 & k_2 \end{bmatrix} \begin{Bmatrix} z_1 \\ z_2 \end{Bmatrix} = \begin{Bmatrix} F_0 \\ 0 \end{Bmatrix} \sin \omega t \qquad (11.4)$$

Since we are interested in the steady-state magnitudes of z_1 and z_2, we can derive the frequency responses of the system directly, as follows. Since there is no damping, z_1 and z_2 must be of the form:

$$\begin{Bmatrix} z_1 \\ z_2 \end{Bmatrix} = \begin{Bmatrix} \underline{z}_1 \\ \underline{z}_2 \end{Bmatrix} \sin \omega t, \quad \text{and also:} \quad \begin{Bmatrix} \ddot{z}_1 \\ \ddot{z}_2 \end{Bmatrix} = -\omega^2 \begin{Bmatrix} \underline{z}_1 \\ \underline{z}_2 \end{Bmatrix} \sin \omega t \qquad (11.5)$$

Substituting Eq. (11.5) into Eq. (11.4), and dividing through by $\sin \omega t$:

$$-\omega^2 \begin{bmatrix} m_1 & 0 \\ 0 & m_2 \end{bmatrix} \begin{Bmatrix} \underline{z}_1 \\ \underline{z}_2 \end{Bmatrix} + \begin{bmatrix} (k_1 + k_2) & -k_2 \\ -k_2 & k_2 \end{bmatrix} \begin{Bmatrix} \underline{z}_1 \\ \underline{z}_2 \end{Bmatrix} = \begin{Bmatrix} F_0 \\ 0 \end{Bmatrix} \qquad (11.6)$$

The amplitudes $\underline{z}_1$ and $\underline{z}_2$ are real, and their sign indicates whether they are in phase (positive) or out of phase (negative) with F_0, since there are no other possibilities in an undamped system.

Equation (11.6) can be manipulated into the form:

$$\frac{\underline{z}_1 k_1}{F_0} = \frac{\underline{z}_1}{z_s} = \frac{1 - \Omega_2^2}{\left(1 - \Omega_1^2\right)\left(1 - \Omega_2^2\right) - \mu \Omega_1^2} \qquad (11.7a)$$

and

$$\frac{\underline{z}_2 k_1}{F_0} = \frac{\underline{z}_2}{z_s} = \frac{1}{\left(1 - \Omega_1^2\right)\left(1 - \Omega_2^2\right) - \mu\Omega_1^2} \tag{11.7b}$$

where $\underline{z}_1$ is the displacement of the main mass, m_1; $\underline{z}_2$ is the displacement of the auxiliary mass, m_2; $z_s = F_0/k_1$, the static displacement of m_1 when force F_0 is applied to it; $\Omega_1 = \omega/\omega_1$ where ω is the excitation frequency, and $\omega_1 = \sqrt{k_1/m_1}$; $\Omega_2 = \omega/\omega_2$ where $\omega_2 = \sqrt{k_2/m_2}$ and $\mu = m_2/m_1$.

It can be seen immediately from Eq. (11.7a) that $\underline{z}_1$ must be zero when the numerator $1 - \Omega_2^2$ becomes zero, i.e. when $\omega = \omega_2$, that is *when the excitation frequency is the same as the natural frequency of the absorber*, and this is the important point to remember.

The amplitude of the absorber mass at the same condition can be found from Eq. (11.7b). If $\omega = \omega_2$, this equation reduces to:

$$\frac{\underline{z}_2 k_1}{F_0} = \frac{1}{-\mu\Omega_1^2} = -\frac{m_1\omega_1^2}{m_2\omega^2} = -\frac{m_1\omega_1^2}{m_2\omega_2^2} = -\frac{k_1}{k_2}$$

Rearranging:

$$\underline{z}_2 = -\frac{F_0}{k_2} \tag{11.8}$$

From Eq. (11.8), and referring to Fig. 11.5, the force applied to m_1 by the absorber spring, k_2, is

$$k_2 \underline{z}_2 \sin \omega t = -F_0 \sin \omega t$$

which exactly balances the applied force $F_0 \sin \omega t$. This only happens, of course, when $\omega = \omega_2$ *exactly*. The displacement $\underline{z}_2$ of the auxiliary mass, m_2, as given by Eq. (11.8), will often be quite large, and must be allowed for in the design.

Figure 11.6 is a plot of $|\underline{z}_1|/z_s$ versus Ω_1, from Eq. (11.7a), with $\omega_1/\omega_2 = \Omega_2/\Omega_1 = 1$, and the mass ratio, $\mu = m_2/m_1 = 0.2$. Here $|\underline{z}_1|/z_s$ is the absolute value of $\underline{z}_1/z_s$, since the sign (i.e. the phase) of the latter is not of interest in this case.

Figure 11.6 illustrates some of the characteristics of the dynamic absorber. Since we have set $\omega_1 = \omega_2$ in this case, $\Omega_1 = \omega/\omega_1 = 1$ corresponds to the point where $\omega = \omega_2$, and complete cancelation of the displacement, $\underline{z}_1$, of the main mass, m_1, is obtained at that point, as we have already deduced.

For values of Ω_1 higher and lower than 1, the displacement of the main mass can become very large, and Ω_1 must not be allowed to stray into these regions. These are simply regions where the excitation frequency, ω, is close to one or other of the natural frequencies of the combined system. The latter are given by the determinant of Eq. (11.6), i.e. they are the two values of $\underline{\omega}$ satisfying:

$$\begin{vmatrix} \left(k_1 + k_2 - \underline{\omega}^2 m_1\right) & -k_2 \\ -k_2 & k_2 - \underline{\omega}^2 m_2 \end{vmatrix} = 0 \tag{11.9}$$

Alternatively, we can say that the two values of $\underline{\omega}^2$ are the eigenvalues of the 2-DOF system represented by Eq. (11.6).

In the example shown in Fig. 11.6 , with $\mu = m_2/m_1 = 0.2$, these natural frequencies occur at approximately 0.80 and 1.25 times the natural frequency of the absorber alone, ω_2.

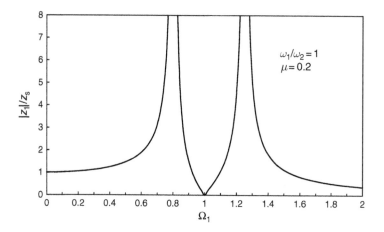

Fig. 11.6 Non-dimensional response of the main system when the dynamic absorber is tuned to the same frequency as the main system. Mass ratio = 0.2.

If the absorber mass is reduced, for example to save weight, both the natural frequencies of the system become closer to the natural frequency of the absorber. This is illustrated by Fig. 11.7, which is a plot similar to Figure 11.6, except that the mass ratio μ has been changed to 0.05, i.e. the absorber mass is 5% of the main mass.

As can be seen from Fig. 11.7, the frequency band within which vibration reduction of the main mass, m_1, occurs is considerably reduced. With $\mu = 0.05$, in fact, the natural frequencies become about 0.90 and 1.12 times that of the absorber, ω_2.

In the cases discussed above, it was assumed that the absorber was tuned to the same frequency as the main system, i.e. $\omega_1 = \omega_2$. This is not necessary, and the absorber will also work when these frequencies are different. However, cancelation of the motion of the main mass will only occur when the excitation frequency, ω, is

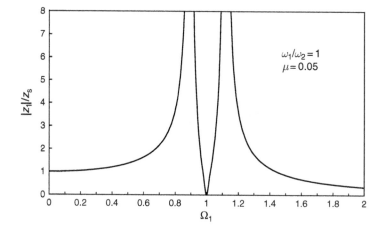

Fig. 11.7   Repeat of Figure 11.6 with mass ratio = 0.05.

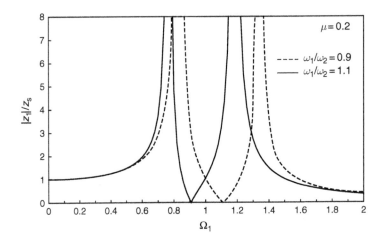

Fig. 11.8   Non-dimensional response of the main system when the dynamic absorber is tuned to 0.9 and 1.1
times its natural frequency. Mass ratio = 0.2.

equal to the absorber natural frequency, ω_2. To demonstrate this, $|\underline{z}_1|/z_s$ is plotted in
Fig. 11.8 with two different values of the ratio ω_1/ω_2 as follows:

Case 1 (dashed line): $\omega_1/\omega_2 = 0.9$, i.e. the natural frequency of the main system is 0.9
times that of the absorber. Since $|\underline{z}_1|/z_s = 0$ when $\omega = \omega_2$, this point occurs on the
graph when:

$$\Omega_1 = \frac{\omega}{\omega_1} = \frac{\omega_2}{\omega_1} = \frac{1}{0.9} = 1.11 \text{ approximately.}$$

Case 2 (solid line): now $\omega_1/\omega_2 = 1.1$, and the natural frequency of the main system is
1.1 times that of the absorber. Since $|\underline{z}_1|/z_s = 0$ still occurs when $\omega = \omega_2$, the point of
zero motion occurs at $\Omega_1 = 1/1.1 = 0.91$ approximately.

This demonstrates the fact that the cancelation of the motion of the main system is
largely independent of its natural frequency, and the important thing is that the
natural frequency of the absorber must be equal to the excitation frequency.

In practice, a dynamic absorber will create a zone of zero or small response at the
point in a structure where it is installed, provided only that its natural frequency is
very close to the excitation frequency. Its use in helicopters is assisted by the fact that
the rotor speed is usually maintained practically constant in flight, and there is often a
large response at the 'blade-passing frequency', i.e. the rotation speed multiplied by
the number of blades. Many ingenious ways of creating dynamic absorbers in helicopters without adding weight have been used, for example by mounting the main
battery on springs. Den Hartog [11.1] has described the application of a dynamic
absorber to a hair clipper vibrating synchronously at 120 Hz, i.e. twice the frequency
of the electricity supply.

11.2.1 The Centrifugal Pendulum Dynamic Absorber

The dynamic absorber, as described above, will only work, of course, at a single
frequency, and is useless if the excitation frequency departs significantly from the

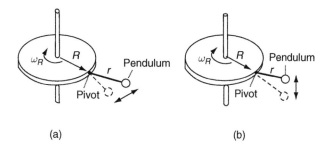

Fig. 11.9   Centrifugal pendulum dynamic absorbers.

natural frequency of the absorber. However, in rotating systems it is possible to replace the auxiliary mass and spring by a centrifugal pendulum, which can be designed to have the remarkable property of automatically tuning itself to any single multiple of the rotation speed. The principle of operation can be seen from the schematic models shown in Fig. 11.9.

The disk shown in Fig. 11.9 at (a) represents part of a machine rotating at a mean angular velocity, ω_R, but superimposed on this velocity are variations, due to vibration, at a fixed multiple, n, of ω_R. A pendulum, of length r, is pivoted at radius R, on the disk, with freedom to move as shown.

If the pendulum was suspended in the usual way, hanging downwards, and acted upon by gravity, its natural frequency, ω_N would be

$$\omega_N = \sqrt{\frac{g}{r}} \tag{11.10}$$

where g is the acceleration due to gravity. However, in this case the centripetal acceleration at angular speed ω_R is $\omega_R^2 R$, and the natural frequency of the pendulum is

$$\omega_N = \sqrt{\frac{\omega_R^2 R}{r}} = \omega_R \sqrt{\frac{R}{r}} \tag{11.11}$$

Thus the natural frequency of the pendulum is a constant multiple, $\sqrt{R/r}$, of the rotation angular speed, ω_R, as required.

To suppress a torsional vibration at n times the rotation speed requires that $\omega_N = n\omega_R$, and this can create design problems: for example, with $n = 5$, which is not unusual, we require $\omega_N = 5\omega_R$, and from Eq. (11.11):

$$\frac{R}{r} = 25$$

This means that, typically, r has to be very small, requiring special designs such the 'bifilar' pendulum absorber [11.1, 11.2].

The pendulum arrangement shown in Fig. 11.9 at (b) can be used to reduce *linear* vibration along the axis of rotation, rather than of angular vibration about the rotation axis, as discussed above. In this case two or more pendulums are used for balance.

The main applications of the centrifugal pendulum are to suppress torsional vibration of the crankshafts of reciprocating engines and rotor vibration on helicopters.

11.3 The damped vibration absorber

The dynamic absorber described above can only work when the excitation is at a single frequency, which is either constant or, if pendulums are used, proportional to the rotation speed. The damped vibration absorber does not have these limitations, and can be made to work with sinusoidal, periodic, and even random, inputs.

The model considered now is shown schematically in Fig. 11.10. The main mass and spring are m_1 and k_1 as before, but the added auxiliary system now consists of m_2 and k_2 with the viscous damper, c, in parallel with k_2. Strictly, there should also be a damper in parallel with k_1, representing the damping in the main system, but this is small and usually ignored in a preliminary analysis.

The practical problem to be solved is as follows. The main system, m_1 and k_1, represents a vibration mode of a structure or machine part, with excessive response, z_1, due to the applied force, F_0, which may now have any form, including broad-band random. The aim is to choose m_2, k_2 and c so as to reduce the response, z_1, as much as possible, although it will never be zero. The maximum size of m_2 will often be decided by other considerations, and in practice we can fix it, as before, as a fraction, μ, of the main mass. The problem then reduces to finding the optimum values of k_2 and c, and this can be done non-dimensionally.

The equations of motion of the system shown in Fig. 11.10 are

$$\begin{bmatrix} m_1 & 0 \\ 0 & m_2 \end{bmatrix} \begin{Bmatrix} \ddot{z}_1 \\ \ddot{z}_2 \end{Bmatrix} + \begin{bmatrix} c & -c \\ -c & c \end{bmatrix} \begin{Bmatrix} \dot{z}_1 \\ \dot{z}_2 \end{Bmatrix} \begin{bmatrix} (k_1 + k_2) & -k_2 \\ -k_2 & k_2 \end{bmatrix} \begin{Bmatrix} z_1 \\ z_2 \end{Bmatrix} = \begin{Bmatrix} F_0 \\ 0 \end{Bmatrix} \quad (11.12)$$

Substituting rotating complex vectors in the usual way, $F_0 = \underline{F}_0 e^{i\omega t}$, and

$$z_1 = \underline{z}_1 e^{i\omega t}; \qquad z_2 = \underline{z}_2 e^{i\omega t};$$
$$\dot{z}_1 = i\omega \underline{z}_1 e^{i\omega t}; \qquad \dot{z}_2 = i\omega \underline{z}_2 e^{i\omega t};$$
$$\ddot{z}_1 = -\omega^2 \underline{z}_1 e^{i\omega t}; \qquad \ddot{z}_2 = -\omega^2 \underline{z}_2 e^{i\omega t}.$$

Equation (11.12) becomes, after dividing through by $e^{i\omega t}$,

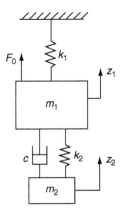

Fig. 11.10 Schematic diagram illustrating a damped vibration absorber.

$$\left(-\omega^2 \begin{bmatrix} m_1 & 0 \\ 0 & m_2 \end{bmatrix} + i\omega \begin{bmatrix} c & -c \\ -c & c \end{bmatrix} + \begin{bmatrix} (k_1 + k_2) & -k_2 \\ -k_2 & k_2 \end{bmatrix} \right) \left\{ \begin{matrix} z_1 \\ z_2 \end{matrix} \right\} = \left\{ \begin{matrix} F_0 \\ 0 \end{matrix} \right\} \quad (11.13)$$

For the present purpose we only require the ratio:

$$\frac{|\underline{z}_1| k_1}{|F_0|} = \frac{|\underline{z}_1|}{z_s} \quad (11.14)$$

where $|\underline{z}_1|$ is the magnitude of the steady-state displacement z_1, for sinusoidal excitation at frequency ω, and z_s is the static displacement of the mass m_1 when a force F_0 is applied.

Equation (11.13) can be manipulated [11.1] into the non-dimensional form:

$$\frac{|\underline{z}_1|}{z_s} = \left[\frac{(2\gamma\Omega_1)^2 + (\Omega_1^2 - R^2)^2}{(2\gamma\Omega_1)^2 (\Omega_1^2 - 1 + \mu\Omega_1^2)^2 + [\mu R^2 \Omega_1^2 - (\Omega_1^2 - 1)(\Omega_1^2 - R^2)]^2} \right]^{\frac{1}{2}} \quad (11.15)$$

where:

$$\mu = \frac{m_2}{m_1} = \frac{(\text{absorber mass})}{(\text{main mass})}$$

$$\Omega_1 = \frac{\omega}{\omega_1} = \frac{(\text{excitation freq}).}{(\text{natural frequency of main system}) = \sqrt{k_1/m_1}}$$

$$R = \frac{\omega_2}{\omega_1} = \frac{(\text{natural frequency of absorber}) = \sqrt{k_2/m_2}}{(\text{natural frequency of main system}) = \sqrt{k_1/m_1}}$$

$$\gamma = \frac{c}{2m_2\omega_1} = \text{damping parameter}.$$

Note: the definition of the damping parameter, γ, is that used by Den Hartog [11.1]. It is based on the mass of the absorber, m_2, and the natural frequency of the *main* system, ω_1. A similar parameter based on m_2 and ω_2 is sometimes used [11.2]. This is $1/R$ times the parameter γ used here.

Any practical installation can now be represented non-dimensionally by the four quantities: μ, Ω_1, R and γ. In Fig. 11.11, $|\underline{z}_1|/z_s$ is plotted against Ω_1, with R and μ fixed at 1 and 0.1, respectively. Three values of the damping parameter, μ, are shown to illustrate the curious feature that at two points, A and B, known as 'fixed points', the values are independent of the damping. Knowing that these points exist, it is possible, first, to find the value of R that makes them of equal height, which turns out to be [11.1]

$$R_{\text{opt}} = \frac{1}{1 + \mu} \quad (11.16)$$

where R_{opt} is the optimum value of R, irrespective of the damping.

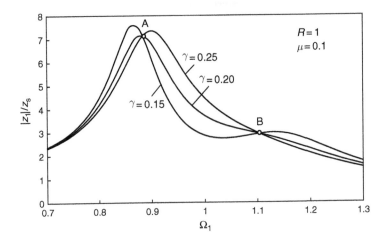

Fig. 11.11   Non-dimensional response of the main system showing the fixed points' A and B.

With R equal to R_{opt}, the optimum value of the damping parameter, γ_{opt} can then be shown [11.1] to be very close to:

$$\gamma_{opt} = \sqrt{\frac{3\mu}{8(1+\mu)^3}} \tag{11.17}$$

So the absorber design giving the lowest overall frequency response of the main mass for any given value of the mass ratio, μ, is found by applying Eqs (11.16) and (11.17).

Example 11.2

In Fig. 11.10, the mass m_1 and spring k_1 represent a single-DOF system with low inherent damping. Its response z_1, due to the applied force, F_0, is found to be excessive. The obvious solution of adding a damper in parallel with k_1 is found not to be feasible in this case, and it is proposed to add a damped vibration absorber consisting of m_2, k_2 and c, where the mass of m_2 is one-tenth that of m_1.
(a) Define the properties of the vibration absorber non-dimensionally, and plot the non-dimensional magnitude of the displacement, $|z_1|/z_s$, versus $\Omega_1 = \omega/\omega_1$. Discuss the improvement compared with the original system.
(b) If the mass of m_1 is 100 kg, and the natural frequency of the original single-DOF system is 10 Hz, define the optimum properties of the absorber dimensionally.

Solution

Part (a):
The value of the mass ratio, $\mu = m_2/m_1 = 0.1$.

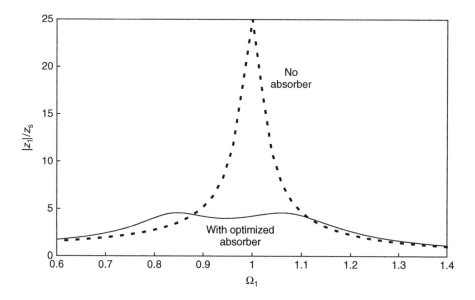

Fig. 11.12   Non-dimensional response of a single-DOF system fitted with an optimized damped vibration
absorber (mass ratio = 0.1), compared with that of the untreated system.

From Eq. (11.16) the optimum value of R is

$$R_{\text{opt}} = \frac{1}{1+\mu} = \frac{1}{1+0.1} = 0.909 \tag{A}$$

From Eq. (11.17), with $\mu = 0.1$, the optimum value of the damping parameter $\gamma = c/2m_2\omega_1$ is

$$\gamma_{\text{opt}} = \sqrt{\frac{3\mu}{8(1+\mu)^3}} = 0.168 \tag{B}$$

Substituting the above values of μ, R_{opt} and, γ_{opt}, into Eq. (11.15), the plot shown as a solid line in Fig. 11.12 was obtained using a standard spreadsheet. The maximum magnification of the system, with the optimized damper, is about 4.6.

An assessment of the improvement effected by adding the damper can be made by plotting the magnification of the original single-DOF system, without the added damper, for comparison. This is given by Eq. (4.12), which with the present notation is

$$\frac{|z_1|}{z_s} = \frac{1}{\sqrt{\left(1-\Omega_1^2\right)^2 + \left(2\underline{\gamma}\Omega_1\right)^2}} \tag{C}$$

where $\underline{\gamma}$ is the damping coefficient of the original single-DOF system before adding the auxiliary damper. Taking $\underline{\gamma} = 0.02$, the original magnification curve is shown as a dashed line in Fig. 11.12. In comparing the two curves in Fig. 11.12 it should be remembered that the damping in the original single-DOF system was neglected when deriving the response for the system with the added damper, giving a small but conservative error. Ignoring this, the maximum magnification at the worst excitation frequency in each case is seen to be reduced by a factor of at least 5.

Part (b):
To define the properties of the vibration absorber dimensionally, we have

$$m_1 = 100\,\text{kg}; \quad \mu = 0.1; \quad m_2 = \mu m_1 = 10\,\text{kg}$$

$$\omega_1 = 2\pi \times 10 = 20\pi\,\text{rad/s}; \quad \omega_2 = R_{\text{opt}}\omega_1 = 0.909 \times 20\pi$$

$$\omega_2 = \sqrt{\frac{k_2}{m_2}}; \quad k_2 = m_2\omega_2^2 = 10(0.909 \times 20\pi)^2 = 32\,620\,\text{N/m}$$

$$\gamma_{\text{opt}} = \frac{c}{2m_2\omega_1}; \quad c = 2\gamma_{\text{opt}}m_2\omega_1 = 2 \times 0.168 \times 10 \times 20\pi = 211\,\text{N/m/s}$$

Thus the auxiliary damper has

Mass $= m_2$ $\qquad = 10\,\text{kg};$
Spring stiffness $= k_2$ $\quad = 32\,620\,\text{N/m};$
Damper constant $= c =$ $211\,\text{N/m/s}.$

11.3.1 The Springless Vibration Absorber

If the spring k_2 shown in Fig. 11.10 is not present, the absorber consists of a mass and damper only. Particular examples of springless devices are the Lanchester and Houdaille vibration absorbers described below.

Equation (11.15) still works with a springless absorber, with all quantities having the same definition as before. The frequency ratio $R = \omega_2/\omega_1$ becomes zero, since ω_2 is now zero.

The expression for the optimum damping parameter, γ_{opt}, of a springless absorber can be shown [11.1] to be

$$\gamma_{\text{opt}} = \frac{1}{\sqrt{2(2+\mu)(1+\mu)}} \tag{11.18}$$

and the maximum value of $|z_1|/z_s$ can be shown to be

$$\frac{|z_1|}{z_s} = 1 + \frac{2}{\mu} \tag{11.19}$$

and to occur at

$$\Omega_1 = \sqrt{\frac{2}{2+\mu}} \tag{11.20}$$

Example 11.3

(a) Find the optimum damping for a springless vibration absorber where the absorber mass is one-half of the mass of the main system. Use Eq. (11.15) to plot the non-dimensional response of the optimized system, and compare it with non-optimized cases.

(b) Compare the optimized response above with that for an absorber with the same mass ratio, but having an optimized spring as well as a damper.

Solution

Part (a):

The mass ratio $\mu = m_2/m_1 = 0.5$ in this case. The optimum value of the damping parameter, γ_{opt}, is given by Eq. (11.18) as:

$$\gamma_{opt} = \frac{1}{\sqrt{2(2+\mu)(1+\mu)}} = 0.365 \tag{A}$$

The response curve is plotted from Eq. (11.15) in Fig. 11.13, using $\gamma_{opt} = 0.365$, with $R = 0$ and $\mu = 0.5$. Curves for the non-optimized values of the damping parameter, $\gamma = 0.2$ and $\gamma = 0.8$, are seen to give increased maximum response.

Part (b):

For an absorber with the same mass ratio, $\mu = 0.5$, except that a spring is used, and both spring and damper are optimized, the optimum value of R is given by Eq. (11.16):

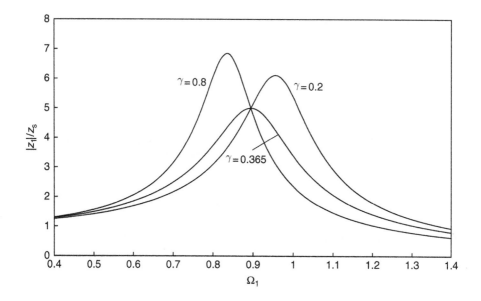

Fig. 11.13 Non-dimensional response of a springless vibration absorber with various damping parameters.

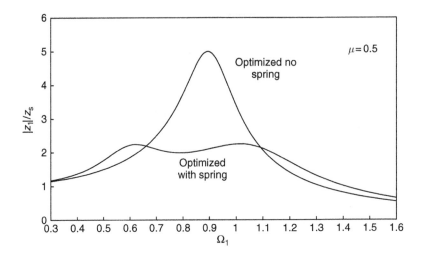

Fig. 11.14 Comparison of the non-dimensional response of optimized vibration absorbers with and without a spring. Mass ratio = 0.5.

$$R_{opt} = \frac{1}{1 + \mu} = 0.6666 \tag{B}$$

and the optimum damping is given by Eq. (11.17) with $\mu = 0.5$:

$$\gamma_{opt} = \sqrt{\frac{3\mu}{8(1 + \mu)^3}} = 0.236 \tag{C}$$

The resulting response curve is plotted from Eq. (11.15) in Fig. 11.14, together with the optimized response with a springless absorber. A system with a tuned spring is seen to give considerably lower (i.e. better) maximum response than a springless design.

Lanchester and Houdaille dampers

These devices are often used on the crankshafts of reciprocating engines to reduce stresses due to torsional vibration, and are particular applications of the springless absorber discussed above. The original Lanchester damper consisted of a disk attached to the crankshaft, similar to a modern automobile brake disk, with a small flywheel, free to rotate concentrically with it, except for spring-loaded friction pads between them, providing Coulomb damping.

If the friction pads are replaced by completely enclosing the small flywheel in a viscous fluid, the device is more usually known as a Houdaille damper. The device is installed at the point of maximum torsional vibration response, which is usually at the end of the crankshaft opposite to that of the main flywheel, and can also serve as a pulley for driving accessories.

Considering only the case of viscous damping, the system described above can still be represented schematically by Fig. 11.10, without k_2, and Eq. (11.15) still applies, with the following modifications:

(a) In the case of an engine crankshaft, the whole assembly is rotating, and thus has a rigid rotation mode as well as various twist modes of the crankshaft. As far as the damper installation is concerned, the steady rotation can be ignored, and each of the crankshaft twist modes can be considered individually, lumping their equivalent inertias and stiffnesses at the damper location. In practice, the main engine flywheel often has enough inertia to be considered fixed.

(b) The non-dimensional displacement response $|z_1|/z_s$ must be re-defined in angular terms, for example as $|\theta_1|/\theta_s$, where $|\theta_1|$ is the magnitude of the twist angle at the point on the main system where the damper is to be installed, and θ_s is the displacement for an applied static moment equal to the magnitude of sinusoidal forcing. In practice $|\theta_1|/\theta_s$ is likely to be inferred from measurements.

(c) Clearly, since k_2 is zero, the frequency ratio R is zero.

(d) The mass ratio, $\mu = m_2/m_1$, is interpreted as the inertia ratio I_2/I_1, where I_2 is the moment of inertia if the damper flywheel and I_1 is the inertia of the crankshaft referred to the damper location in the vibration mode concerned.

References

11.1 Den Hartog, JP (1956). *Mechanical Vibrations*. Fourth edition. McGraw-Hill.

11.2 Harris, CM and Piersol, AG eds (2002). *Harris' Shock and Vibration Handbook*. McGraw-Hill.

12 Introduction to Self-Excited Systems

Contents

Self-excited systems are those in which the motion of the system itself produces sufficient excitation to sustain an oscillation. There are many such phenomena, and in this chapter we introduce a few of the more important practically. Those discussed are the following:

(a) Vibration induced by friction, which can occur when the rubbing surfaces of items such as tires on runways, brakes and clutches have a coefficient of friction which tends to decrease with increasing velocity.

(b) Aircraft flutter, a branch of aeroelasticity, where the aerodynamic driving forces are derived from the airstream by the elastic motion of the flying surfaces, i.e. the wings, fins, stabilizers, controls, etc.

(c) 'Shimmy' of aircraft landing gear, which still tends to occur with cantilevered landing gear units, particularly nose units, when the nose-wheel is free to castor. It is a phenomenon similar to classical flutter, involving two degrees of freedom, which are rotation of the wheel or wheels in a castoring or steering sense, and lateral bending of the leg. The driving force is derived from tire side forces.

12.1 Friction-induced vibration

The essential mechanism of vibration induced by friction is illustrated by the system shown in Fig. 12.1(a). This shows a spring, mass and damper, where the mass m slides on a surface, which is moving to the right at constant velocity v_1. The argument is unchanged, however, if the surface is fixed and the spring–mass system is moving.

12.1.1 Small-Amplitude Behavior

The equation of motion of the system is

$$m\ddot{x} + c\dot{x} + kx = F \tag{12.1}$$

where m, c, k and x are defined by Fig. 12.1(a). The force F is the friction force applied to the mass by the moving base, and P is the normal force, considered constant, acting between the rubbing surfaces.

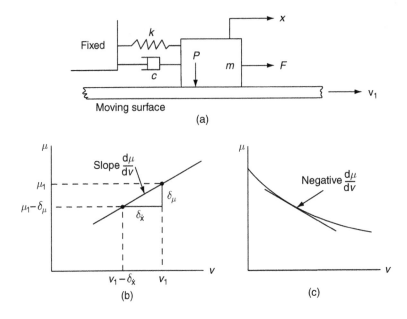

Fig. 12.1   Model for friction-induced vibration.

There is assumed to be a relationship between the coefficient of friction, μ, of the rubbing surfaces and the rubbing velocity, v, as shown in Fig. 12.1(b). Since the base is moving at constant velocity v_1, and the corresponding coefficient of friction at that velocity is μ_1, the system reaches static equilibrium with:

$$F_1 = \mu_1 P \tag{12.2}$$

If now a disturbance applies a small positive increment of velocity, $\delta\dot{x}$, to the mass, as shown in Fig. 12.1(b), this momentarily *reduces* the rubbing velocity, which becomes $v_1 - \delta\dot{x}$. The coefficient of friction then becomes $\mu_1 - \delta\mu$, and the friction force becomes

$$F_1 - \delta F = (\mu_1 - \delta\mu)P \tag{12.3}$$

It can be seen from Fig. 12.1(b), however, that

$$\delta\mu = \delta\dot{x}\left(\frac{d\mu}{dv}\right) \tag{12.4}$$

where $(d\mu/dv)$ is the local slope of the μ versus v curve. Therefore from Eqs (12.3) and (12.4):

$$F_1 - \delta F = \mu_1 P - \delta\dot{x}\left(\frac{d\mu}{dv}\right)P \tag{12.5}$$

Subtracting Eq. (12.2) from Eq. (12.5) removes the steady quantities, leaving the incremental values:

$$\delta F = \delta\dot{x}\left(\frac{d\mu}{dv}\right)P \tag{12.6}$$

We now see that $(d\mu/dv)P$ gives the change in force for a unit change in velocity, and is therefore equivalent to a viscous damping coefficient. The actual damper, c, in response to the same positive velocity change $\delta\dot{x}$, produces a force $c\delta\dot{x}$, which acts in the same sense as δF. Therefore, for small perturbations about the steady-state position, we can write Eq. (12.1) as:

$$m\ddot{x} + \left[c + \left(\frac{d\mu}{dv} \right) P \right] \dot{x} + kx = 0 \qquad (12.7)$$

It is usually more convenient to write Eq. (12.7) in the non-dimensional form:

$$\ddot{x} + 2(\gamma_0 + \gamma_F)\omega_n \dot{x} + \omega_n^2 x = 0 \qquad (12.8)$$

where γ_0 is the original non-dimensional viscous damping coefficient of the system, always positive, given by:

$$\gamma_0 = \frac{c}{2m\omega_n} \qquad (12.9)$$

and γ_F is an effective non-dimensional viscous damping coefficient, caused by friction, which may be positive or negative, depending on the sign of $d\mu/dv$, and is given by:

$$\gamma_F = \frac{P}{2m\omega_n} \left(\frac{d\mu}{dv} \right) \qquad (12.10)$$

If now $d\mu/dv$ should be negative, as shown in Fig. 12.1(c), γ_F will be negative, and the overall non-dimensional damping coefficient of the system, $(\gamma_0 + \gamma_F)$, will be reduced, and may become negative. From Eq. (2.28), the free response of Eq. (12.8) is

$$x = e^{-\gamma\omega_n t}(A \cos \omega_d t + B \sin \omega_d t) \qquad (12.11)$$

where $\gamma = (\gamma_0 + \gamma_F)$, in this case, and $\omega_d = \omega_n \sqrt{1 - \gamma^2}$, which for practical purposes can be taken as ω_n, the undamped natural frequency. The constants A and B depend upon the initial conditions, and in practice any small disturbance will be sufficient to start an oscillation. If then the total non-dimensional damping coefficient, $(\gamma_0 + \gamma_F)$, is negative, it can be seen from Eq. (12.11) that the exponent $\gamma\omega_n t$ becomes positive, and x will always consist of an exponentially growing oscillation, until limited in some way.

12.1.2 *Large-Amplitude Behavior*

Once started, how the oscillation builds up depends upon the shape of the μ versus v curve. The amplitude will limit if the energy derived from the friction force over a complete cycle becomes equal to the energy dissipated by the damper. Limiting will always occur when the peak velocity due to the oscillation, $\dot{x}_{max}$, exceeds the mean rubbing velocity, v_1, by more than a small margin, although it can occur sooner. This can be seen as follows. As the oscillation builds up, the velocity, $\dot{x}_{max}$, the maximum velocity of the mass in the positive direction, may eventually reach the value v_1, and the rubbing velocity will instantaneously be zero at some point in the cycle. If $\dot{x}_{max}$ tries to increase beyond this point, the rubbing velocity changes sign, and the motion is then opposed, over part of the cycle, by the whole of the friction force, μP, which gives a high level of positive damping. Therefore, in practice, the velocity of the oscillation will limit at a value not exceeding $\dot{x}_{max} = v_1$, approximately. Assuming that the waveform remains sinusoidal, this corresponds to a displacement limit, x_{max}, given by:

$$x_{max} = \frac{v_1}{\omega_n} \tag{12.12}$$

This normally sets an amplitude limit for any friction-induced vibration. It should be noted that it applies at the rubbing surface.

12.1.3 Friction-Induced Vibration in Aircraft Landing Gear

The conventional cantilever type of aircraft landing gear, when fitted with brakes, can exhibit two main forms of friction-induced vibration:
(1) When the brakes are working normally, if the coefficient of friction between the rubbing surfaces in the brakes tends to decrease as the velocity increases, negative damping may result, producing the vibration known as 'brake judder' or 'brake chatter', at low frequencies, or 'brake squeal' at high frequencies.
(2) When the brakes lock completely, perhaps because the anti-skid device, if fitted, does not always operate down to very low speeds, the friction characteristics between the tires and the runway can then provide the negative damping instead. Measurements of tire friction show that the negative slope of μ versus v, necessary for instability, tends to occur when the runway is wet, and the speed is low.
In both cases, the oscillations of the gear, in the fore-and-aft direction, can be quite pronounced, and lead to possible fatigue damage.

When a case of apparent brake judder is encountered, it should first be checked that this is not simply due to the operation of the anti-skid device. The operation of these devices is usually cyclic, and if the frequency of operation, or possibly a harmonic of it, coincides with the fore and aft natural frequency of the gear, a large response is possible. This should not be confused with true brake judder, which is caused by the friction characteristics of the rubbing surfaces.

Cantilevered landing gear units can be modeled by lumped parameter systems, consisting of one, two or more degrees of freedom. These will always include the fore-and-aft bending mode of the gear, but may also include other freedoms, such as incremental wheel and tire rotation.

The following example is based upon an actual investigation of skidding oscillations on a main landing gear.

Example 12.1

The main landing gear sketched in Fig. 12.2 is fitted with two wheels, on a live axle, and has two sets of disk brakes. These were sometimes found to lock when the brakes were applied firmly at very low speeds, resulting in a skid, during which severe fore-and-aft vibration of the gear occurred if the runway was wet.

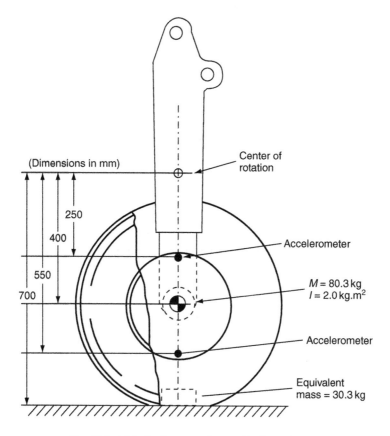

Fig. 12.2   Main landing gear discussed in Example 12.1.

The two accelerometers shown, fitted at the top and bottom of one brake back-plate, were used to monitor the vibration. From the readings, it was found that the mode shape could be represented by rotation of the lower parts of the gear, essentially the wheels, tires and brakes, about a point 0.70 m above the ground line. The mass properties were represented by a mass $M = 80.3$ kg, at the axle, and a moment of inertia about the axle $I = 2.0$ kg m^2. The equivalent mass, m, referred to the ground line was then given by:

$$M(0.4)^2 + I = m(0.7)^2 \tag{A}$$

From Eq. (A), the equivalent mass, $m = 30.3$ kg.

A vibration test on the unit had previously shown the non-dimensional damping coefficient to be about 0.05 of critical under static conditions.

In one measured trial, at a ground speed of 1.5 m/s, the oscillation, at 29 Hz, was found to grow at a rate corresponding to a damping coefficient of -0.04 of critical, and to limit at an amplitude of about ± 6 mm, as measured at the lower accelerometer. The vertical static load on the landing gear was 45 000 N.

(a) Explain why the skidding oscillations only occurred when the runway was wet, and why they grew at a rate corresponding to about −0.04 of critical damping.
(b) Explain why the oscillation was found to limit at about ± 6 mm, at the lower accelerometer, at a ground speed of 1.5 m/s.

Solution

Part (a):

From Eq. (12.10), the effective non-dimensional damping coefficient due to the friction force only is

$$\gamma_F = \frac{P}{2m\omega_n}\left(\frac{d\mu}{dv}\right) \tag{B}$$

where all the quantities are referred to the ground line. We now need the slope of the coefficient of friction versus speed curve. This was not measured, but Fig. 12.3 shows averaged published data for tires on both wet and dry concrete. At speeds below about 10 m/s, $(d\mu/dv)$ is seen to be equal to about −0.02 per m/s for wet concrete, but close to zero for dry concrete.

In Eq. (B), for the system operating on wet concrete at low speed, numerical values are

$P = 45\,000$ N, the static vertical load on the gear;
$m = 30.3$ kg, the equivalent mass of the system referred to the ground line;
$\omega_n = 2\pi f_n = 2\pi \times 29 = 182$ rad/s, the natural frequency;
$(d\mu/dv) = -0.02$, the slope of the μ versus v curve at low speed.

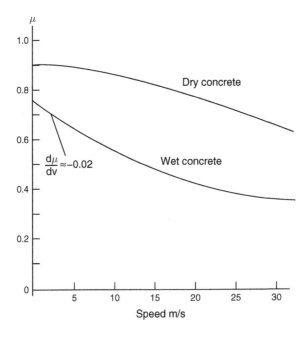

Fig. 12.3   Averaged coefficients of friction for wet and dry concrete runways.

Substituting these values into Eq. (B) gives $\gamma_F = -0.082$. However, the damping coefficient measured under static conditions, γ_0, was $+0.05$, so the predicted net damping coefficient is -0.032, close to the measured value, -0.04.

In dry conditions at low speed, $(\mathrm{d}\mu/\mathrm{d}v) \approx 0$, and the net damping coefficient remains at about the original value of 0.05, explaining why divergent oscillations did not occur.

Part (b):

From Eq. (12.12) the amplitude would be expected to limit when:

$$x_{max} = \frac{v_1}{\omega_n} \tag{C}$$

With $v_1 = 1.5\,\mathrm{m/s}$, and $\omega_n = 182\,\mathrm{rad/s}$, Eq. (C) gives $x_{max} = 0.0082\,\mathrm{m}$, or 8.2 mm. This is the vibration displacement defined at the ground. From the dimensions in Fig. 12.2, it can be seen that this corresponds to:

$$\frac{0.55 \times 8.2}{0.70} = 6.4\,\mathrm{mm}$$

at the lower accelerometer, which is close to the measured value of 6 mm.

12.2 Flutter

In flutter, the energy required to excite the system is derived from an airstream by the motion of the system itself. Many types of structural vibration involving aero-dynamic forces can be described as flutter, affecting structures as diverse as chimneys, suspended cables and even complete suspension bridges. However, the discussion here will be limited to the most common application, aircraft flutter.

The most important type is *classical flutter*. As is well known, this can be very destruc-tive, and much of the development time of a new or highly modified aircraft is devoted to its elimination. Classical flutter can be distinguished from other types by the fact that it requires at least two degrees of freedom, and has a definite onset speed, at a given air density, below which the system is stable. Routine prediction methods for classical flutter are now reasonably accurate, except possibly at transonic speeds, although, in addition, two types of test, a ground vibration test and flight tests, are still regarded as essential.

Before discussing classical flutter in more detail, two similar phenomena should be mentioned:

(1) A single-DOF form of flutter, more usually known as *buzz*, can affect aircraft control surfaces such as rudders, elevators and ailerons. The most common buzz phenomenon occurs in a fairly narrow range of Mach number, typically just below the speed of sound. The mechanism of this form of buzz usually involves the formation of a local shock wave on the control surface, which promotes an area of flow separation. This changes the pressure distribution, moving the control slightly, which, in turn, moves the shock, and this cycle repeats indefinitely. Another form of buzz can occur under supersonic conditions.

Buzz cannot be predicted reliably by normally available analytical methods, but empirical criteria for its avoidance exist. These require the control to have a high rotational stiffness, which can usually be achieved by powered actuators, provided the connecting linkages are also stiff.

(2) *Stall flutter*, as the name implies, can occur when a wing, or other flying surface, a
helicopter rotor blade or a compressor blade in a jet engine, operates close to the
stall. Two factors appear to be involved: the negative slope of the lift curve, just
beyond the stall, tending to reduce the aerodynamic damping, and the flow
separation associated with the stall. In practice, such phenomena are usually
dealt with empirically, often using wind tunnel models.

Traditionally, the several different classical flutter mechanisms that can occur
with a conventional aircraft had to be examined separately: for example, the
bending-torsion flutter of a wing, including the aileron, was examined with the
wing cantilevered from the fuselage, which was considered fixed. This permitted
the problem to be studied using as few as two or three degrees of freedom.
Similarly, empennage, or 'tail', flutter, could be considered using symmetric and
anti-symmetric models consisting of the cantilevered rear fuselage, with vertical
and horizontal stabilizers, rudder and elevators.

Modern developments in structural analysis and unsteady aerodynamics now allow
most flutter investigations to be carried out using the normal modes of the whole
aircraft as the degrees of freedom, the modes being divided into separate symmetric
and anti-symmetric sets. The basic mechanisms may still be recognized, for example,
as mainly wing bending-torsion flutter, or mainly a flutter of the empennage, but the
contribution from the rest of the structure is then correctly allowed for.

To illustrate how flutter equations are set up, we first consider a very simple model,
representing the bending-torsion flutter of an aircraft wing. We then look briefly at
how larger sets of equations are solved, and finally, how the flutter clearance of a
modern aircraft is organized.

12.2.1 The Bending-Torsion Flutter of a Wing

Consider the straight wing shown in Fig. 12.4(a) , divided into streamwise strips, as
shown. Structurally, it is represented by a single beam, AB, free to bend and twist,
cantilevered from the fuselage at A. This hypothetical beam is assumed to lie along the
'flexural axis' of the actual wing, i.e. the line of points where a vertical shear load
would produce bending but no twist. The strips of the wing are individually rigid, have
mass and rotational inertia, and define the external shape of the wing. They are
assumed to have the same values of z and α as the underlying beam. The external
aerodynamic force and moment, F and M, acting in the same sense as z and α,
respectively, are shown for one strip.

Aerodynamic forces
Although the following in no way represents current practice, it indicates how
rough aerodynamic forces can be derived from the motion of a wing, based only on
the simple idea that an airfoil develops a lift force proportional to its angle of attack
with respect to the airstream.

Considering the strip shown in Fig. 12.4(a), we know that statically, and at low
frequencies, the non-dimensional lift coefficient, c_L, of an airfoil, is related to its angle
of attack, α, as shown in Fig. 12.4(b).

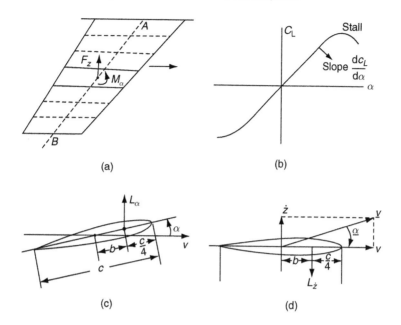

Fig. 12.4   Simple illustration of aerodynamic forces on a moving airfoil.

The lift coefficient, c_L, is defined by:

$$L = \frac{1}{2}\rho v^2 s c_L \tag{12.13}$$

where

L is the lift, ρ the air density, v the airspeed and s in this case the area of the strip.

For subsonic flow, the lift, L, acts at a point $c/4$ from the leading edge, where c is the chord. For supersonic flow, the lift tends to act nearer to $c/2$, but we will now consider only the subsonic case.

From Fig. 12.4(b), over the normal working range of the airfoil, c_L is linearly related to α, but at extreme positive and negative values of α the flow breaks down, and the airfoil stalls. Over the linear range, the slope, $(dc_L/d\alpha)$, which will now be written c_{L_α}, depends upon the aspect ratio of the wing, and for high aspect ratio it can be shown to approach the value 2π per radian. The lift due to unit α, say L_α, is then:

$$L_\alpha = \frac{1}{2}\rho v^2 s c_{L_\alpha} \alpha \tag{12.14}$$

Equation (12.14) is represented by Fig. 12.4(c).

Now suppose, as shown in Fig. 12.4(d), that $\alpha = 0$, but the wing section is translating upwards with positive velocity $\dot{z}$. Since it is also traveling to the right at velocity v, its instantaneous direction of travel will be given by the resultant of the two velocity vectors, $\dot{z}$ and v, which is the vector $\underline{v}$, at angle $\underline{\alpha}$. The airfoil now has a negative angle of attack, $-\underline{\alpha}$, with respect to the airstream, where, for small $\underline{\alpha}$, $\underline{\alpha} = \dot{z}/v$. The lift force $L_{\dot{z}}$ due to $\dot{z}$, acting at the $c/4$ point is therefore:

$$L_{\dot{z}} = \frac{1}{2}\rho v^2 s c_{L_\alpha}(-\underline{\alpha}) = -\frac{1}{2}\rho v^2 s c_{L_\alpha}\frac{\dot{z}}{v} = -\frac{1}{2}\rho v s c_{L_\alpha}\dot{z} \tag{12.15}$$

The total lift, L, acting at the $c/4$ point, will then be, from Eqs (12.14) and (12.15):

$$L = L_\alpha + L_{\dot{z}} = \frac{1}{2}\rho v^2 s c_{L_\alpha} \alpha - \frac{1}{2}\rho v s c_{L_\alpha} \dot{z} \qquad (12.16)$$

Resolving L as a force F, and a moment, M, at the reference center, which is taken as the twist axis of the wing, since $F = L$ and $M = bL$, we have

$$\left\{ \begin{matrix} F \\ M \end{matrix} \right\} = \rho v^2 \begin{bmatrix} 0 & \frac{1}{2}sc_{L_\alpha} \\ 0 & \frac{1}{2}bsc_{L_\alpha} \end{bmatrix} \left\{ \begin{matrix} z \\ \alpha \end{matrix} \right\} + \rho v \begin{bmatrix} -\frac{1}{2}sc_{L_\alpha} & 0 \\ -\frac{1}{2}bsc_{L_\alpha} & 0 \end{bmatrix} \left\{ \begin{matrix} \dot{z} \\ \dot{\alpha} \end{matrix} \right\} \qquad (12.17)$$

where b is the distance of the quarter chord point from the twist axis.

Simple flutter equations

To create a simple set of flutter equations roughly representing bending-torsion flutter of a wing, we can now consider a simple model similar to that used by Theodorsen and Garrick in 1940 [12.1]. A section of the wing, at about $\frac{3}{4}$ of the semi-span, as shown in Fig. 12.5, is taken as representing the bending and torsion behavior of the wing as a whole. The springs k_z and k_α represent the bending and torsional stiffnesses. The mass of the wing section, located at the mass center, is m and I is the mass moment of inertia about the mass center, which is a distance a forward of the stiffness axis.

The equations of motion of the system can be found by applying Lagrange's equations, taking z and α as generalized coordinates, and F and M as the corresponding external generalized 'forces'.

Also including some viscous damping due to the structure, this gives

$$\begin{bmatrix} m & ma \\ ma & I + ma^2 \end{bmatrix} \left\{ \begin{matrix} \ddot{z} \\ \ddot{\alpha} \end{matrix} \right\} + \begin{bmatrix} c_z & 0 \\ 0 & c_\alpha \end{bmatrix} \left\{ \begin{matrix} \dot{z} \\ \dot{\alpha} \end{matrix} \right\} + \begin{bmatrix} k_z & 0 \\ 0 & k_\alpha \end{bmatrix} \left\{ \begin{matrix} z \\ \alpha \end{matrix} \right\} = \left\{ \begin{matrix} F \\ M \end{matrix} \right\} \qquad (12.18)$$

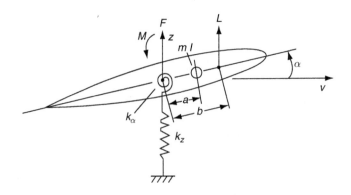

Fig. 12.5 Simple 2-DOF model representing classical flutter.

The aerodynamic force and moment, $\{^F_M\}$, on the right side of Eq. (12.18), are given by Eq. (12.17), and we can write

$$
\begin{bmatrix} m & ma \\ ma & I+ma^2 \end{bmatrix} \begin{Bmatrix} \ddot{z} \\ \ddot{\alpha} \end{Bmatrix} + \begin{bmatrix} c_z & 0 \\ 0 & c_\alpha \end{bmatrix} \begin{Bmatrix} \dot{z} \\ \dot{\alpha} \end{Bmatrix} + \begin{bmatrix} k_z & 0 \\ 0 & k_\alpha \end{bmatrix} \begin{Bmatrix} z \\ \alpha \end{Bmatrix}
$$
$$
= \rho v^2 \begin{bmatrix} 0 & \frac{1}{2}sc_{L_\alpha} \\ 0 & \frac{1}{2}bsc_{L_\alpha} \end{bmatrix} \begin{Bmatrix} z \\ \alpha \end{Bmatrix} + \rho v \begin{bmatrix} -\frac{1}{2}sc_{L_\alpha} & 0 \\ -\frac{1}{2}bsc_{L_\alpha} & 0 \end{bmatrix} \begin{Bmatrix} \dot{z} \\ \dot{\alpha} \end{Bmatrix} \tag{12.19}
$$

It is usual to take the aerodynamic forcing terms over to the left side of the equations, i.e.,

$$
\begin{bmatrix} m & ma \\ ma & I+ma^2 \end{bmatrix} \begin{Bmatrix} \ddot{z} \\ \ddot{\alpha} \end{Bmatrix} + \rho v \begin{bmatrix} \frac{1}{2}sc_{L_\alpha} & 0 \\ \frac{1}{2}bsc_{L_\alpha} & 0 \end{bmatrix} \begin{Bmatrix} \dot{z} \\ \dot{\alpha} \end{Bmatrix} + \rho v^2 \begin{bmatrix} 0 & -\frac{1}{2}sc_{L_\alpha} \\ 0 & -\frac{1}{2}bsc_{L_\alpha} \end{bmatrix} \begin{Bmatrix} z \\ \alpha \end{Bmatrix}
$$
$$
+ \begin{bmatrix} c_z & 0 \\ 0 & c_\alpha \end{bmatrix} \begin{Bmatrix} \dot{z} \\ \dot{\alpha} \end{Bmatrix} + \begin{bmatrix} k_z & 0 \\ 0 & k_\alpha \end{bmatrix} \begin{Bmatrix} z \\ \alpha \end{Bmatrix} = 0 \tag{12.20}
$$

Equation (12.20) are the simple flutter equations for the system shown in Fig. 12.5. Although considerably over-simplified, they nevertheless illustrate some general principles common to all such equations. It can be seen that
(1) the whole aerodynamic damping matrix has the factor ρv;
(2) the whole aerodynamic stiffness matrix has the factor ρv^2;
(3) the eigenvalues and eigenvectors of the system are, in general, complex, since the aerodynamic damping matrix is not small or 'proportional' as discussed in Section 6.4.4 and cannot be made diagonal by any real transformation applied to the whole system.

The last point above makes the solution of flutter equations more complicated than for the systems considered in earlier chapters. This is briefly discussed in Section 12.2.2.

Static divergence

Although it is not flutter, the phenomenon of *static divergence* should be mentioned, since it can be equally dangerous. It can be explained by reference to Fig. 12.5 and Eq. (12.20), as follows.

In Eq. (12.20), since we are considering only static forces, $\ddot{z}$, $\ddot{\alpha}$, $\dot{z}$ and $\dot{\alpha}$ are all zero. Considering only the second equation, representing the freedom α, this reduces to:

$$
\rho v^2 \left(-\frac{1}{2}bsc_{L_\alpha} \right) \alpha + k_\alpha \, \alpha = 0 \tag{12.21}
$$

For b positive, i.e. for the lift acting forward of the twist axis, $\rho v^2 \left(-\frac{1}{2}bsc_{L_\alpha} \right)$ is seen to be a negative aerodynamic stiffness, which will subtract from the positive structural stiffness, k_α. The total stiffness about the twist axis is therefore $\bar{k}_\alpha$, say, where:

$$
\bar{k}_\alpha = k_\alpha - \frac{1}{2}\rho v^2 bsc_{L_\alpha} \tag{12.22}
$$

For subsonic conditions, where b does not change appreciably with v, there is a speed v at which the total stiffness becomes zero, and the slightest input will cause α to grow indefinitely until failure occurs. This is known as the *divergence speed*, given by putting $\bar{k}_\alpha = 0$ in Eq. (12.22), i.e.,

$$v_d = \sqrt{\frac{k_\alpha}{\frac{1}{2}\rho bsc_{L_\alpha}}} \qquad (12.23)$$

In practice, the reduction in torsional stiffness becomes noticeable at speeds much lower than the divergence speed, and safe flight is possible only up to a fraction of v_d.

Although divergence is a static phenomenon, the easiest way to detect it is by the reduction in natural frequency as the divergence speed is approached.

12.2.2 Flutter Equations

Analysis of the structure, initially using global coordinates, transforming to modal coordinates, followed by insertion of the aerodynamic forces, leads, in general, to flutter equations of the form:

$$[A]\{\ddot{q}\} + \rho v[B]\{\dot{q}\} + \rho v^2[C]\{q\} + [D]\{\dot{q}\} + [E]\{q\} = 0 \qquad (12.24)$$

where:

$[A]$ is the mass or inertia matrix of the structure, $[B]$ the aerodynamic damping matrix, $[C]$ the aerodynamic stiffness matrix, $[D]$ the damping matrix of the structure, viscous in this case, and $[E]$ the structural stiffness matrix.

Equation (12.24), which is quite general, is seen to be of the same form as Eq. (12.20), which was derived from a very simple model.

It should be mentioned that aerodynamic forces proportional to acceleration are also possible, but are usually small enough to ignore.

The aerodynamic matrices $[B]$ and $[C]$ are functions of two parameters, the reduced frequency, $\omega c/v$ and the Mach number v/a, where ω is the flutter frequency in rad/s, v the true air speed, c a reference length, not necessarily the mean chord and a the speed of sound.

Solution methods

The usual problem to be solved is to find the lowest airspeed v, at which flutter occurs, for any given value of the air density, ρ, which is dependent on altitude. This may have to be done for a range of structural variations and aircraft configurations.

One way to investigate the flutter behavior of the system represented by Eq. (12.24) is known as the p method. Historically, this was not the first method to be widely used: that was the k method, developed by the US Air Materiel Command in 1942.

In the p method, Eq. (12.24) is solved directly for its roots, or eigenvalues, which, as already stated, will be complex. Substituting the following into Eq. (12.24):

$$\{q\} = \{\underline{q}\}e^{\lambda t}, \quad \{\dot{q}\} = \lambda\{\underline{q}\}e^{\lambda t}, \quad \{\ddot{q}\} = \lambda^2\{\underline{q}\}e^{\lambda t}$$

and dividing through by $\{q\}e^{\lambda t}$ gives

$$[A]\lambda^2 + ([D] + \rho v[B])\lambda + ([E] + \rho v^2[C]) = 0 \qquad (12.25)$$

The complex roots, λ, of Eq. (12.25) can be found by standard methods. They may be real, and negative or positive, indicating static stability or instability, respectively. However, most roots will occur as complex conjugate pairs, of the form $\lambda = \sigma \pm i\omega$, and the response then contains the factor $e^{\sigma t}e^{i\omega t}$ or $e^{\sigma t}(\cos \omega t + i \sin \omega t)$. Divergent flutter, at frequency ω, is then indicated by a positive value of σ, and freedom from flutter by a negative value of σ. The flutter boundary is indicated by $\sigma = 0$.

It is usual to plot the variation of the apparent non-dimensional damping coefficient $\gamma = -\sigma/\omega$ and the flutter frequency in Hz, $f = \omega/2\pi$, versus v, at fixed values of the air density, ρ, representing altitude, for each mode. Then, in this method, a flutter speed is indicated by the value of v at which the damping coefficient, γ, changes from positive to negative.

The p method uses complex eigenvalues and eigenvectors throughout, and the second-order equations of motion are expressed as first-order equations of twice the size, as briefly discussed in Section 7.5.

In the early 1940s, the large-scale computation of complex eigenvalues was not possible, and the k method, developed at that time, avoided this by solving the equations only at a flutter boundary, where the motion is harmonic. This was achieved by combining the aerodynamic forces into a single complex matrix, and introducing fictitious hysteretic damping, proportional to the stiffness matrix. This added damping is then negative below the flutter speed and positive above it. Later, the p–k method, combining some of the advantages of the p and k methods, was introduced. More details of methods for solving flutter equations, and the development of methods for calculating unsteady aerodynamic forces are given by Hodges and Pierce [12.2].

Notation systems

The notation system used above is known as RAE notation, after the (British) Royal Aircraft Establishment (RAE), where the p method of solution was developed, and it is still used in Great Britain. It was used above for illustration purposes, on account of its simplicity. In practice, other notation systems are likely to be encountered, and as an example, the system used in MSC/NASTRAN [12.3] for the p–k method of solution can be related to the RAE system as follows.

The matrices representing the mass, the viscous damping of the structure and the structural stiffness are, in fact, the same, i.e.,

$$[A] = [M_{hh}]; \quad [D] = [B_{hh}]; \quad [E] = [K_{hh}]$$

where $[A]$, $[D]$ and $[E]$ are in RAE notation and $[M_{hh}]$, $[B_{hh}]$ and $[K_{hh}]$ are the corresponding matrices in MSC/NASTRAN notation.

The aerodynamic matrices are related by:

$$[B] = -\frac{1}{2}\frac{v}{\omega}[Q^I_{hh}]; \quad [C] = -\frac{1}{2}[Q^R_{hh}]$$

where $[Q_{hh}^R]$ and $[Q_{hh}^I]$ are the real and imaginary parts, respectively, of the complex aerodynamic matrix used in the MSC/NASTRAN version.

Substituting the above in Eq. (12.24), noting that the reduced frequency, k, used in the *p-k* method is $k = \omega\bar{c}/2v$; and also substituting:

$$\{q\} = \{u_h\}; \quad \{\dot{q}\} = p\{u_h\}; \quad \{\ddot{q}\} = p^2\{u_h\}$$

gives the flutter equations used in MSC/ NASTRAN [12.3] for the *p–k* method:

$$\left\{ [M_{hh}]p^2 + \left([B_{hh}] - \frac{1}{4}\rho\bar{c}v[Q_{hh}^I]\frac{1}{k}\right)p + \left([K_{hh}] - \frac{1}{2}\rho v^2[Q_{hh}^R]\right) \right\}\{u_h\} = 0 \quad (12.26)$$

where p corresponds to λ in Eq. (12.25).

NASTRAN is a registered trade mark of the National Aeronautics and Space Administration. MSC/NASTRAN is marketed by the MacNeal-Schwendler Corporation.

12.2.3 An Aircraft Flutter Clearance Program in Practice

The formal clearance of a new aircraft for flutter is a complex operation involving the participation of many specialists. The essential steps in the process are usually as follows. This description is intended only to indicate how the various calculation and test activities involved in a flutter clearance program are coordinated.

Analysis of the structure

A finite element model of the complete aircraft structure is set up in global coordinates. This usually consists of two separate half-aircraft models, representing the symmetric and anti-symmetric behavior, with appropriate constraints at the plane of symmetry. Typically, up to several thousand grid points will be used, with three displacement and three rotation coordinates at each. The type of elements used in the FE analysis will depend upon the type of structure.

Calculation of normal modes

Symmetric and anti-symmetric normal modes, of the undamped structure, are then calculated from the FE model. Each mode is characterized by its natural frequency and mode shape. The modes are, of course, real, since no damping has been incorporated at this stage. Figure 12.6 shows sketches of a few of the calculated whole-aircraft symmetric normal modes, in increasing order of natural frequency, of a typical conventional aircraft. The displacements of the left side of the aircraft are not shown; these are, of course, mirror images of those on the right. The anti-symmetric modes, which are not shown, would have equal and opposite displacements on the two sides. The relative signs of the displacements of corresponding grid points on opposite sides of a symmetrical structure, in symmetric and anti-symmetric modes, are defined in Section 8.5.

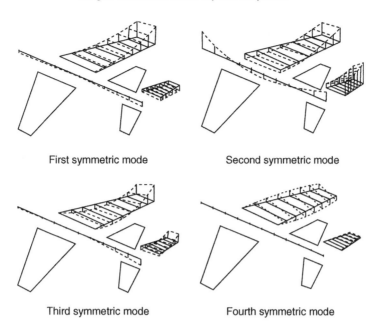

First symmetric mode   Second symmetric mode

Third symmetric mode   Fourth symmetric mode

Fig. 12.6   Sketches of the first four symmetric normal modes of a typical conventional aircraft.

Normal modes are essentially modes of the complete structure, with all parts responding to some extent in each mode. However, it is often possible to identify each of the normal modes with a particular component, or components, of the aircraft. Thus, in Fig. 12.6, the first mode shown is essentially the symmetric fundamental wing bending mode. In the second normal mode, the fuselage first symmetric bending and the first horizontal stabilizer mode are both evident. The third and fourth modes shown can be identified with the wing second bending and wing torsion mode, respectively.

Comparison of calculated and measured modes

Although the normal modes of the complete aircraft will usually have been calculated long before a prototype aircraft exists, it is not until they have been verified, by comparing them with those measured in the *ground vibration test* (GVT), that they are regarded as a valid representation of the structure. A procedure for measuring the normal modes of a typical aircraft is briefly described in Section 13.1.2.

Calculation of the unsteady aerodynamic forces

The aerodynamic force matrices are calculated from the whole-aircraft normal mode displacements, using appropriate computer codes. Many methods for computing the unsteady forces are available, such as the doublet lattice method for subsonic flow, and the Mach box method for supersonic flow. Bodies such as the fuselage and engine nacelles can be incorporated. The grid points required by the unsteady aerodynamics program will be different from those used for the structural model, and spline fits are extensively used to interpolate from one to the other.

Flutter calculations

Having assembled the flutter equations, in terms of normal mode coordinates, by combining the structural model with the unsteady aerodynamic forces, they can be solved, for example by the k or p–k method, to identify possible flutter mechanisms, such as bending-torsion flutter of the wing, flutter involving control surfaces, flutter of the empennage, etc.

Flight flutter tests

Flight flutter tests are always carried out, first because the possible flutter mechanisms identified by the flutter calculations must be verified, and secondly, in the case of transonic aircraft, because this flight regime can still introduce unpredictable flutter behavior.

Flight flutter testing essentially involves measuring the natural frequencies and damping coefficients of the normal modes in flight, while the air speed and Mach number are increased in steps. The expected variation of the damping coefficients and natural frequencies, as speed is increased, will have been predicted theoretically, and these trends are now compared with measured values. It is essential to compare natural frequencies as well as damping coefficients, because although flutter occurs when a damping coefficient changes from positive to negative, the coalescence of the natural frequencies as speed is increased, well before this occurs, can give important information about the mechanism involved.

The measurement of the natural frequencies and damping coefficients in flight requires the response to some form of excitation to be measured, for which accelerometers are almost always used. It also requires the aircraft to be excited over a suitable range of frequencies. Telemetry is frequently used to relay the data to a ground station for analysis. Excitation can be via the normal control surfaces, such as by 'stick jerks' applied by the pilot, or by injecting suitable inputs, such as swept sinusoids, or pseudo-random series, into the control system. Special devices, such as small oscillating airfoils, are sometimes used. Methods used in the past, now less popular, have included pyrotechnic devices, and inertia exciters. The required natural frequencies and damping coefficients are usually obtained by the cross-spectral density method discussed in Section 10.6.2, but several other methods are used.

Flight flutter testing is regarded as slightly hazardous, and steps are always taken to minimize this a far as possible. In principle, a flutter boundary can be predicted by extrapolating the trend of a series of damping and frequency measurements made at speeds well below the flutter speed.

12.3 Landing gear shimmy

Conventional cantilevered aircraft landing gear units are still sometimes found to vibrate violently due to a phenomenon known as shimmy. The name appears to be derived from the 'shimmy-shake', a popular dance of the 1920s. Typically, the gear

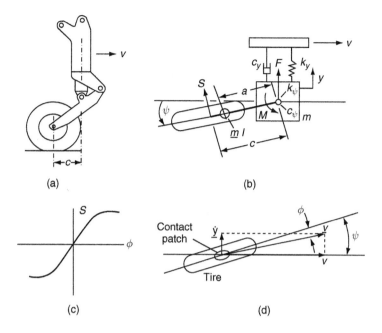

Fig. 12.7 Shimmy analysis of an aircraft landing gear.

vibrates laterally in its cantilever bending mode, and rotation of the castoring wheel assembly about a vertical axis is another essential degree of freedom. The driving force is derived from side forces developed by the tire on the runway. Shimmy is rarely immediately catastrophic, but can lead to structural failures from fatigue, if allowed to persist.

The phenomenon can be illustrated by the following simple analysis. Figure 12.7(a) shows a typical aircraft nose landing gear. On simple aircraft, the fork and wheel assembly may be free to castor at all times. More complex aircraft may have two modes of operation, a freely castoring mode, engaged during touch down, to allow the nose wheel to align itself with the runway, and a steered mode, using a hydraulic motor. Shimmy is more likely to occur when the wheel assembly castors freely.

In Fig. 12.7(b) the leg bending mode is represented by mass m, stiffness k_y, and viscous damper c_y. The fork and wheel assembly is represented by the trailing arm, with its mass, $\underline{m}$, at the mass center, and moment of inertia I about the mass center. A rotational stiffness, k_ψ, is included, but would be set to zero when the gear is free to castor. The rotational damper, c_ψ, assumed viscous, is discussed below.

The mass center of the trailing assembly is distance a behind the pivot axis.

The equations of motion for the system without damping are obtained by applying Lagrange's equations in the usual way:

$$\frac{d}{dt}\left(\frac{\partial T}{\partial \dot{q}_i}\right) + \left(\frac{\partial U}{\partial q_i}\right) = Q_i \quad i = 1, 2. \tag{12.27}$$

where the generalized coordinates are taken as $q_1 = y$ and $q_2 = \psi$, and the generalized forces are $Q_1 = F$ and $Q_2 = M$, where F and M are the tire force and moment, referred to the pivot axis, as shown in Fig. 12.7(b).

The kinetic energy of the system, T, is

$$T = \frac{1}{2}m\dot{y}^2 + \frac{1}{2}\left[\underline{m}(\dot{y} - a\dot{\psi})^2\right] + \frac{1}{2}I\dot{\psi}^2 \qquad (12.28)$$

and the potential energy, U, is

$$U = \frac{1}{2}k_y y^2 + \frac{1}{2}k_\psi \psi^2 \qquad (12.29)$$

Applying Eq. (12.27) gives the undamped equations of motion:

$$\begin{bmatrix} m + \underline{m} & -\underline{m}a \\ -\underline{m}a & I + \underline{m}a^2 \end{bmatrix} \begin{Bmatrix} \ddot{y} \\ \ddot{\psi} \end{Bmatrix} + \begin{bmatrix} k_y & 0 \\ 0 & k_\psi \end{bmatrix} \begin{Bmatrix} y \\ \psi \end{Bmatrix} = \begin{Bmatrix} F \\ M \end{Bmatrix} \qquad (12.30)$$

The damping in the leg bending mode c_y should be obtained from a test, if possible.

The damping in the castor mode, c_ψ, is found to be important in preventing shimmy, and special arrangements are often made to increase it, using one or more of the following methods.
(a) On simpler aircraft, friction disks can be incorporated into the pivot;
(b) A hydraulic damper may be used;
(c) If the gear incorporates a hydraulic steering motor, this may have two modes, as mentioned above, a steered mode, and a castoring mode. In the latter case, the motor can usually be arranged to provide passive hydraulic damping when castoring.
(d) Tire friction can also be used to increase the damping in the castor mode. If the gear has two wheels, connecting them by a live axle can have this effect. In the case of a single wheel, it is possible to obtain 'anti-shimmy' tires incorporating two contact areas.

Friction methods will tend to provide Coulomb damping, and hydraulic methods square-law damping, both of which are non-linear. These can conveniently be expressed as equivalent viscous dampers, where the value is a function of the vibration amplitude, as discussed in Section 5.7.

Adding the damping terms, here assumed linear, Eq. (12.30) becomes

$$\begin{bmatrix} m + \underline{m} & -\underline{m}a \\ -\underline{m}a & I + \underline{m}a^2 \end{bmatrix} \begin{Bmatrix} \ddot{y} \\ \ddot{\psi} \end{Bmatrix} + \begin{bmatrix} c_y & 0 \\ 0 & c_\psi \end{bmatrix} \begin{Bmatrix} \dot{y} \\ \dot{\psi} \end{Bmatrix} + \begin{bmatrix} k_y & 0 \\ 0 & k_\psi \end{bmatrix} \begin{Bmatrix} y \\ \psi \end{Bmatrix} = \begin{Bmatrix} F \\ M \end{Bmatrix} \quad (12.31)$$

We now require the tire force F and moment M. These can be provided by theoretical or empirical methods. *String theory*, developed by von Schlippe, has been used, and empirical forces based on measurements under oscillatory conditions can be found in the literature. The following very simple approach, known as a point contact theory, will be used here.

From steady yawed rolling tests on tires [12.4], it is found that the side force, S, is as sketched in Fig. 12.7(c), and is independent of speed. The force S is found to act a small distance behind the center of the tire contact patch, and at distance c behind the pivot axis. For a given vertical load on the tire, S is proportional to the *slip angle*, ϕ, for small values of ϕ. The slope, $dS/d\phi$ is known as the *cornering power*, N.

The slip angle ϕ is defined by Fig. 12.7(d). The path of the aircraft, assumed to be in a straight line, is in the direction of the vector v. The tire contact patch, assumed small, has a yaw angle, ψ, with respect to this line. However, the contact patch also has instantaneous lateral velocity, $\underline{\dot{y}}$, due to $\dot{y}$ and $\dot{\psi}$, given by:

$$\underline{\dot{y}} = \dot{y} - c\dot{\psi} \tag{12.32}$$

so it is instantaneously moving in the direction of $\underline{v}$. The slip angle, ϕ, for small angles, is therefore given by:

$$\phi = \psi - \frac{\underline{\dot{y}}}{v} = \psi - \frac{1}{v}(\dot{y} - c\dot{\psi}) \tag{12.33}$$

Thus, the side force, S, is given by:

$$S = \frac{dS}{d\phi}\phi = N\phi = N\psi - \frac{1}{v}N\dot{y} + \frac{1}{v}cN\dot{\psi} \tag{12.34}$$

Now $F = S$ and $M = -cS$, so from Eq. (12.34):

$$\left\{ \begin{array}{c} F \\ M \end{array} \right\} = \left[\begin{array}{cc} 0 & N \\ 0 & -cN \end{array} \right] \left\{ \begin{array}{c} y \\ \psi \end{array} \right\} + \frac{1}{v} \left[\begin{array}{cc} -N & cN \\ cN & -c^2N \end{array} \right] \left\{ \begin{array}{c} \dot{y} \\ \dot{\psi} \end{array} \right\} \tag{12.35}$$

and Equation (12.31) becomes

$$\left[\begin{array}{cc} m+\underline{m} & -\underline{m}a \\ -\underline{m}a & I+\underline{m}a^2 \end{array} \right] \left\{ \begin{array}{c} \ddot{y} \\ \ddot{\psi} \end{array} \right\} + \frac{1}{v} \left[\begin{array}{cc} N & -cN \\ -cN & c^2N \end{array} \right] \left\{ \begin{array}{c} \dot{y} \\ \dot{\psi} \end{array} \right\} + \left[\begin{array}{cc} 0 & -N \\ 0 & cN \end{array} \right] \left\{ \begin{array}{c} y \\ \psi \end{array} \right\}$$
$$+ \left[\begin{array}{cc} c_y & 0 \\ 0 & c_\psi \end{array} \right] \left\{ \begin{array}{c} \dot{y} \\ \dot{\psi} \end{array} \right\} + \left[\begin{array}{cc} k_y & 0 \\ 0 & k_\psi \end{array} \right] \left\{ \begin{array}{c} y \\ \psi \end{array} \right\} = 0 \tag{12.36}$$

Equations (12.36) are the shimmy equations, which are seen to have similarities to a set of flutter equations. Using this analogy, Eqs. (12.36) could be written in a form similar to Eqs. (12.24):

$$[A]\{\ddot{q}\} + \frac{1}{v}[B]\{\dot{q}\} + [C]\{q\} + [D]\{\dot{q}\} + [E]\{q\} = 0 \tag{12.37}$$

where in the case of shimmy, $[A]$ is the inertia matrix; $[B]$ the tire damping matrix; $[C]$ the tire stiffness matrix; $[D]$ the damping matrix for the structure and $[E]$ the structural stiffness matrix.

Equations (12.37) could now be solved as if they were a set of flutter equations, for their complex roots, investigating the effect of changing parameters, such as speed, the trail of the castoring wheel, damping, etc., as required. Traditionally, however, a much simpler method, based on E.J. Routh's criteria of stability, published in 1877, has been used. This allows the stability, or otherwise, of any set of equations of motion to be determined, but the roots are not actually found. The method was used in some early flutter investigations, and remains useful in shimmy, where there are usually only two degrees of freedom, and the mechanism and frequency are not in question. It can be illustrated as follows.

The roots, or eigenvalues, of Eq. (12.37), which, in general, will be complex are given by the determinant:

$$\left| [A]\lambda^2 + \frac{1}{v}[B]\lambda + [C] + [D]\lambda + [E] \right| = 0 \tag{12.38}$$

If the matrices are of size 2×2, Eq. (12.38) can be written as the quartic equation:

$$\lambda^4 + A_3\lambda^3 + A_2\lambda^2 + A_1\lambda + A_0 = 0 \tag{12.39}$$

The four roots, $\lambda_1 - \lambda_4$, can be real, or occur as complex conjugate pairs. For stability, the real parts of all the roots must be negative or zero. For a quartic stability equation such as Eq. (12.39), the Routh criteria state that for the system to be stable, i.e. free from shimmy:
(a) All coefficients $A_3 - A_0$ must be positive, and
(b) the following must be true: $A_1A_2A_3 > A_1^2 + A_0A_3^2$.

Published criteria for the avoidance of shimmy are usually based on this method, rather than roots solutions.

References

12.1 Scanlan, RH and Rosenbaum, R (1951). *Introduction to the Study of Aircraft Vibration and Flutter.* Macmillan. (Reprinted by Dover Publications Inc. 1968.)
12.2 Hodges, DH and Pierce, GA (2002). *Introduction to Structural Dynamics and Aeroelasticity.* Cambridge University Press.
12.3 Rodden WP, ed. 1987. *Handbook for Aeroelastic Analysis, Volume 1 – MSC/NASTRAN Version 65.* The MacNeal-Schwendler Corporation.
12.4 Smiley, RF and Horne, WB (1958). *Mechanical Properties of Pneumatic Tires with Special Reference to Modern Aircraft Tires.* NACA Report 4110.

1 3 Vibration testing

Contents

Vibration testing in structural dynamics work can be classified into three main groups.

(1) First, in *modal testing*, the aim is to test a system or structure to obtain its vibration characteristics, from which its underlying equations of motion can be found. This 'identifies the system', leading to the alternate name of *system identification*. Before structural analysis had reached its present state of development, this was, in fact, the only way in which the equations of motion of complex structures, such as complete aircraft, could be determined with sufficient accuracy for, say, a flutter investigation to be made. At the present time, although it is possible to model virtually any structure by mathematical analysis alone, significant errors are still encountered, and a comparison between the predicted and the actual properties remains essential in critical cases. Therefore, current practice, particularly in aerospace, is to vibration test the structure as soon as at least one specimen is reasonably complete, in a ground vibration test, GVT, and compare the results with predictions. In the event of significant differences, the mathematical model must be changed.

(2) The second group is *environmental vibration testing*. Here, the aim is to demonstrate that an item of equipment, typically electronic equipment installed in or on a vehicle, is capable of withstanding its vibration environment for the duration of its service life. The vendor of such equipment usually has to demonstrate this to the purchaser, by subjecting at least one specimen to the equivalent of the actual vibration environment. These tests are usually of only a few hours duration, and the test level is often increased to compensate. The tests can be based on standard specifications, such as the MIL series, or, when possible, on actual measurements of vibration levels.

(3) The third group is *vibration fatigue testing*. Items such as skin panels, exposed to high acoustic pressures, have been tested in noise generating facilities for many years. To test complete fins, stabilizers, wings, etc. exposed to high buffet or acoustic loads in this way, however, would require extremely large and expensive facilities, and such items are therefore usually tested by applying sequences of static loads, as part of the normal fatigue test. Although still regarded as novel,

367

fatigue tests of these large components by the use of shakers are occasionally carried out.

In this chapter these three main types of test are introduced, followed by a brief discussion of some of the hardware used in vibration testing.

13.1 Modal testing

Modal testing can be applied to any resonant structure, from a simple single-DOF system, where only the natural frequency, damping coefficient and the shape of the single mode is required, to a complete vehicle, such as an aircraft, space-craft or land vehicle, with many normal modes.

When only a rough idea of the natural frequencies, damping coefficients and approximate mode shapes of a simple system is required, the modal testing procedure can be correspondingly simple. The natural frequencies and damping coefficients are first found, by applying a sinusoidal force from a shaker, and plotting the amplitude and phase of the response, using an accelerometer, for a suitable range of frequencies. To find the approximate mode shapes, the single accelerometer can then be moved around to different locations on the structure, while exciting each mode at its natural frequency, and recording the amplitude and phase at each location.

At a more advanced level, modal testing can be used to find the natural frequencies, damping coefficients and vibration modes, of large, complex, structures. All aircraft and space-craft, and many other structures, are now modeled mathematically, typically using the finite element method, followed by transformation into normal modes. In the case of aircraft, such models are absolutely essential, to enable the flutter and dynamic loads analyses to be carried out, and they must be shown to be accurate. In all cases, therefore, the mathematical model has to be 'validated' by carrying out a GVT. This aims to measure the natural frequencies, and the shapes, of the normal modes of the actual structure, for comparison with those predicted by the mathematical model. It also gives the damping coefficients of the normal modes, which usually cannot be obtained theoretically.

A simple approach to large-scale modal testing, as generally used for finding the undamped normal modes of aircraft structures, is briefly presented here. A wider view of modal testing in general, including rotating structures, which are not considered here, is given by Ewins [13.1].

13.1.1 Theoretical Basis

Suppose that the structure, of which the modes are to be measured, is linear, and could be represented by a *hypothetical* set of equations of the form:

$$[M]\{\ddot{z}\} + [C]\{\dot{z}\} + [K]\{z\} = \{f\} \tag{13.1}$$

where

$\{z\}$ is a vector of actual displacement coordinates at local level, $\{f\}$ a vector of corresponding external forces, $[M]$ the corresponding inertia matrix, $[K]$ the corresponding stiffness matrix and $[C]$ the corresponding damping matrix.

Assuming sinusoidal excitation, let:

$$\{f\} = \{F\}e^{i\omega t}$$

$$\{z\} = \{Z\}e^{i\omega t}$$

$$\{\dot{z}\} = i\omega\{Z\}e^{i\omega t},$$

$$\{\ddot{z}\} = -\omega^2\{Z\}e^{i\omega t}. \tag{13.2}$$

Substituting Eqs (13.2) into Eq. (13.1), and dividing through by $e^{i\omega t}$ gives

$$-\omega^2[M]\{Z\} + i\omega[C]\{Z\} + [K]\{Z\} = \{F\} \tag{13.3}$$

where $\{Z\}$ and $\{F\}$ represent the displacements and applied forces, respectively, and are regarded as complex vectors for the present.

Now suppose that it is possible to adjust the external forces, $\{F\}$, experimentally, in such a way that

$$\{F\} = i\omega[C]\{Z\} \tag{13.4}$$

If ω is one of the undamped natural frequencies of the system, then it is actually possible to satisfy Eq. (13.4) for any damping matrix $[C]$ when the vectors $\{Z\}$ and $\{F\}$ are real: they do not have to be complex. Physically, this means that all the displacements are either in phase, or exactly out of phase, with each other, as are all the applied forces, and there is a $\pm 90°$ phase relationship between the forces and the displacements. The applied forces exactly balance the damping forces, and cancel each other out in Eq. (13.3), leaving the equation of motion of the undamped system:

$$\left([K] - \omega^2[M]\right)\{Z\} = 0 \tag{13.5}$$

Thus, the structure vibrates as if it has no damping, and the modes excited are the normal modes of the undamped system. This is usually just what is required, because the modes of the structure will also have been calculated without damping, and can be compared directly with the measured modes.

Although described above for viscous damping, the idea of canceling out the damping terms by applied forces is valid for any form of damping, provided it is linear, and so includes the case of hysteretic as well as viscous damping.

In practice, a test method based on this approach is found to work well with aircraft structures, even with a relatively small number of applied forces. Its practical application is described below.

13.1.2 Modal Testing Applied to an Aircraft

Figure 13.1 shows a typical set up for modal testing a small jet aircraft.

Several exciters will be required on each major component of the aircraft, such as the fuselage, wings, fins, stabilizers and any discrete masses such as engines or wing pods. The rated force output of each exciter should be related to the mass of the part of the structure to which it is attached, and all the possible freedoms of each

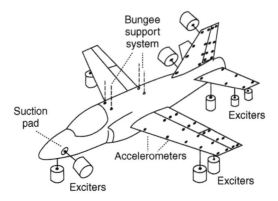

Fig. 13.1   Modal testing applied to a small jet aircraft.

component should be allowed for, for example the vertical bending and torsion modes of a wing will always be required, but fore and aft wing bending modes may or may not be in the frequency range of interest.

Supporting the test aircraft is important, and it should be tested in the condition for which the modes are required. The important case is usually in flight, and this can be simulated, in the case of a small aircraft, by suspending it from a suitable arrangement of bungee cords, with the landing gear retracted. Larger aircraft are usually supported on their landing gear, each unit of which, with or without wheels, rests on special spring supports. In either case, the rule of thumb is that the highest rigid mode natural frequency, of which there are six, should be less than one-third of the frequency of the lowest expected vibration mode in the airborne condition.

The equipment used to excite the specimen will typically consist of the following elements. Some of these are described in more detail in Section 13.4. Electro-dynamic exciters are assumed, but electro-hydraulic devices are also available.

(1) A *signal generator*, providing voltages representing the time functions required to excite the specimen. This function may, in fact, be provided by the computer used to control the test.

(2) *Power amplifiers*, converting these signals to the heavy currents required to drive the exciters.

(3) The *exciters*, of which several are shown in Fig. 13.1.

The equipment required for measuring the forces applied to the specimen, and the resulting responses, will typically consist of:

(1) *Force transducers* connected between each exciter and the specimen. These are usually of the piezo-electric type.

(2) *Accelerometers* measuring the response of the structure. Some of these are shown in Fig. 13.1. They are usually of the piezo-electric type, and several hundred are required for a typical aircraft modal test.

(3) *Signal conditioning amplifiers* to convert the outputs of the force and acceleration transducers to analog voltages acceptable to the analysis equipment.

(4) The *analysis equipment*, which will usually be based on a general purpose computer. This must be preceded by suitable equipment for digitizing the analog voltages from the force and acceleration transducers, i.e. A-to-D converters.

The controlling computer will usually handle the signal analysis, and this will not be described in detail. The essential feature, however, is deriving the relationships, in complex or amplitude/phase form, between the master oscillator signal, driving the exciters, and both the force transducer and accelerometer outputs, a process for which a 'resolver' was once used. Any other relationships are, of course, easily obtainable from these. This is essentially a simple form of Fourier analysis, as discussed in Chapter 9, except that only the fundamental Fourier components of the response, i.e. those at the excitation frequency, are required. If $x(t)$ represents either a steady-state measured force signal, or an accelerometer response signal, it will be of the form, ignoring any overall scaling:

$$x(t) = a \cos 2\pi ft + b \sin 2\pi ft \qquad (13.6)$$

where a and b are the fundamental Fourier force or acceleration components to be found and f is the excitation *and* response frequency.

then

$$a = 2\langle x(t) \cos 2\pi ft \rangle \qquad (13.7a)$$

and

$$b = 2\langle x(t) \sin 2\pi ft \rangle \qquad (13.7b)$$

where the sin and cos waveforms in Eqs (13.7a) and (13.7b) are derived from the master oscillator. The averaging is carried out over an integer number of cycles. Having found the Fourier components a and b, the real and imaginary components of the force or acceleration signal relative to the master oscillator are given by Re $= a$ and Im $= -b$, from which any required amplitude and phase relationships are easily computed.

Test procedure

If the structure is symmetrical, as in the case of most aircraft, advantage is always taken of this fact to simplify the test, since the modes will either be symmetric or antisymmetric, as discussed in Section 8.5. Two completely separate tests are carried out, one with symmetric excitation, and one with antisymmetric excitation, and each of these may have to be repeated with several different fuel states and external store configurations.

Generally, a symmetrically placed pair of exciters should be operated as a single entity, i.e. the forces should either be the same or equal and opposite, at all times.

The following procedure for exciting the undamped normal modes of the aircraft is based on the idea of canceling the damping by applying external forces, as discussed in Section 13.1.1.

(1) The approximate natural frequencies of the system, and a rough estimate of the corresponding mode shapes, are first found by measuring a series of frequency response functions, using any suitable exciter locations and forces. These will usually be in the form of inertance, i.e. acceleration per force. They are typically plotted in the form of amplitude and phase versus frequency, and also as 'circle plots', as discussed in Chapter 4. Although many techniques for extracting natural frequencies and damping coefficients from FRFs have been devised since, the circle plot method, presented by Kennedy and Pancu [13.2] in 1947, remains one of the best.

(2) Knowing the natural frequencies, it is possible to excite each undamped normal mode in turn. This requires adjustment of the individual forces, while, at the same time, adjusting the excitation frequency to maintain a 90° relationship between the forces and, initially, one of the responses. When the correct condition is obtained, the acceleration responses will be either in phase, or exactly out of phase with each other, while all being at 90° to the forces.

These criteria are usually considered satisfied when all the response vectors lie within about 10° from the ideal, and the mode is then assumed to be a normal mode of the underlying undamped structure. If this condition cannot be achieved, it is probable that the exciters are incorrectly located, and another set of positions should be tried. In this iterative procedure, it will be found that it is just as important *not* to excite the other modes as it is to excite the desired mode.

Having found the optimum distribution of applied forces, each normal mode is measured, by recording the amplitude and phase of the response at every accelerometer location, while the structure is excited at resonance.

The adjustment procedure to excite the undamped normal modes described above is somewhat tedious to carry out manually, particularly when there are several modes close in frequency, and a number of schemes for automating the process have been devised. These do not always work in practice, but practical experience suggests that a software loop to control the excitation frequency, so that there is always a 90° phase angle between the forces and a selected acceleration response, is extremely valuable, and makes manual adjustment of the forces relatively easy, provided a clear display showing the instantaneous phase relationships is provided. This is similar in principle to a *phase locked loop* often used in electronics to keep resonant circuits tuned to a carrier frequency.

Equations of motion derived from measured data

Modal testing, as described above, can provide the natural frequencies, damping coefficients and normal mode shapes of the structure. While these may provide sufficient information to judge whether a mathematical model representing the same structure is reasonably accurate, they are insufficient to allow the equations of motion of the structure to be written, since they provide no way of calculating the generalized masses of the normal modes. Therefore modal testing an aircraft structure traditionally used to include a further activity: measurement of the generalized masses. These were measured in one of two ways:

(1) small known masses were added to the structure, sufficient in size to change the natural frequencies, but not the mode shapes. From the changes in natural frequencies, the generalized masses could be calculated, or:

(2) while the structure was vibrating at resonance in a normal mode, each exciter force was modified by adding a small, proportional, quadrature force component, of appropriate sign. This represented an increase or decrease of stiffness, again allowing the generalized mass to be found from the change in natural frequency.

These methods do not appear to be used now, and the modern approach is to use the *calculated* mass distribution, with the measured modes, to calculate the generalized masses. This looks like cheating, but is probably justified in these days of computerized mass control. This is carried out in the same way as for calculated modes, i.e. the generalized mass of mode i, $\underline{m}_{ii}$, is given by:

$$\underline{m}_{ii} = \{z\}_i^T [M]\{z\}_i \tag{13.8}$$

where $\{z\}_i$ is the measured mode shape for mode i and $[M]$ is the corresponding mass matrix.

Orthogonality check

If the measured normal modes represent the undamped structure, they should be orthogonal with respect to the mass matrix, and the cross-inertia between any two modes, i and j, should be zero, as discussed in Section 6.3.1. Thus for two perfect measured normal modes, we should have

$$\underline{m}_{ij} = \{z\}_i^T [M]\{z\}_j = 0 \tag{13.9}$$

where $\{z\}_i$ and $\{z\}_j$ are two different measured normal mode shapes and $[M]$ is the corresponding *calculated* mass matrix.

It is usual to calculate both the generalized masses from Eq. (13.8) and the cross-inertias from Eq. (13.9). The resulting matrix, in normal mode coordinates, can be conveniently normalized by expressing it in orthonormal form, i.e. by scaling the diagonal terms to unity. The non-diagonal terms should then be small compared with the diagonal terms, and this is a valuable test of the quality of the measured modes. This is, of course, carried out separately for the symmetric and the antisymmetric modes.

13.2 Environmental vibration testing

Electronic and other devices carried in aircraft, spacecraft, etc., can now be so vital to the correct functioning, or even the survival, of the vehicle, that environmental testing of at least one sample of each piece of equipment is usually specified in a contract between the purchaser and the supplier. This has created the science and industry of *environmental engineering*, which covers all environmental conditions, including high and low temperatures and pressures, shock, steady acceleration, resistance to acids and alkalis, and mold growth, as well as vibration. Here we shall discuss only the vibration aspects of environmental testing, but it should be remembered that it may occasionally be necessary to combine vibration testing with other environmental conditions, such as high or low temperatures, or low or zero atmospheric pressure.

In general, the vibration behavior of equipment items is not modeled mathematically in detail, and vibration qualification relies almost entirely on testing.

13.2.1 Vibration Inputs

Originally, all vibration testing was sinusoidal, and was carried out in accordance with standard specifications, measurements in service being practically unknown at that time. The early specifications required a resonance search, and having found the major resonances, they were each excited by a sinusoidal dwell test for a specified period. Sometimes the test was supplemented by a sinusoidal sweep test. As electronic equipment became miniaturized, it became progressively more difficult to find the resonances, and it was possible for them to be missed, resulting in inadequate testing. This led to increased use of broad-band random testing, when appropriate, and the resonance search is now not usually specified.

Random testing is now the most widely used method, but it must be supplemented by swept sinusoids in the case of equipment installed in helicopters, which generate periodic vibration at several multiples of the main and tail rotor rotation frequencies. Sinusoidal testing is also still usually specified for equipment installed on engines, including jet engines, and, in combination with random vibration, to represent gunfire.

Vibration inputs are now based on standard specifications, or on measurements. The latter option is often possible, because the life-cycle of most avionics equipment tends to be quite short, and in practice a large proportion of new equipment is actually installed in aircraft that have been in service for some time. When equipment vibration tests are based on measurements, a standard test specification is normally still used, but with 'tailored' adjustments to the actual test levels.

Although equipment in service is subjected to vibration in three axes simultaneously, it is invariably tested by exciting for one-third of the total test time in each direction.

Standard specifications

Standard national specifications for environmental vibration tests are issued by several countries, but the MIL standards issued by the US government are the most comprehensive and widely used. Method 514.5 of MIL-STD-810F [13.3], not only presents detailed test procedures, but also includes comprehensive background material on vibration mechanisms and test methods, and is essential reading. The specification covers equipment installed in all aircraft types, and transportation of equipment by land, sea and air.

13.2.2 Functional Tests and Endurance Tests

A typical environmental vibration test consists of:
(1) a functional test; and
(2) an endurance test.

The aim of the functional test is simply to ensure that the equipment functions correctly in the presence of vibration. This test, which need only last a few minutes, is usually carried out before, after and halfway through the endurance test. The vibration level used should be equal to the highest that the equipment will experience during operation, but need not be scaled up in any way, as may be the case for the endurance test, discussed below. The vibration level for the functional test may therefore be lower than that of the endurance test. Vibration can cause a piece of equipment to malfunction without necessarily damaging it, and in such cases it may be found to function perfectly when the vibration is removed.

The endurance test is usually regarded as a fatigue test, and scaling procedures often have to be used to condense long periods of vibration in service to the shorter time allowed for the test. This is usually achieved by scaling up the lower levels of vibration to that of the highest, shortening their duration according to an assumed S–N curve (see Section 10.8.2). If the test time obtained in this way exceeds the standard duration specified for the test, usually a few hours in each of the three axes, a further factor may be applied. For scaling purposes, a standard S–N curve of the form:

$$NS^{\alpha} = \text{constant} \tag{10.94}$$

is assumed, where N is the number of cycles to failure at stress level, S, and α is a constant dependent on the material, usually lying in the range 4–8. Now N can be assumed proportional to the time, T, for which the vibration is applied, and S can be taken as proportional to the vibration level, g. Thus for sinusoidal vibration, with $\alpha = 6$, Eq. (10.91), can be written as:

$$T_0 g_0^6 = T_1 g_1^6 \quad \text{or} \quad \frac{g_0}{g_1} = \left(\frac{T_1}{T_0}\right)^{\frac{1}{6}} \tag{13.10}$$

where g_0 is the original sinusoidal vibration level, of duration T_0 and g_1 is the equivalent, increased level, of shorter duration, T_1.

If the vibration levels are instead expressed as power spectral densities, W_0 and W_1, and $\alpha = 8$ for random vibration, Eq. (13.9) becomes,

$$\frac{W_0}{W_1} = \left(\frac{T_1}{T_0}\right)^{\frac{1}{4}} \tag{13.11}$$

noting that the PSD ratio is proportional to the square of the RMS amplitude ratio.

Equations (13.10) and (13.11) appear in Annex B of MIL-STD-810F [13.3]. It should be noted that this standard uses different values of the index, α, for sinusoidal and random vibration.

Vibration levels for endurance tests should only be scaled in such a sense as to increase the level and shorten the time, due to the possible existence of a fatigue limit (see Fig. 10.24). This is conservative, in that it will over-estimate rather than under-estimate the required test level. It is never permissible to scale the amplitude down and the test time up, due to the danger of the test level being taken below the fatigue limit.

13.2.3 Test Control Strategies

The traditional test method for equipment items of moderate size is the *acceleration input control* strategy. In this method, the equipment is attached to a vibration table, or other fixture, and a purpose-designed *controller* is used to maintain the acceleration input, at the equipment attachment points, at the desired value, usually a specified power spectral density (PSD) function. The controller achieves this by comparing the output from one or more monitoring accelerometers, located close to the attachment points, with the desired value, and adjusting the drive to the shaker system accordingly. The desired input spectrum and the controller's attempt to reproduce it are usually displayed, for comparison. The equipment is vibrated in each of the three axes in sequence, to the specified levels, for the specified periods. Since the table moves in only one axis, the test item must either be rotated relative to the table for the other two axes or a slip table can be used. For larger items, a test fixture excited by one or more standard shakers is often used instead of a vibration table.

For larger, more flexible, items of equipment, controlling the acceleration input only at the points where it is attached to the vehicle can be very unsatisfactory, resulting in vibration levels at some locations within the equipment being too high, and some too low. Several test procedures have been developed to rectify this situation. The *acceleration limit strategy* [13.3] can be used when it is known from measured data that some locations within the equipment respond excessively, due to resonance. The input spectrum is then modified in appropriate narrow frequency bands to limit these responses to the measured values. The *acceleration response control strategy* [13.3] is a somewhat similar method, used for vibration testing aircraft externally mounted stores, when flight measurements at several points within the store are available. The excitation is still controlled at the points where the store is attached to the test fixture, but in this case is initially set to an arbitrary low level, which is experimentally increased in selected frequency bands until the flight measured spectra are reproduced as closely as possible, erring on the high side, if necessary. Several other test strategies [13.3] are also used. The aim, in all cases, is that the response at the majority of locations in the equipment should equal, or slightly exceed, those measured in service. Of course, these improved strategies require comprehensive vibration measurements in service, and it is now generally accepted that in expensive development programs, the cost of these can easily be justified.

13.3 Vibration fatigue testing in real time

Although aircraft structural failures due to buffet loads have been known for many years, some relatively recent developments in aerodynamics, as applied to highly maneuverable combat aircraft, appear to have increased buffet-induced vibration, so that fin, stabilizer and even wing structure can be at risk. Typically, the problem occurs in flight at high angles of attack, where vortices shed by the wing can excite fins and stabilizers to high enough response levels in flexural modes to threaten their structural integrity. Such components therefore have to be fatigue tested specifically for the buffet loads. This tends to be beyond the capability of acoustic test facilities, and it is usually carried out by static rather than dynamic loading, in the same way as for the quasi-static maneuver loads.

Fatigue testing in real time, using exciters, offers several advantages:
(a) It is relatively cheap, compared with acoustic methods.
(b) By exciting at the correct range of frequencies, the structure responds in its normal modes, which means that vibration stresses are automatically correctly distributed.
(c) If the damping of the modes is light, as is usually the case, the fact that the system is excited at resonance means that exciter forces are many times smaller than the static forces that would be needed to produce equivalent local stresses, with less risk of unwanted local damage at the points where the loads are applied.
(d) There is a large saving of test time compared with static loading.

In the case of conventional aircraft fins and stabilizers, it is usually found that only one or two of the modes have to be excited to significant levels.

The practical application of this test method is quite difficult. The response levels required are, by definition, extremely high, in the order of 50 to several hundred *g* at the tip of a fin or stabilizer, and this stretches the capability of electrodynamic exciters to the limit, not so much due to the forces required, but because the moving mass of the exciters becomes important at such high levels of response, and hydraulic exciters will probably be essential in extreme cases.

The excitation required is, in this case, random, and it is easily shown that exciter force must not be wasted by setting the excitation bandwidth too wide. For each mode, an excitation bandwidth equal to about ten half-power bandwidths of the mode, centered on the natural frequency, appears to be wide enough to avoid distorting the response time history, without wasting exciter power.

13.4 Vibration testing equipment

A wide range of measuring and excitation equipment can be used in vibration testing, including some very advanced devices employing lasers. The following discussion introduces only the more common devices used in most everyday testing situations. Most vibration test work, in fact, is carried out using electrodynamic or hydraulic exciters to apply vibration forces to systems under test; force transducers to measure the applied forces and accelerometers to measure vibration response. Strain gages are also used to measure vibration response and loads, but will not be discussed here.

13.4.1 Accelerometers

As discussed in Example 4.3, most accelerometers are based on the principle that in a single-DOF spring, mass, damper system, excited via the base, the displacement of the spring is proportional to the acceleration at the base up to a certain frequency. When the damping is low, as in most accelerometers used in vibration testing, this frequency is often taken as 20% of the natural frequency, since the error is then limited to about 0.5 dB, or about 6%. Thus the natural frequency is the most important parameter in defining the upper limit of the useful frequency response, and this is independent of the type of accelerometer.

Accelerometers used in vibration work are usually of two types: piezoelectric and piezoresistive. In the piezoelectric type, which are the more common, the inertia force produced by a small mass, when acted upon by acceleration, is applied to a piezoelectric element, which both acts as a spring and generates an electric charge proportional to the acceleration. A piezoelectric accelerometer must be used with a suitable pre-amplifier, which can be a charge amplifier or a voltage amplifier. The former type is generally preferred, since the sensitivity of the combined system is largely unaffected by the length of the connecting coaxial cable, a problem with voltage amplifiers. The frequency response of piezoelectric accelerometers cannot extend down to 'DC', or constant acceleration. Charge amplifiers are better than voltage amplifiers in this respect, and can be used down to a fraction of 1 Hz.

Piezoresistive accelerometers work on a different principle, in that the inertia force of the mass acts upon an actual spring element, of which the displacement is measured, essentially by strain gages made of semiconductor material, and require a low voltage DC supply. The main advantage of piezoresistive accelerometers over the piezoelectric type is that they can be used when response down to DC is required.

A fairly recent development in accelerometer design is to incorporate some of the amplifier electronic components into the same casing as the accelerometer itself. A point to watch when ordering accelerometers of this type is that sometimes the sensitivity is fixed in the design, and can only be changed by replacing the whole unit.

13.4.2 Force Transducers

Force transducers are mainly used in vibration work to measure the force applied to a structure by an exciter. They usually employ the piezoelectric principle, and the preamplifiers used are similar to those used for accelerometers. Some older types of force transducer measure load by strain gages fitted to metal rods or rings.

13.4.3 Exciters

Vibration exciters or 'shakers' are of two main types, electrodynamic and electrohydraulic. The former are the more common, and are used for most modal testing, except for the very largest structures. They are also used for most environmental testing, except, again, for the largest equipment items. Electrohydraulic exciters are expensive, and tend to be used only where electromagnetic units would be unsuitable, such as when very large displacements, or a combination of steady and oscillatory forces are required. They will not be discussed further here: a good introduction to electrohydraulic exciters is given in [13.4, Chapter 25]. A wealth of practical information on all forms of exciters is, of course, available from manufacturers' literature.

Electromagnetic exciters and power amplifiers

Although other types exist, the moving coil electromagnetic exciter is the type commonly encountered. They have excellent linearity and frequency range. The principle of operation is very simple, and is identical to that of a moving coil loudspeaker, found in every radio and hi-fi unit, but on a much larger scale. The moving coil, which may be from a few millimeters to more than a meter in diameter, is wound on a non-ferrous former, and constrained to move only axially in the annular air gap of a large magnet. Permanent magnets are used in the smaller sizes, and electromagnets, supplied with DC, on larger units. This makes the units relatively heavy, but this is not necessarily a disadvantage in use, since a large mass is often desirable to react the oscillatory forces produced. The moving coil is usually kept central in the air gap of the magnet by flexures, free to bend in the axial direction, but stiff in the radial direction. Linear ball bearings have also been used for this purpose.

Electromagnetic exciters are available in a wide range of peak force outputs, from less than 1 N, to several thousand newtons. The stroke varies accordingly, but exciters with a total possible movement exceeding about 40 mm are quite rare. The output force is usually limited by the heat generated in the moving coil, and can be maximized by forced cooling.

Like a loudspeaker, an amplifier is required to provide the power required to drive an electrodynamic exciter, and its characteristics are important in ensuring overall satisfactory operation. The power amplifier is similar, in principle, to those used in ordinary hi-fi units, but very much larger, usually incorporating liquid-cooled output transistors.

In the remainder of this section, two of the important, yet often not well-understood effects that can be encountered in the operation of electrodynamic exciters are discussed.

The back-EMF effect in electromagnetic exciters

It is commonly observed that when a moving coil electromagnetic exciter is used to shake a resonant system, the exciter current tends to decrease as the amplitude of the system increases, for example when it passes through a resonance. It may also be found that the damping of the system appears to increase due to the presence of the exciter. The effect is important, and has considerable impact on the design of the amplifier used to drive the exciter, and the conduct of vibration tests. It can be explained, as follows, using the laws of electromagnetism; however, it is stressed that the following is a somewhat simplified explanation of the effect.

Figure 13.2 shows the moving coil of an electro-magnetic exciter, considered mass-less. The DC electromagnet (or permanent magnet) producing the magnetic field is not shown. The exciter drive rod, rigid, in the axial sense, is connected, to the mass, m, which together with the spring, k, and damper, c, represents one mode of the system being excited.

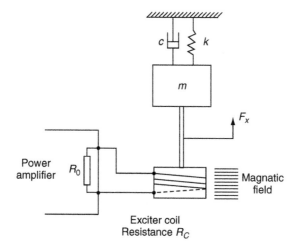

Fig. 13.2

If a current is passed through the coil by the power amplifier, the instantaneous force applied by the coil to the mass is given by:

$$F = BLI \tag{13.12}$$

where:

F is the force, (newtons), B the flux density of the magnetic field (T), L the total length of wire in the coil (m) and I the instantaneous value of the current in the coil (A).

If the coil is moving, it generates a back-EMF given by:

$$E_B = BL\dot{x} \tag{13.13}$$

where:

E_B is the back-EMF (V), $\dot{x}$ the velocity (m/s) and B and L are as defined above.

From Ohm's law, the back-EMF produces a 'back-current', I_B (in A), given by dividing E_B by the total resistance of the circuit, which consists of the resistance of the coil and the output resistance of the driving amplifier in series, i.e.,

$$I_B = \frac{E_B}{R_C + R_O} \tag{13.14}$$

where:

R_C is the resistance of the exciter coil (Ω) and R_O is the output resistance of the amplifier (Ω).

Strictly, these quantities are impedances rather than resistances, but will approximate to resistances at low frequencies, since the coil is air-cored and has relatively low inductance.

The actual force produced by the exciter is still as given by Eq. (13.12), but now I is reduced from the value, I_0, say, that would be obtained in the absence of a back-EMF effect, that is with $\dot{x} = 0$, by an amount I_B, so:

$$F = BLI = BL(I_0 - I_B) \tag{13.15}$$

Substituting Eqs (13.13) and (13.14) into Eq. (13.15) gives

$$F = BLI_0 - \frac{(BL)^2}{R_C + R_0}\dot{x} \tag{13.16}$$

Equation (13.16) clearly shows how the force F actually generated is reduced from the intended value, BLI_0, by a magnetically generated amount proportional to the velocity of the response, $\dot{x}$.

The constants required in Eq. (13.16), incidentally, should all be available from the manufacturer of the exciter: R_C and R_0 are defined above, and BL is the force per current ratio for the exciter in newtons/ampere.

To see how the damping of the system is affected, we first write the equation of motion of the complete system, ignoring the mass of the exciter armature and any stiffness or damping due to the exciter coil suspension. This is simply:

$$m\ddot{x} + c\dot{x} + kx = F \tag{13.17}$$

Combining Eqs (13.16) and (13.17) gives

$$m\ddot{x} + \left[c + \frac{(BL)^2}{R_C + R_0}\right]\dot{x} + kx = BLI_0 \qquad (13.18)$$

Equation (13.18) shows that the back-EMF effect can also be interpreted as an increase in the damping of the system.

For a given exciter, BL and R_C are fixed by the design, but R_0, a property of the *amplifier*, can be changed to obtain desired characteristics, as follows.

(a) It can be seen from Eqs (13.16) and (13.18) that changing R_0 has a powerful effect on the back-EMF phenomenon.

(b) Secondly, R_0 can be changed fairly easily by changing the type of *negative feedback* (NFB) used in the design of the amplifier. NFB is used in virtually all power amplifiers to reduce distortion, and consists of feeding back a voltage derived from the output to the input.

(c) If the NFB is made proportional to the output *voltage* of the amplifier, it can be shown that R_0 is reduced, and it can be seen from Eq. (13.16) that the back-EMF effect is increased. This increases the tendency for the force to 'dip' at resonance, and increases the damping of the structure.

(d) If the NFB is made proportional to the output *current* of the amplifier, it can be shown that R_0 is increased. The back-EMF effect is then reduced, or eliminated, as is the 'dipping' of the force, and the damping added to the structure.

What is desirable depends upon the application. An amplifier designed for use with moving-coil loudspeakers should have a low output resistance, because this tends to damp cone resonances, which are undesirable for high-fidelity sound reproduction. Similarly, shakers used with vibration test rigs can often be used to damp unwanted rig resonances, as well as to excite the system. On the other hand, in modal testing, the last thing we want to do is to damp the system being tested, and the power amplifiers used require high output impedance. In general, a large degree of control is possible, and in the best designs the NFB mode of the amplifier, and hence the behavior of the associated exciter can be changed by a simple switch. At least one manufacturer claims that the damping effect can be eliminated, or made large, whichever is desired, in this way.

It should be noted that the effect described above does not necessarily affect the accuracy of damping measurements, provided the correct procedure is used. FRFs should be found, for example, by dividing the response of the structure by the force actually applied at each frequency. However, using an amplifier with low output resistance for modal testing can practically reduce the applied force to zero at resonance.

The effect of armature mass

Another effect, that can change the force applied to a structure by an exciter, is caused by its moving mass. Like the back-EMF effect described above, the force actually applied to the structure tends to vary when passing through a resonance. However, this is a purely mechanical effect due to the inertia of the moving parts of the exciter. Any type of exciter can be affected by this phenomenon, but an electro-dynamic type will be assumed in the following analysis.

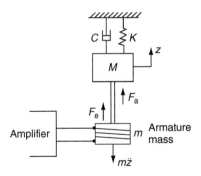

Fig. 13.3

Figure 13.3 represents a single-DOF system, consisting of M, K and C, which are the equivalent properties of a single mode being excited, lumped at the point where the exciter is attached. The mass of the moving parts of the exciter is represented by m, and this is assumed to be rigidly connected to the main mass, and therefore has precisely the same motion as the latter. The electromagnetic force generated by the exciter coil is F_e, but the force actually transmitted to the 'structure' is F_a, which differs from F_e by the inertia force $m\ddot{z}$. It is assumed, in this analysis, that there is no back-EMF effect to confuse the issue.

The equations of motion are

$$M\ddot{z} + C\dot{z} + Kz = F_e - m\ddot{z} \tag{13.19}$$

and

$$F_a = F_e - m\ddot{z} \tag{13.20}$$

Equation (13.19) is conveniently written in the form:

$$(M + m)\ddot{z} + C\dot{z} + Kz = F_e \tag{13.21}$$

Equation (13.21) can be written in the form of a FRF, giving the complex inertance, $\ddot{z}/\underline{F}_e$, using the expression derived as Eq. (C) in Example 4.2, i.e.:

$$\frac{\ddot{z}}{\underline{F}_e} = \frac{-\Omega^2 \left[\left(1 - \Omega^2 \right) - i(2\gamma\Omega) \right]}{(M+m) \left[\left(1 - \Omega^2 \right)^2 + (2\gamma\Omega)^2 \right]} \tag{13.22}$$

where

$$\Omega = \frac{\omega}{\omega_n}; \quad \omega_n = \sqrt{\frac{K}{M+m}}; \quad \text{and} \quad \gamma = \frac{C}{2\omega_n(M+m)}$$

where ω is the excitation frequency. It should be noted that ω_n is strictly the natural frequency of the system with the exciter mass added.

For compactness, Eq. (13.22) will be written as:

$$\frac{\ddot{z}}{\underline{F}_e} = \frac{1}{M+m}(a + ib) \tag{13.23}$$

Expressing Eq. (13.20) in the form of complex vectors also:

$$\underline{F}_a = \underline{F}_e - m\underline{\ddot{z}} \tag{13.24}$$

From Eqs (13.23) and (13.24):

$$\frac{\underline{F}_a}{\underline{F}_e} = 1 - \frac{m}{M+m}(a+ib)$$

or for practical purposes:

$$\frac{\underline{F}_a}{\underline{F}_e} = 1 - \frac{m}{M}(a+ib) \tag{13.25}$$

If we are only interested in the relative magnitudes of F_a and F_e:

$$\frac{|\underline{F}_a|}{|\underline{F}_e|} = \sqrt{\left(1 - \frac{ma}{M}\right)^2 + \left(\frac{mb}{M}\right)^2} \tag{13.26}$$

Equation (13.26) can be used to investigate how the actual force applied to the structure, F_a, varies with frequency when passing through a resonance, assuming that F_e remains constant. This is seen to be dependent on the damping coefficient, and the ratio m/M. This ratio can be estimated roughly in the following way.

First, the weight (not the mass) of the moving parts of most electrodynamic exciters usually lies between 1 and 2% of the *rated* thrust, i.e.,

$$mg = \lambda F_R \tag{13.27}$$

where:

m is the mass of the moving parts, g the acceleration due to gravity, F_R the rated peak force output of the exciter and λ a constant, usually between 0.01 and 0.02 depending upon the design of the exciter.

Secondly, assuming that an exciter is running at its rated force output, the maximum acceleration at resonance, $\ddot{x}_{max}$, at the point on the structure where it acts, is easily shown to be given approximately by:

$$\ddot{x}_{max} = \frac{F_R}{2\gamma M} \tag{13.28}$$

where M is the effective mass of the structure at the exciter position and γ is the fraction of critical damping of the mode concerned.

Combining Eqs (13.27) and (13.28):

$$\frac{m}{M} = \frac{2\gamma \lambda \ddot{x}_{max}}{g} = 2\gamma \lambda a_g \tag{13.29}$$

where $\ddot{x}_{max}/g = a_g$ is the peak acceleration in g units at the point on the structure where the exciter acts. For an exciter *running at its rated output*, the value of m/M is thus determined by the damping coefficient, γ, the ratio of the weight of the moving parts of the exciter to its rated thrust, λ, and the acceleration response a_g at the exciter location. We can now estimate the value of m/M under some typical working conditions from Eq. (13.29), as follows.

(1) In a typical modal test, taking $\gamma = 0.025$ and $\lambda = 0.02$, then with a_g about 1g, from Eq. (13.29):

$$\frac{m}{M} \approx 2 \times 0.025 \times 0.02 \times 1 = 0.001$$

(2) If the application is a severe environmental test, or fatigue test, the peak response a_g at the exciter station may well be as high as 20g, and taking $\gamma = 0.025$ and $\lambda = 0.02$ as before:

$$\frac{m}{M} \approx 2 \times 0.025 \times 0.02 \times 20 = 0.02$$

These two cases probably represent the extreme values likely to be encountered following best practice, which is to choose an exciter so that it is running close to its rated output. It can be seen, however, that in case (2), m/M could be much larger if the exciter was used well below its rated capacity, or if the damping coefficient was higher.

The modulus ratio $|\underline{F}_a|/|\underline{F}_e|$ is plotted in Fig. 13.4, from Eq. (13.26), for the two cases above.

It can be seen from Fig. 13.4 that assuming that electrodynamic exciters are used close to their rated outputs, the variation in output force with frequency is much greater at high levels of excitation than at low levels. Provided the effect is allowed for, by actually measuring the force applied, it is generally not a problem in modal testing. However, in high-level vibration tests, the effect can be quite significant, possibly causing the force applied to vary up to $\pm 50\%$ or so around resonance in extreme cases.

It should be remembered that in environmental tests, the controller used to define the spectrum shape can, in principle, be used to correct the effect, up to a point.

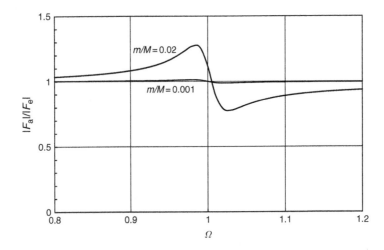

Fig. 13.4 Variation of the force actually applied to a test structure by an exciter due to its own moving mass.

References

13.1 Ewins, DJ (2000). *Modal Testing: Theory, Practice, and Application.* Research Studies Press Ltd.

13.2 Kennedy, CC and Pancu, CDP (1947). Use of vectors in vibration measurement and analysis. *J.Aero. Sci.* 14, 603–25.

13.3 Anon (2000). *MIL-STD-810F, METHOD 514.5.* US Government publication.

13.4 Harris, CM and Piersol, AG eds, (2002). *Harris' Shock and Vibration Handbook.* McGraw-Hill.

Appendix A

A Short Table of Laplace Transforms

$f(t)$	$F(s)$	Notes
$\delta(t)$	1	Unit impulse or Dirac function.
$H(t)$	$\dfrac{1}{s}$	Unit step or Heaviside function.
e^{-at}	$\dfrac{1}{s+a}$	
te^{-at}	$\dfrac{1}{(s+a)^2}$	
t^n	$\dfrac{n!}{s^{(n+1)}}$	n must be a positive integer. $n! = $ factorial n
$\sin \omega t$	$\dfrac{\omega}{s^2 + \omega^2}$	
$\cos \omega t$	$\dfrac{s}{s^2 + \omega^2}$	
$e^{-at} \sin \omega t$	$\dfrac{\omega}{(s+a)^2 + \omega^2}$	
$e^{-at} \cos \omega t$	$\dfrac{s+a}{(s+a)^2 + \omega^2}$	
$e^{-at}(1 - at)$	$\dfrac{s}{(s+a)^2}$	
$1 - \cos \omega t$	$\dfrac{\omega^2}{s(s^2 + \omega^2)}$	
$\dfrac{1}{\omega_d} e^{-\gamma \omega_n t} \sin \omega_d t$ where $\omega_d = \omega_n \sqrt{1 - \gamma^2}$	$\dfrac{1}{s^2 + 2\gamma \omega_n s + \omega_n^2}$	

Appendix B
Calculation of Flexibility Influence Coefficients

Flexibility influence coefficients for simple beams in bending or torsion can be calculated easily using the following methods. The bending case is considered first.

The potential energy, U_B, due to bending, in a beam, is

$$U_B = \int \frac{M^2}{2EI} \, dx \tag{B1}$$

where the integration is carried out over the whole length of the beam, and
M is the bending moment as a function of x;
x is the distance along the beam;
E is the Young's modulus and
I is the second moment of area of the cross-section.
Since

$$M = EI \frac{d^2 y}{dx^2} \tag{B2}$$

where $\frac{d^2 y}{dx^2}$ is the curvature. Eq. (B1) can also be written in the form:

$$U_B = \frac{1}{2} \int EI \left(\frac{d^2 y}{dx^2} \right)^2 dx \tag{B3}$$

If a load P is applied to the beam, Castigliano's first theorem states that the resulting displacement, at the same point, and in the same direction, is y_P, where:

$$y_P = \frac{\partial U_B}{\partial P} = \int \frac{M}{EI} \cdot \frac{\partial M}{\partial P} \cdot dx \tag{B4}$$

or

$$y = \int \frac{Mm}{EI} \cdot dx \tag{B5}$$

where $m = \frac{\partial M}{\partial P}$ is the moment per unit load, and can be interpreted as the bending moment function due to a dummy unit load at the point where the displacement, y, is required.

Equation (B5) can be used to calculate flexibility influence coefficients for simple beams in bending. Taking the cantilevered beam shown in Fig. B1 as an example, suppose we require the flexibility influence coefficient α_{21}, which is defined as the displacement y at x_2 due to a unit load at x_1. Figure B1 shows M, the bending moment

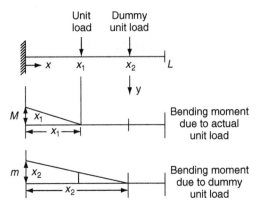

Fig. B1

function due to the actual unit load at x_1, and m, the bending moment due to a dummy unit load at x_2. The equations are

$$M = (x_1 - x) \quad 0 < x < x_1$$
$$M = 0 \qquad\quad x_1 < x < L$$
(B6)

and

$$m = (x_2 - x) \quad 0 < x < x_2$$
$$m = 0 \qquad\quad x_2 < x < L$$
(B7)

Applying Eq. (B5):

$$y = \int_0^{x_1} \frac{Mm}{EI} \cdot \mathrm{d}x = \int_0^{x_1} \frac{(x_1 - x)}{EI}(x_2 - x) \cdot \mathrm{d}x$$
(B8)

This is sometimes known as the area-moment method, because the last integral can be seen to be equal to the first moment of the area under the function M/EI, in this case $(x_1 - x)/EI$, between $x = 0$ and $x = x_1$, about the point where the displacement is required, x_2 in this case.

If EI varies along the beam in an arbitrary way, the integration in Eq. (B8) is necessarily numerical. However, in the case of constant EI, the areas involved have simple shapes, and the area moments can be evaluated by inspection.

Alternatively, Eq. (B5) can be evaluated directly.

The method described above works equally well for unit bending moments and unit angular displacements (or slopes). In this case the moment functions M and m are rectangles of unit height. It can also be used in the case of an influence coefficient which is an angular displacement due to an applied force, or the linear displacement due to an applied moment, in which case one moment function is triangular, and the other rectangular.

For beams subjected to pure torsion, Eq. (B5) becomes

$$\phi = \int \frac{Tt}{GJ} \mathrm{d}x$$
(B9)

where T is the torque as a function of x due to an applied moment and t is the torque function due to a dummy unit torque applied at the point where the angular displacement ϕ is required. When using Eq. (B9) to find influence coefficients, both T and t have the value 1 or 0.

Example B1

Use the area-moment method to derive the flexibility matrix for the cantilever beam shown in Fig. B2 at (a), where EI is constant over the length of the beam.

Solution

The flexibility influence coefficient matrix is defined by:

$$\begin{Bmatrix} z_1 \\ z_2 \end{Bmatrix} = \begin{bmatrix} \alpha_{11} & \alpha_{12} \\ \alpha_{21} & \alpha_{22} \end{bmatrix} \begin{Bmatrix} F_1 \\ F_2 \end{Bmatrix} \tag{A}$$

where z_1, z_2, F_1 and F_1 are defined by Fig. B2(a), and

$$[\alpha] = [K]^{-1} = \begin{bmatrix} \alpha_{11} & \alpha_{12} \\ \alpha_{21} & \alpha_{22} \end{bmatrix} \tag{B}$$

is the flexibility influence coefficient matrix. The quantity α_{12}, for example, is the displacement z_1 when F_1 is zero and F_2 is a unit load.

Figure B2 at (b) shows the bending moment m_1 due to a unit load at node 1, and Fig. B2 at (c) shows the bending moment m_2 due to a unit load at node 2.

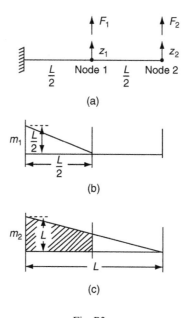

Fig. B2

The individual terms of Eq. (A) are found as follows:

α_{11} is the displacement z_1, at node 1, due to a unit load F_1 also at node 1.
Then

$$\alpha_{11} = \int_0^{\frac{L}{2}} \frac{\left(\frac{L}{2}-x\right)}{EI}\left(\frac{L}{2}-x\right) \cdot dx \qquad (C)$$

but this is equal to the area moment of the triangle in Fig. B2 at (b) about node 1:

$$\alpha_{11} = \frac{1}{EI} \times \left[\frac{1}{2}\left(\frac{L}{2}\right)^2\right] \times \left[\frac{2}{3}\left(\frac{L}{2}\right)\right] = \frac{L^3}{24EI} \qquad (D)$$

α_{12} is the displacement z_1 at node 1 due to a unit load F_2 at node 2. This is given by the area moment of the shaded area in Fig. B2(c) about node 1. This can be found by splitting it into a triangle and a rectangle, and multiplying these areas by the distances of their centroids from node 1. It should be noted that α_{12} is not given by the area moment of the whole triangle in Fig. B2(c) about node 1. Thus:

$$\alpha_{12} = \frac{1}{EI}\left\{\left[\frac{1}{2}\left(\frac{L}{2}\right)^2 \times \frac{L}{3}\right] + \left[\left(\frac{L}{2}\right)^2 \times \frac{L}{4}\right]\right\} = \frac{5L^3}{48EI} \qquad (E)$$

It can similarly be shown that:

$$\alpha_{21} = \alpha_{12} = \frac{5L^3}{48EI} \qquad (F)$$

and

$$\alpha_{22} = \frac{L^3}{3EI} \qquad (G)$$

The final flexibility matrix is

$$[\alpha] = [K]^{-1} = \begin{bmatrix} \alpha_{11} & \alpha_{12} \\ \alpha_{21} & \alpha_{22} \end{bmatrix} = \frac{L^3}{EI}\begin{bmatrix} \frac{1}{24} & \frac{5}{48} \\ \frac{5}{48} & \frac{1}{3} \end{bmatrix} \qquad (H)$$

Appendix C
Acoustic Spectra

Spectra representing sound pressure histories are usually presented in a special way, expressing the power in *decibels* (dB) in octave-based frequency bands, rather than as PSD plots. This method of presentation is largely due to two factors: (1) it was originally designed for studies involving human hearing, which has a naturally logarithmic response, and (2) it was based on the use of analog filter methods, which are more conveniently set up using bands consisting of a fixed fraction of an octave, rather than of a fixed number of hertz.

This method of presentation, although designed for audio work, using analog methods, has survived into the digital age, and is still widely used in purely structural applications. Structural dynamics work may therefore involve converting from the 'acoustics' method to the 'spectral density' method, and vice versa. The 'acoustic' method of presentation is now discussed, with an example showing a typical conversion to the PSD method.

Decibels

The *decibel* is one-tenth of the original unit, the bel (B). This was found inconvenient for practical use, and was divided into 10 decibels. The decibel scale alone, without reference to a standard level, is simply a way of expressing the factor by which an oscillatory quantity, such as voltage, force, pressure, etc., changes. It is defined in terms of power (mean square) values. So, for example, if an RMS voltage, v_1, changes to v_2, the change expressed in dB is N, say, where:

$$N(\text{dB}) = 10 \log_{10}\left(\frac{v_2^2}{v_1^2}\right) \quad \text{or} \quad \frac{v_2^2}{v_1^2} = 10^{N/10} \tag{C1}$$

Equation (C1) can be written as:

$$N(\text{dB}) = 20 \log_{10}\left(\frac{v_2}{v_1}\right) \quad \text{or} \quad \frac{v_2}{v_1} = 10^{N/20} \tag{C2}$$

As an example, if $v_1 = 1\,\text{V RMS}$, but changes to $v_2 = 2\,\text{V RMS}$, the change expressed in dB is

$$N(\text{dB}) = 10 \log_{10}\left(\frac{2^2}{1^2}\right) \quad \text{or} \quad 20 \log_{10}\left(\frac{2}{1}\right) = +6.02\,\text{dB} \tag{C3}$$

This is referred to as an increase of 6 dB, as indicated by the positive sign. It should be noted that a change from 2 V to 4 V RMS would also be an increase of 6 dB.

Table C1

N (dB)	-40	-20	-6	-10	-3	0	3	6	10	20	40
Power ratio	10^{-4}	0.01	≈0.25	0.1	≈0.5	1	≈2	≈4	10	100	10^4
RMS ratio	0.01	10	≈0.5	≈0.316	≈0.708	1	≈1.41	≈2	≈3.16	10	100

The changes in dB corresponding to some simple multiples of RMS and mean square levels are shown in Table C1, to sufficient accuracy for most purposes.

When used to express sound pressure levels, the decibel is used in a completely different way. The RMS sound pressure is then defined as being N dB above a reference level, p_{REF}, which is fixed at the assumed threshold of human hearing, taken as an *RMS value* of 20 μPa, i.e. 20×10^{-6} N/m^2 (or 2.90×10^{-9} lbf/in.^{2}).

The *sound pressure level* (SPL), N (dB) is then given by:

$$N(\text{dB}) = 10\log_{10}\left(\frac{p^2}{p_{REF}^2}\right) = 20\log_{10}\left(\frac{p}{p_{REF}}\right) \tag{C4}$$

where p is the RMS value of the pressure concerned, and p_{REF} is the RMS reference level defined above. The dB figure given by Eq. (C4) can be an *overall sound pressure level*, (OASPL), implying a broad band of frequencies, the sound pressure level in a particular frequency band, or a single sinusoidal tone.

As an example, the OASPL close to the nozzle of a rocket motor or jet engine can be as high as 1 lbf/in.^{2} RMS (= 6895 Pa RMS). This would be referred to as an OASPL of:

in British units:

$$N(\text{dB}) = 20\log_{10}\left(\frac{1}{2.90 \times 10^{-9}}\right) = 170.7\,\text{dB}$$

or, in SI units:

$$N(\text{dB}) = 20\log_{10}\left(\frac{6895}{20 \times 10^{-6}}\right) = 170.7\,\text{dB}$$

Octaves

Acoustic spectra are given as dB in bands, with band centers usually spaced at a given fraction of an octave, although dB levels in 1 Hz bands are also used. An octave, as in music, is an interval over which the frequency doubles. Very often, the center frequencies of the bands are spaced at $\frac{1}{3}$-octave intervals, but other fractions, or even whole octaves, can be used. Taking the $\frac{1}{3}$-octave system as an example, and starting at 10 Hz, the band centers, f_c, are as shown in Table C2. These are awkward numbers, but the rounded, 'standard', values shown are usually close enough for practical purposes.

Table C2

Center frequency (Exact) f_c (Hz)	Center frequency (Standard value) (Hz)
$10 \times 2^0 = 10.000$	10
$10 \times 2^{1/3} = 12.599$	12.5
$10 \times 2^{2/3} = 15.874$	16
$10 \times 2^1 = 20.000$	20
$10 \times 2^{4/3} = 25.198$	25
$10 \times 2^{5/3} = 31.748$	32
etc.	etc.

If the center frequencies are spaced at $\frac{1}{3}$-octave intervals, the bandwidth associated with each may be regarded as extending from $\frac{1}{6}$ of an octave below to $\frac{1}{6}$ of an octave above, f_c. Thus:

$$\text{Bandwidth} = \left(2^{\frac{1}{6}}f_c - 2^{-\frac{1}{6}}f_c\right) = 0.2315 f_c \tag{C5}$$

Example C1

The noise measured in the vicinity of a jet engine is shown in Table C3, expressed as dB re. 20 µPa in standard third-octave bands.
(a) Find the overall sound pressure level (OASPL) of the noise in dB re. 20 µPa.
(b) Express the spectrum given in Table C3 in the form of power spectral density, (PSD), in Pa²/Hz, suitable for use in a structural response calculation.

Table C3

Band center (Hz)	50	64	80	100	125	160	200	250	320	400	500
SPL in band (dB re.20 µPa)	144	146	148	150	153	155	156	156	154	153	152

Solution

Part (a)
To add dB levels, they must first be expressed as mean square levels. In Table C4, the RMS level, p, in each band is given in the third column, using Eq. (C4) in the inverted form:

$$p = p_{REF} 10^{\frac{N}{20}}$$

where N is the dB figure in the second column and $p_{REF} = 20\,\mu\text{Pa}$. Mean square values, p^2, are given in the fourth column. These are summed, giving an overall mean square value of 8.596×10^6 Pa², and, taking the square root, an overall RMS value of 2932 Pa. Equation (C4) then gives

$$\text{OASPL(dB)} = 20\log_{10}\frac{p}{p_{REF}} = 20\log_{10}\frac{2932}{20 \times 10^{-6}} = 163.3\,\text{dB re. } 20\,\mu\text{Pa}$$

Table C4

Band center f_c (Hz)	S.P.L. in band (dB re.20 μPa)	RMS pressure p (Pa)	Mean SQ. pressure p^2 (Pa^2)	Bandwidth (Hz)	PSD (Pa^2/Hz)
50	144	317	100×10^3	11.58	8 680
64	146	399	159×10^3	14.82	10 750
80	148	502	252×10^3	18.52	13 630
100	150	632	400×10^3	23.15	17 280
125	153	893	798×10^3	28.93	27 580
160	155	1 125	$1\ 265 \times 10^3$	37.04	34 150
200	156	1 262	$1\ 592 \times 10^3$	46.30	34 390
250	156	1 262	$1\ 592 \times 10^3$	57.88	27 510
320	154	1 002	$1\ 005 \times 10^3$	74.08	13 560
400	153	893	798×10^3	92.60	8 620
500	152	796	634×10^3	115.80	5 480
Overall	163.3	2 932	8.596×10^6		

Part (b)

The bandwidth of each third octave band is given by Eq. (C5), i.e. $0.2315 f_c$, and is entered in the fifth column. The PSD in each frequency band is then given by dividing the mean square pressure in the fourth column by the bandwidth in the fifth column.

The last column gives the spectral density in Pa^2/Hz, as required for a structural response calculation.

Index

Printed and bound by CPI Group (UK) Ltd, Croydon, CR0 4YY

03/10/2024

01040334-0012